# Studies in Rhythm Engineering

**Series Editors**

Anirban Bandyopadhyay, Senior Researcher, ANCC, National Institute for Materials Science, Tsukuba, Ibaraki, Japan

Kanad Ray, Amity School of Applied Sciences, Amity University Rajasthan, Jaipur, Rajasthan, India

Chi-Sang Poon, Department of Health Sciences and Technology, Massachusetts Institute of Technology, Cambridge, MA, USA

This is a multi-disciplinary book series, ranging from astrophysics to biology, chemistry, mathematics, geophysics and materials science. Its primary scope is the fundamental science and associated engineering wherever cyclic and rhythmic oscillations are observed.

Time neither being an entity nor a process is unmeasurable and undefined, although a clock only measures the passage of time. The clock drove recurring processes are observed in the biological rhythms, astrophysical and geophysical environments. Always, the clocks are nested, arranged in a geometric shape to govern a phenomenon in nature. These clocks are made of atoms, molecules, their circuits, and complex networks. From biology to astrophysics, the clocks have enriched the science and engineering and the series would act as a catalyst and capture the forthcoming revolution of time cycles expected to unfold in the 21st Century. From the cyclic universe to the time crystal, the book series makes a journey through time to explore the path that time follows.

The series publishes monographs and edited volumes.

More information about this series at http://www.springer.com/series/16136

Tanusree Dutta · Manas Kumar Mandal
Editors

# Consumer Happiness: Multiple Perspectives

Springer

*Editors*
Tanusree Dutta
Indian Institute of Management Ranchi
Ranchi, Jharkhand, India

Manas Kumar Mandal
Indian Institute of Technology Kharagpur
Kharagpur, West Bengal, India

ISSN 2524-5546 ISSN 2524-5554 (electronic)
Studies in Rhythm Engineering
ISBN 978-981-33-6376-2 ISBN 978-981-33-6374-8 (eBook)
https://doi.org/10.1007/978-981-33-6374-8

© The Editor(s) (if applicable) and The Author(s), under exclusive license to Springer Nature Singapore Pte Ltd. 2021

This work is subject to copyright. All rights are solely and exclusively licensed by the Publisher, whether the whole or part of the material is concerned, specifically the rights of translation, reprinting, reuse of illustrations, recitation, broadcasting, reproduction on microfilms or in any other physical way, and transmission or information storage and retrieval, electronic adaptation, computer software, or by similar or dissimilar methodology now known or hereafter developed.

The use of general descriptive names, registered names, trademarks, service marks, etc. in this publication does not imply, even in the absence of a specific statement, that such names are exempt from the relevant protective laws and regulations and therefore free for general use.

The publisher, the authors and the editors are safe to assume that the advice and information in this book are believed to be true and accurate at the date of publication. Neither the publisher nor the authors or the editors give a warranty, expressed or implied, with respect to the material contained herein or for any errors or omissions that may have been made. The publisher remains neutral with regard to jurisdictional claims in published maps and institutional affiliations.

This Springer imprint is published by the registered company Springer Nature Singapore Pte Ltd.
The registered company address is: 152 Beach Road, #21-01/04 Gateway East, Singapore 189721, Singapore

# Preface

The relentless search of happiness by humans is sought in different ways. Scientific discussion on happiness for long was considered a forte of Philosophers. Other disciplines seldom delved into this. But today, not only science but neuroscience, marketing and other varied fields have started delving into it and have developed a keen interest. Consumers these days are seeking to buy from brands not only for what they are but also for the emotional experiences that they deliver. Today's consumer seeks happiness rather than satisfaction. They are willing to spend money to seek happiness. This is the reason why brands are willing to walk the extra mile in generating durable happiness.

For brands to sustain in the long run, 'happiness' has emerged as one of the biggest unmet challenges of the consumer world. When consumers share their stories, they are not sharing the satisfaction derived from a store but an emotional experience in a cognitive way. Happy consumers seek repeated happy interaction with the brand, make purchases and recommend it to others. It helps to promote loyalty and not to forget the revenue that it generates.

Thus, we are 'BUYING' our way to happiness. If we are buying our way to happiness, then it is important to identify the factors of happiness. Does happiness of a consumer reside in product disposability, i.e. being able to buy a product, use it and dispose it without liability or is it durability. Is a consumer happy because of the product experience or product quality? Do advertisements have to address the consumers mind or mindset to generate positive experience? Are consumers happy with product information (cognitive) or product experience (emotion)? Searching an answer for such and other questions in the quest to consumer happiness forms the content of this book.

'Keeping consumer happy' and 'understanding consumer happiness' are not one and the same. While the first raises the question of 'what', the latter addresses the 'how' of consumer happiness Most available research, empirical studies, books, marketing initiatives and more are directed towards keeping a consumer happy, and very few have made an attempt to understand consumer happiness. Initiatives undertaken to keep a consumer happy are 'momentary changes', because continuation of the same leads to habituation. On the other hand, understanding consumer happiness has an enduring effect. And a platform that combines information about the understanding of consumer happiness and measure to keep consumer happy helps in the systematic and comprehensive understanding of enduring consumer happiness. This book has been conceptualized on this line of thinking and thus divided into two parts.

The first part is customized towards understanding the multiple perspectives of happiness. The first chapter in this part is dedicated towards understanding the "Behavioural Perspective" of happiness. Is the expression and experience of happiness universal? Studies on happiness conducted over the past 20 years suggest that the definition of happiness has evolved over time. Do we seek happiness in all that we do or are there some activities that generate happiness more than others? Is there a cultural perspective to it?

The second chapter is tilted "Cross-Cultural Perspective". It addresses the question of universality vs specificity. Is everybody happy or are there differences that exist among cultures? Happiness indices suggest huge difference across societies having different cultures worldwide. While some countries rank higher in global assessment, some countries are lower down the rank. Is the difference due to collectivist vs individualistic culture? Or are there any other reason?

The third chapter in this part questions the role of technology in inducing happiness. The widely accepted belief is that technology has a negative impact on subjective wellbeing. The author questions this popular belief and triggers a critical thinking on connectivity vs isolation: Does technology play a role in generating happiness or does it make us self sufficient?

The fourth chapter discusses the relation between socioeconomic status on consumer happiness. Does socioeconomic status moderate the relation between happiness and satisfaction? Does socioeconomic status influence consumer preference and consumer happiness?

Knowledge about the different perspectives and theories has a wide range of benefits. It informs us about how the brain works, interprets and reacts. This theoretical understanding helps us to move beyond the trial and error methods towards a more scientific underpinning of adoption of measures that would generate long-lasting happiness in consumers.

The second part of the book is dedicated towards understanding consumer happiness, i.e. keeping consumer happy. This segment has ten chapters.

The first and second chapters in this part depart from the functioning of the conscious brain and try to understand the impact of subliminal messages. While one chapter is dedicated on understanding the impact of subliminal messaging on sports, the other is dedicated towards understanding the impact of subliminal messaging on consumer behaviour. What really makes us think is whether the power of subliminal is more effective than message that address out conscious processing? The power of subconscious? Which among the two generate more positive experience and how ethical it is to use messages designed for the subconscious?

The third chapter is on understanding consumer happiness from the perspective of cognitive dissonance and customer experience. The chapter is titled as "Customer Happiness: The Role of Cognitive Dissonance and Customer Experience". Creating a positive customer experience is one of the key ingredients of consumer happiness. Cognitive dissonance on the other hand generates unhappiness and disengagement. This chapter takes an unique perspective of cognitive dissonance, customer experience, their link and possible outcomes in order to enhance customer happiness.

A product needs to be heard, to be seen to be known to increase sales. And who else but the act of advertisement helps to reach out to the masses. Advertisements connect with the consumers and are an essential aspect of marketing. We address this in the fourth chapter which is "Advertising: A New Visual World (Re-conceptualisation of Advertising Through Creative Design)". The author argues for creative ways of advertising and deliberates whether celebrity endorsement works better than presentation of plain facts? Are creative experiments better than traditional ones? Do they impress upon the brain differently? Are they related to consumer happiness in any way?

In order to reach out the consumer in a meaningful way, it is important to develop an understanding about the cognitive biases which implicitly play an important role in consumer happiness and decision-making. The next two chapters, namely fifth chapter titled "Consumer Happiness and Decision Making: The Way Forward", and the sixth chapter titled "Paying a Price to Get a Value: Choose Wisely", enlist the different biases that play a role on consumer decision-making. The authors of these chapters argue that if price, scarcity, loss activate the brain and triggers certain behaviours does playing up to the cognitive bias help negate the effect of these triggers?

"Personality Metatraits, Neurocognitive Networks, and Reasoning Norms for Creative Decision-Making" is the seventh chapter of the book. This chapter deviates from the routine thought process and discusses about the creative thinking process. Creative decision-making is undoubtedly a function of individual personality traits, processed either sequentially or simultaneously at the individual cognitive level. Creative decision-making helps in the development of innovative products, and this helps to reach out the consumer in a meaningful way.

The next two chapters highlight the integration of technology in the field of consumer behaviour. While the eighth chapter is on the use of technology to develop recommender system to assist consumer decision-making, the ninth chapter describes the use of technology in creating augmented and virtual reality to enhance consumer experience and marketing strategies. These two chapters highlight the synergy between the use of technology to enhance consumer experience.

The closure of the book marks the beginning of new thoughts for research and application to enhance consumer behaviour. The last chapter of the book is titled "The Path Less Traversed: Neuroscience and Robots in Nudging Consumer Happiness". The chapter maps the future of consumer behaviour and ends with the note on how customer satisfaction can be enhanced in the days to come.

|  |  |
|---|---|
| Ranchi, India | Tanusree Dutta |
| Kharagpur, India | Manas Kumar Mandal |

# Contents

**Understanding Happiness: Different Perspectives**

Behavioral Perspective .................................................. 3
Gina Pipoli de Azambuja and Gustavo Rodríguez-Peña

The Cultural Perspective: Are Some Societies Happier Than
Others? .................................................................. 23
Vivian Eternod

Happiness Digital Technology and Social Networks ................... 43
Francisco Mochón

Socioeconomic Status and Consumer Happiness ....................... 69
Lucia Savadori and Austeja Kazemekaityte

**Consumer Happiness: Neuroscience Perspective**

Subliminal Messaging and Application in Sports: Moving Beyond
the Conscious ........................................................... 89
Özge Ercan

Subliminal Messaging: Moving Beyond Consciousness ............... 101
Ratul Sur

Customer Happiness: The Role of Cognitive Dissonance
and Customer Experience ............................................. 117
Anil V. Pillai

Advertising: A New Visual World (Re-Conceptualization
of Advertising Through Creative Design) ............................ 127
Aleksandra Krajnović

Consumer Happiness and Decision Making: The Way Forward ....... 153
Tanusree Dutta and Manas Kumar Mandal

Paying a Price to Get a Value: Choose Wisely ....................... 163
Vijay Victor and Elizabeth Dominic

**Personality Metatraits, Neurocognitive Networks, and Reasoning Norms for Creative Decision-Making** .............................. 179
Paul Hangsan Ahn and Lyn M. Van Swol

**Recommender Systems Beyond E-Commerce: Presence and Future** .... 203
Alexander Felfernig, Thi Ngoc Trang Tran, and Viet-Man Le

**An Overview of How VR/AR Applications Assist Specialists in Developing Better Consumer Behavior and Can Revolutionize Our Life** ........................................................... 231
Rocsana Bucea-Manea-Țoniş, Elena Gurgu, Oliva Maria Dourado Martins, and Violeta Elena Simion

**The Path Less Traversed: Neuroscience and Robots in Nudging Consumer Happiness** ............................................. 255
Elena Gurgu and Rocsana Bucea-Manea-Țoniş

**Author Index** ...................................................... 285

# Editors and Contributors

## About the Editors

**Dr. Tanusree Dutta** is presently working in Indian Institute of Management, Ranchi. She was associated with Rekhi Centre of Excellence for the Science of Happiness, Indian Institute of Technology, Kharagpur (on lien from Indian Institute of Management Ranchi). She has research publications in peer-reviewed national and international journals which have been cited in manuscripts and books of national and international repute. She has co-authored a book titled 'Neuromarketing in India: Understanding the Indian Consumer' and has co-edited a book titled 'Bias in Human Behavior'.

**Dr. Manas Kumar Mandal** is currently serving as Distinguished Visiting Professor at Indian Institute of Technology - Kharagpur. He is also Adjunct Professor at the National Institute of Advanced Studies, IISC, Bangalore. He was formerly Distinguished Scientist and Director-General – Life Sciences in DRDO [2013–2016]. Prior to this, he was Chief Controller R&D (Life Sciences), DRDO. He was also Director, Defense Institute of Psychological Research, for about nine years. During 2003, he was Fulbright Visiting Lecturer at Harvard University, USA.

## Contributors

**Paul Hangsan Ahn** Department of Communication Arts, University of Wisconsin-Madison, Madison, WI, USA

**Rocsana Bucea-Manea-Țoniș** Doctoral School, University of Physical Education and Sports, Bucharest, Romania

**Gina Pipoli de Azambuja** Faculty of Business Science, Universidad del Pacífico, Lima, Perú;
Business Sciences, United Partners for Excellence in Research (UPER), Lima, Perú

**Elizabeth Dominic** Saintgits Institute of Management, Kerala, India

**Tanusree Dutta** Indian Institute of Management, Ranchi, India

**Özge Ercan** Faculty of Sports Sciences, Sinop University, Sinop, Turkey

**Vivian Eternod** Business School, Tecnologico de Monterrey, San Pedro Garza García, Mexico

**Alexander Felfernig** Institute of Software Technology, Applied Artificial Intelligence, TU Graz, Graz, Austria

**Elena Gurgu** Department of Economic Science, Spiru Haret University, Bucharest, Romania;
Springer Nature Singapore Pte Ltd., Singapore, Singapore

**Austeja Kazemekaityte** University of Trento, Trento, Italy

**Aleksandra Krajnović** Department of Economics and Management, University of Zadar, Zadar, Croatia

**Viet-Man Le** Institute of Software Technology, Applied Artificial Intelligence, TU Graz, Graz, Austria

**Manas Kumar Mandal** Dept of Humanities and Social Sciences, Indian Institute of Technology, Kharagpur, India

**Oliva Maria Dourado Martins** Department of Business Science, Institute Polytechnic of Tomar, Tomar, Portugal

**Francisco Mochón** UNED, Madrid, Spain

**Anil V. Pillai** Pune, India

**Gustavo Rodríguez-Peña** Business Sciences, United Partners for Excellence in Research (UPER), Lima, Perú

**Lucia Savadori** Department of Economics and Management, University of Trento, Trento, Italy

**Violeta Elena Simion** Department of Veterinary Medicine, Spiru Haret University, Bucharest, Romania

**Ratul Sur** University of Kalyani, Kalyani, India

**Thi Ngoc Trang Tran** Institute of Software Technology, Applied Artificial Intelligence, TU Graz, Graz, Austria

**Lyn M. Van Swol** Department of Communication Arts, University of Wisconsin-Madison, Madison, WI, USA

**Vijay Victor** Department of Economics, CHRIST (Deemed to be University), Bangalore, Karnataka, India

# Understanding Happiness: Different Perspectives

# Behavioral Perspective

**Gina Pipoli de Azambuja and Gustavo Rodríguez-Peña**

**Abstract** The objective of this chapter is to understand happiness and identify the factors that make consumers happy. The literature review examined studies on the happiness of different parts of the world in the last 20 years. The research method was the literature review, and the approach was a theoretical research study. The main findings of the chapter are presented: Happiness has many definitions from ancient Greeks like Aristotle who considered happiness to be the cornerstone of a good life, a life well-lived. During the last two decades, various definitions of happiness have emerged. In this context, recent studies have found that there are four main activities that can make happy humans: (1) physical activity, (2) social experiences, (3) tourism, and (4) altruism. However, the concept of happiness can vary according to culture. For example, Americans define happiness in terms of pleasure whereas Asian culture in terms of marital status and standard of living. The chapter examines happiness approaches and highlights the special case of Latin America, to illustrate, it was measure subjective well-being and live satisfaction in a Peruvian sample.

## Introduction

The human race wants to be happy (Diener, 2000), and we pursue happiness in different ways, for example: increase our incomes, have successful careers to achieve prestige and money, purchase luxury cars, travel to exotic destinies, and others. However, are we happy? Scientific evidence has demonstrated that tactics based on changing life circumstances like career development and incomes not necessarily imply greater well-being (Kristin & Lyubomirsky, 2014).

Happiness has begun to study by psychologist and economist since represent an agenda issue in the world. First, according to World Health Organization (2017), it

---

G. P. de Azambuja (✉)
Faculty of Business Science, Universidad del Pacífico, Lima, Perú
e-mail: pipoli_gm@up.edu.pe

G. P. de Azambuja · G. Rodríguez-Peña
Business Sciences, United Partners for Excellence in Research (UPER), Lima, Perú

© The Author(s), under exclusive license to Springer Nature Singapore Pte Ltd. 2021
T. Dutta and M. K. Mandal (eds.), *Consumer Happiness: Multiple Perspectives*,
Studies in Rhythm Engineering,
https://doi.org/10.1007/978-981-33-6374-8_4

is estimated that 300 million or 4.4% of the world's population suffering depression, and it is expected that depression will be the leading cause of disability in 2030. This way, some psychologists are studying the causes and solutions to prevent and combat depression: The understanding of feelings such as happiness can provide critical solutions against depression. Second, some economists are focusing on studying the well-being in societies, according to Layard (2006), the theory behind public economics needs a radical reform: Human well-being should be the cornerstone of public policies. Consequently, the economist and psychologist are studying the role of happiness in our civilization. When did happiness begin to have a more significant role?

After World War II, there was an increase in per-capita income in several countries around the world. However, human happiness measured by several techniques had not increased. In consequence, several economists and psychologists paid attention to happiness in the world. Mainly, Easterlin (1973, 1974) was one of the leading scholars who studied the relationship between subjective well-being and material living standard. The author found that there is a trend that humans with higher income are happier than others.

Notably, economists have been paid attention to happiness since the Easterlin paradox can suppose public policy implication: Can economic growth promote happiness in current society? This issue represented one of the main topics about happiness literature. Hence, economic perspective and interest on the issue of happiness are increasing. For example, some researchers such as Oswald (1997) studied the relation between well-being/utility with some macroeconomic variables like unemployment, inflation, and per-capita income.

Simultaneously, several psychologists have started to study human feelings such as well-being. According to Feldman (2010), many psychologists have studied happiness from a medical perspective: They research the causes and cures of pathologies related to human happiness. In consequence, happiness is not a trivial issue nowadays, and definitely, it started to be a new issue in the academy.

## What is Happiness?

Usually, happiness is understood by ordinary people as the individual perception about our quality of life or life satisfaction degree. Scholars have paid interest to understand what is happiness? There is no total consensus about what happiness is? Indeed, happiness has been a discussion topic since the ancient Greek times.

Aristotle was one of the pioneers to define happiness. Notably, he believed that happiness is the primary human purpose and defined it as well-lived life, which means realizing one's full potential since most people will not realize it (Uchida, Norasakkunkit, & Kitayama, 2004a, 2004b). Hence, Aristotle's happiness concept is not defined as an emotion but a measure of a well-lived life. Similarly, Aristotle highlighted a virtuous life as a happy life. For example, Aristotle said that one of the human virtues is a magnificence that remits to be generous with others to be happy.

Consequently, Aristotle agrees that happiness depends on ourselves and represents a measure of our entire life and how well we lived it (Kraut, 1979).

Kant has a different happiness concept since he defines it as the satisfaction of all our inclinations. However, Kant thought that the more we try to be happy, the more unhappy we will be. He highlighted that sometimes we think that some activities or things can make us happy. However, sometimes we just disappointed. For this reason, one of the main essential conclusions about Kant's happiness perception is that we should not obsess about happiness, but we try to be a good person instead: When we do the right thing, happiness will naturally follow (Guyer, 2000).

Another approach was performed by Socrates, who perceived happiness as the pleasure that we experiment; it implies that happiness does not come from external factors. Consequently, according to Socrates, our happiness should not be based on external things but on how material things are used in our lives (Urstad, 2007). For example: using money to help others, have a professional career to make a better world, and others.

Nietzsche undertook a complete happiness definition: Happiness is a kind of power that humans exert over the world. Hence, we get the power, we feel happy, and when we feel unhappy, we need to attempt to take back control. We can show such power when we moved to the city we want, work at a job we enjoy, spend time with the person that we want, and others. This happiness approach is based on the idea of using our freedom and ability to decide the life that we want (Bett, 2005).

Late seventeenth and eighteenth centuries, it arises a religious perspective of happiness. In this sense, Thomas Aquino represents a unique and subjective happiness definition: He perceives happiness as the result of the relation between humans and God. Aquino thought that happiness is like being drunk on God or have God's grace: He has pleasure as feeling to some degree (Campodonico, 2011).

We highlight that these philosophers had a concept of happiness based on their own experiences, society beliefs, and perceptions. However, it does not mean that their definitions are wrong. Gilbert (2006) suggests that the happiness concept can be subjective and idiosyncratic. In this context, is the concept of the philosophers as mentioned earlier different from modern and contemporary happiness scholars? Below are the definitions of researchers about happiness.

Bentham (1789) conceptualized happiness as the result of the sum of pleasures and pains in our lives. Pleasure denotes enjoyable experiences during pain, aversive life events. Hence, happiness is an overall evaluation of our kind and bad experiences in our lives. However, the author does not deepen the analysis of the specific causes of happiness and how it works in our minds.

Such issues were examined by Schachtel (1954), who proposed that *"happiness is not having what you want, but wanting what you have"*. This perspective stated that happiness is the result of appreciation and value with such things, friends, relations, and others that we have. In this sense, the author proposed that attitudes are more important than facts, and attitudes are shaped by mental and spiritual health: Both are the primal needs of human life. This way, happiness can be considered a decision that is based on the appreciation of our conditions. Also, happiness is considered as the result of the right mental health.

Later, Kitayama et al. (1995, 2000) conceptualized happiness as a positive emotional state independently of the culture, country, society, and circumstances. This point of view highlights the internal factors as key to achieving happiness similarly as postulated Schachtel (1954).

A comparable perspective of happiness is stated by Deci and Ryan (2000), who argue that well-being is the result of the satisfaction of basic human needs such as have food, be healthy, and others like being autonomous and competent. This way, this perspective refers to internal aspects (such as our needs as humans) as Kitayama et al. (1995, 2000). Such ideas correspond to the Self-determination Theory that postulates that happiness can be achieved if we meet our goals and needs.

A modern happiness perspective is leading by Lyubomirsky (2008). She describes happiness as *"the experience of joy, contentment, or positive well-being, combined with a sense that one's life is good, meaningful, and worthwhile."* Hence, modern happiness concepts refer to experiences and satisfaction. Kesebir and Diener (2008, p. 118) proposed that happiness primarily refers to a subjective well-being, as they stated: *"people's evaluations of their lives and encompasses both cognitive judgments of satisfaction and affective appraisals of moods and emotions"* (Kesebir & Diener, 2008, p. 118).

Later, Feldman (2010) proposed that happiness is mainly a psychological state of some mental state. The author proposes that happiness is the result of the human attitude towards life, similarly as a happiness perspective of Schachtel (1954). For example: If you had a bad day, you choose the attitude towards such a day, then you could be happy; however, it depends on you. However, Feldman (2010) argues that most people think that happiness means living well, not necessarily as a psychological state. This way, Feldman focuses more on the human attitude and psychological condition as a basis of happiness than Lyubomirsky (2008) and Kesebir and Diener (2008), who focus more on satisfaction.

What says the neuroscience about happiness? Neuroscientists have researched human pleasure since it is a happiness dimension and represents a critical factor in our well-being sense. Neuroscientists argue that pleasure is not just a sensation or thought, but a result of brain activity in a system denominated "hedonic systems". In fact, scholars have found that all the pleasures seem to involve the same brain system: fundamental pleasures such as food and sex, and altruists (Kringelbach y Berridge, 2010).

Another happiness research approach has focused on study what means happiness to persons. Mogilner, Kamvar, and Aaker (2010) studied 12 million personal blogs that aim to capture the meaning of happiness. The authors examined 12 million sentences posted on blogs that contained the words "I feel" or "I am feeling" and identified a total of 6,302 sentences expressed feeling happy and others. They found that in their youth, people are more likely to relate happiness with excitement, and as they get older, they become more likely to associate happiness with peacefulness.

Delle, Brdar, Freire, and Wissing (2010) studied happiness, meaning in different countries from Europe, Africa, and Oceania. One of their main findings is that happiness is primarily defined as a psychological balance and harmony. Also, words such as family and social relations were mainly associated with happiness and meaningfulness.

In consequence, happiness has many meanings since ancient times. This concept has evolved continuously in the world. It is possible to say that most of the experts define happiness as well-being or subjective well-being, and satisfaction is commonly used to conceptualize happiness. Such definitions can be considered mutual perceptions about happiness. We highlight that contemporary society relates happiness with is pleasure and satisfaction. For example, experiences like eating a tasty dish, listen to favorite music, travel experiences, and others. According to the authors as mentioned earlier, the key to happiness can exploit our full potential, do not too focused on the happiness pursuit, control our own lives, use material things to our well-being, and seek satisfaction in our life.

## Factors that Generate Happiness

In this section, we present the factors that can generate happiness. Mainly, it is analyzed the factors that have been researched and proved by scholars from different parts of the world. A common argument from skeptics is that people's happiness levels are genetically determined and cannot be substantively changed. Despite persuasive evidence from twin studies that well-being levels are strongly influenced by genetics (Lykken & Tellegen, 1996), more recent research suggests that genetic influences on happiness might be weaker than initially thought, with environmental influences explaining a large portion of individual differences in happiness (Røysamb et al. 2003; Stubbe, Posthuma, Boomsma, & de Geus, 2005). Furthermore, research on gene-environment interactions suggests that biology and the environment continuously interact in a dynamic process to influence people's behavior, so even genetic predispositions are not deterministic (Plomin & Spinath, 2004).

Although individual differences in biology and circumstances combine to explain part of the happiness puzzle, a gap exists in the explained variance (Lyubomirsky, Sheldon, & Schkade 2005). For example, people's happiness levels have been found to shift over time. In a 17-year longitudinal study, 24% of participants showed substantive changes in their happiness over time, with 9% changing by over two standard deviations (Fujita & Diener, 2005). Theory and research supporting hedonic adaptation suggest that even substantial changes in people's life circumstances (e.g., marriages, divorces, job layoffs, Luhmann, Hofman, Eid, & Lucas, 2012) do not lead to such sustained shift s in well-being. We argue instead that these observed changes can be explained in part by deliberate ways that people choose to think and behave in their daily lives. How and why these practices and habits can "work" to shift well-being has been the focus of our laboratory's efforts for over a decade (Kristin & Lyubomirsky, 2014).

## Physical Activity

In 2018, Word Health Organization (2018) recognized the critical role of physical activity, such as sports, to improve public health and social well-being. Also, several authors have studied the relationship between physical activity and happiness. There is not a complete consensus about the physical activity-effet on well-being. However, most researchers have shown that physical activity has a positive effect on well-being.

On the one hand, Rasciute and Downward (2010) concluded that physical activities like cycling and walking have a positive effect on well-being while Wang et al. (2012) determined that the higher leisure-time physical activity, the lower levels of unhappiness after two years and four years. Similar results have been found in different target populations. The study of de Souto (2014) and Khazaee-pool, Sadeghi, Majlessi, and Rahimi Foroushani (2014) demonstrated a positive relationship between physical activity and happiness in the case of older adults and Thirdi and Jowkar (2011), in the case of postmenopausal women. Notably, Nelson et al. (2007) suggested doing physical activity 30 min moderate-intensity on five days/week or at least 20 min on three days/week to improve happiness.

Some explanations of such results are that happiness is a positive construct of mental health, which increases resilience to emotional stress or troubles. Then, when physical activity is performed, it positively affects the human factor, which promotes well-being or happiness (Richards et al., 2015). Scientific evidence has demonstrated that physical activity can work as a therapy and prevention of diseases: reduce the risk of several chronic diseases (Warburton et al., 2006), functional declines (Stuck et al., 1999), and can reduce the osteo-articulatory chronic pain (Roddy et al., 2005), menopausal complications (Thirdi & Jowkar, 2011), and others. A different research approach was performed by Rasmussen and Laumann (2014), who studied the long-term effect of physical activity on happiness. Their results show that a higher frequency of physical activity during adolescence implies higher happiness levels in adults.

On the other hand, few researchers have found different results as before presented. In 2000, Mack et al. (2000) found that physical activity does not change the happiness of undergraduate students significantly. Later, Baruth et al. (2011) found that physical activity can improve happiness in the case of adult men; however, not in adult women. Also, Richards et al. (2015) found a weak relation between physical activity and happiness.

There are a few studies that concluded that physical activity could be irrelevant to improve happiness. However, there is a full acceptance of the positive effects of physical activity on happiness. It was shown in different samples like older adults, adolescents, children, and others. Consequently, science shows that physical activity can be a factor in improving happiness.

## *Social Experiences*

During the last 30 years, the happiness of US citizens is declining, on average, in this context, Bartolini, Bilancini, and Pugno (2013) examined the role of social connections and other variables to explain such happiness decreases. Their results demonstrated that the decline of social connections in Americans could explain their happiness decrease. This example can reflect that social experiences are a crucial factor in promoting happiness.

Even more, happiness and social experiences have been studied from different perspectives. First, it was widely studied the effect of social interactions on happiness according to different variables. Phillips (1967) indicated that social interactions exert a positive influence on happiness, even under control for age, social class, and sex. Similarly, according to (Demır & Weitekamp, 2007), social experiences can be reflected in friendships, since friendship quality has a significant and positive effect on happiness above and beyond the personality and number of friends. Also, Hills and Argyle (1998) concluded that the social aspects of religious and musical experiences have a positive relationship with happiness.

Second, from the medicine and psychological perspective, the studies of Broadhead et al. (1983) have discovered that several diseases like depression, anxiety, cancer, and others, can be combated and prevented through social experiences. Also, de Camargos, Ribeiro, Leite de Almeida, and Paiva (2019) concluded that social experiences have a crucial role in generating happiness in both healthy people and cancer survivors.

Third, multicultural approaches were performed by Gallup World Poll (2017) and Shin, Suh, Eom, and Kim (2017). On the one hand, Gallup World Poll studied the relationship between social support and well-being: It was found that in the four highest-ranking countries for life satisfaction (Denmark, Finland, Norway, and the Netherlands), 95% of cases perceived that they had persons who can support in bad times (Diener, Helliwell, & Kahneman, 2010). On the contrary, the authors found that only 55 percent of the studied sample from the lowest-ranking countries (Togo, Burundi, Sierra Leone, and Zimbabwe) have social support (Kristin & Lyubomirsky, 2014).

On the other hand, Shin et al. (2017) showed that individuals of their studied sample (cases from Korea and the USA) who mentioned more common words were more satisfied with their lives. In consequence, common words are well perceived regardless of the culture. It represents a reasonable clue that social experience can contribute to happiness. Fifth, the contribution of Holder and Coleman (2009) was based on the study of the role of social experience in children's happiness and concluded that higher social experiences, higher happiness.

Fifth, a philosophy perspective was carried out by Aaker, Rudd, and Mogilner (2011) who proposed five principles for happiness-maximizing ways to spend time. The first one was related to social experiences: Spend your time with the right people. Authors indicate that social leisure activities generate more happiness than solitary ones, and spend time with significant persons like friends and family is also vital to improve happiness. Similar results have found by authors like Reyes-Garcia et al. (2009) and Lloyd and Auld (2002).

Empirical evidence of other authors like Caprariello and Reis (2013), Diener et al. (1999), Epley and Schroeder (2014) have found that social experiences can foster happiness in humans.

## *Tourism*

The tourism industry is a strategic sector of several countries around the world since it involves several sectors like transportation, accommodation, recreation and entertainment, and others. Indeed, tourism can imply economic growth and development (Seetanah & Fauzel, 2018). Rátz and Michalkó (2011) recognized that tourism is a crucial element of quality of life. In this sense, the role of tourism in human well-being represents a growing research issue. It is widely accepted the benefits of tourism in happiness. Notably, the experiment Nawijn (2012) concluded that vacationers' life satisfaction scores were significantly higher than non-vacationers' life satisfaction scores (Nawijn, 2012).

There is some explanation of this phenomenon. Lent et al. (2005) highlight that tourism improves several life domains; this way, it can increase life satisfaction. McCabe and Johnson (2013) suggested that tourism can promote happiness through leisure and family life domains and Corvo (2011) that tourism can represent the satisfaction of needs such as break the routine. Another explanation corresponds to medical tourism analysis since it is a tourism kind that is directly associated with health and well-being (Connell, 2006). For example, tourism activities like a spa can reduce stress and promote well-being.

Many tourism-linked factors have been analyzed to enhance levels of well-being. However, according to Lyubomirsky, Sheldon, & Schkade (2005), approximately 50% of an individual's happiness is predetermined through heredity, circumstances determine 10%, and 40% is affected by intentional activity. This theory makes it possible for the individual to increase happiness through tourism activity, although just 40% of happiness can control.

A different perspective studied holidays and happiness, and the research showed that spend time with family (46.8%), traveling together as a couple (36.1%), and traveling together with the family (25.7%) represent the three main activities that produce more happiness (Rátz & Michalkó, 2011). Other authors focused on nature in tourism since tourists tend to visit places with strong links with nature since it can involve outdoor activities, play sports, contemplate nature, and others (Corvo, 2011). Consequently, nature-based tourism can foster happiness in humans. Notably, Bimonte and Faralla (2014) found similar results and indicated that their values and ideologies could moderate the effect on happiness.

However, Nawijn (2010) studied the tourism effect on happiness in the long run and found that people who are doing tourism activities are happier. The author demonstrated that memories about vacationing not necessarily have prevalence in people's lives in the long term. Indeed, the effect of tourism activities on happiness is mostly short-lived. Nevertheless, happiness can affect the long term if the travels are repeated often enough (Puczkó, 2012), and memories about the travels can extend the effect on happiness (Liu, 2013).

Also, Rátz and Michalkó (2011) identified that age and financial resources could moderate the effect of tourism on happiness. Similarly, Milman (1998) indicated that happiness could depend on the activity level during the trip: Travelers who perform more activities were much happier. Examples of activities are sightseeing, restaurant dining, shopping, and others.

This way, the studied relation can be defined by both variables. This way, tourism effect on happiness is not necessarily durable in the long-term, and it can depend on age and financial capability.

## *Altruism*

Altruism is the selfless willingness to make voluntary transfers of resources such as money or food to another human. Notably, one of the main characteristics of altruism is the concern for the welfare of the other and not for future rewards. Altruism has been associated with higher well-being. Midlarsky and Kahana (1994) were one of the initial scholars who studied altruism and well-being. They indicated that altruism could improve well-being, self-esteem, morale, and positive affect. Similar results have been found since the altruism-effect on well-being has been demonstrated in adults (Omato y Snyder, 1995), children (Aknin, Kiley, & Dunn, 2012), multicultural contexts (Dunn, Aknin, & Norton, 2008), and others.

Altruism is so beneficial for human well-being, as stated by Post (2005) who concluded that altruism activities foster well-being, health, and longevity when the persons are emotionally compassionate, and they are not overwhelmed by helping tasks. Another effect can be higher life satisfaction, self-esteem, perception of control over life, and others (Choi & Kim, 2011).

A significant contribution was made by Dunn et al. (2008), Appau and Awaworyi (2018) and Lok and Dunn (2020). Authors stated that altruism-effect could vary according to different contexts or factors. Dunn et al. (2008) indicated that altruism could be predicted by how persons spend their money. If persons spend their money more on other people than themselves, they will be happier than others.

Appau and Awaworyi (2018) suggested that the level of altruism-effect depends on the type of charity or volunteering, for example, local or international. Notably, charity to near people (local) produced higher well-being. The authors highlight that if the charity is performed in informal ways, it generates higher well-being, in comparison if the charity would be performed through formal ways. Also, Lok and Dunn (2020) identified three factors that can impact on altruism-effect: (1) ability to perceive the effect of their generosity, (2) social connection with the cause, (3) feeling of freedom to support other persons. Such factors can represent the mechanisms in which altruism makes happy people.

Other variables have been studied to generate happiness or well-being. One of them is age. Another factor to consider is age. That is, might certain age groups gain more from the practice of positive activities? A meta-analysis provided evidence that older participants benefit more from positive activities than younger ones (Sin & Lyubomirsky, 2009). This finding was limited, however, by the small number of studies included in the youngest (less than 17 years old) and oldest (60 years old or more) age ranges. Hence, what the age moderator analysis showed was that young adults (18 to 35 years old) gained less from positive interventions than middle adults (36–59 years old). Furthermore, because the middle adult samples were more likely than the young adult ones to self-select into their respective positive interventions, age was confounded with self-selection.

In sum, the evidence is emerging that positive activities can benefit youth populations. Studies with even younger children are needed, however, to identify the youngest age range for which specific positive activities will still be efficacious. Also, although positive interventions that prompt elderly participants to reflect on their lives within the context of group and individual therapy (e.g., life review therapy and reminiscence) have been found to increase life satisfaction and alleviate depressive symptoms, to our knowledge, self-administered positive activities have not been tested individually among older adults. Interestingly, subjective well-being increases from early adulthood to old age, and this shift is likely to occur because older people are thought to be emotionally wiser—that is, as they increasingly view their time as limited, they compose their social networks, regulate their emotions, and make meaning of their life events in ways that increase and sustain happiness (Kristin & Lyubomirsky, 2014).

## *Other Factors that Promote Happiness*

Other factors can promote happiness. There is no enough quite evidence to understand how these factors can work in happiness. For example, according to Mookerjee y Beron (2005), religion and gender are key factors to promote well-being. Also, Michalos (2008) highlighted the role of education in happiness. Notably, the use of social networks represents a new research issue in happiness studies, as stated by Utz and Beukeboom (2011). Another happiness factors are beauty (Hamermesh & Abrevaya, 2013), age, race (Philipp, 2000), and others. Additionally, there are positive activity interventions that are medical and therapy recommended to improve well-being. For example, it is recommended to writing letters of gratitude, practice optimism, meditate and feel positive emotions to self and others. It is demonstrated that such activities increase the levels of happiness, and even it can decrease depression symptoms (Layous & Lyubomirsky, 2014).

## Culture and Happiness

Is happiness universal or differs depending on the culture? According to Ryff and Keyes (1995), is likely that happiness being universal and similarly valued across cultures. However, it is possible to affirm that humans from everywhere tend to prefer the desirable over the undesirable well-being (Diener, Diener, & Diener, 1995). Nevertheless, it not implies that happiness meaning is the same across cultural variations. Recent evidence suggests that there are three differences in which happiness can vary across cultures: (1) cultural meanings of happiness, (2) motivations underlying happiness, and (3) correlates of happiness.

According to Uchida et al. (2004a, 2004b), emotions are based on culturally shared ideas like sadness and happiness. Mainly, the authors differ in European-American and East Asian regions as two primary cultures that correspond to different perspectives about happiness. There may be within-regional variations on each culture; however, such two regions have historically shared ideas, values, practices, and social institutions.

### *European and North American Region*

European and North American cultures are characterized to have a strong belief in independence and autonomy of the self. Additionally, the self is the basis of their live assessment since the self is understood as the center of thought, action, and motivation. Hence, happiness in this culture is built on positive attributes of the self and can be interpreted as personal attributes that should be pursued and achieved through personal striving: Personal achievement is the basis of happiness. There is scientific evidence that personal accomplishment is associated with happiness, as Emmons (1986) demonstrated.

Similarly, happiness depends on self-esteem. The experiment of Kwan et al. (1997) demonstrates self-esteem was the only predictor of life satisfaction in U.S. culture. In contrast, the authors found that social harmony is the cornerstone of life satisfaction in the Hong Kong region. It does not mean that social relations are not crucial to European-American people, but social relations are a personal choice and form part of the independence of each one. In this context, European and North American persons seek to maximize the experience. Also, Campbell (1981) found a positive association between happiness and self-esteem in North American cultures.

According to Uchida et al. (2004a, 2004b), such a cultural perspective about happiness has been built from history, ideologies and religious ideas. For example, the Protestant worldview shows that humans are predestined to be either selected or doomed. In consequence, the belief in predestination arises, and it generates the desire to be competent and worthy as citizen, father, son, worker, and others. Notably, the Protestant perspective historically has influenced to achieve happiness through personal achievement which is based on the idea of the personal self.

These ideas are broadly dispersed in European-American cultural contexts. It can be shown through different practices and routines like discourses narratives and institutions in which the self and personal achievement is the basis. Notably, empirical evidence has shown that Americans are characterized to actively seek positive emotional experiences in comparison with persons from Asian culture. For example, Americans are highly interested in practice sports and others activities that produce excitement.

## *East Asian Region*

Shin et al. (2017) researched what three words come in a person's mind in association with happiness. The authors found that *family* is the main word that is associated with happiness in the Korean population. Additionally, common words such as relationships and social emotions were frequently found in the Korean population. A different result was found in Americans since smile was the leading word associated with happiness.

A similar result has been widely confirmed by several authors like Suh et al. (1998) who argue that East Asian culture is affected more by social appraisal than inner feelings and Oishi et al. (2008) who affirm that happiness is a social process based on being understood by others. Complementary, Kitayama and Markus (2000) define happiness in East Asian culture as an intersubjective experience that necessarily involves others, such as family members.

In consequence, humans from the East Asian region build their happiness in the measure that they strengthen their current social relations and enrich their social network. East Asian people adjust their behavior in order to fit with others since social validation is value for them. Cialdini and Goldstein (2004) affirm that social validation is a principle related to social norms, in which people observe the behavior of others to decide how to behave, given that they consider that an action is more appropriate when they observe that others react in the same way to a particular situation. Notably, the commitment to social roles and the importance of social obligations configure the right behavior that contributes to well-being (Morling et al., 2002; Weisz et al., 1984).

Indeed, they tend to give social support to others that allow building their happiness (Aknin et al., 2013; Dunn et al., 2008; Thoits & Hewitt, 2001). Also, happiness is embedded in interdependence and connectedness of self with others. Consequently, happiness in the East Asia region depends on social relationship quality and social harmony (Kitayama & Markus, 2000). Reasonably, social relations and trust build represent the locus of thought, action, and motivation. According to Uchida et al. (2004a, 2004b), the interdependence of self and others is so high that the boundary between self and others is blurred. Examples of interdependence can be shown in values like sympathy, compassion, and support.

Opposite to European and North American culture, the pursuit of personal happiness in the East Asia region can damage social relationship quality. For example, professionals who actively seek to move up from a job can generate jealousy by colleagues, which damages social relations. Hence, personal happiness is perceived as imperfect happiness, and it is not desired in such a culture.

Religions and spiritual approaches like Confucianism, Taoism, and Buddhism can explain happiness perception in East Asian culture. Such religions propose that reality is holistic, and everything is connected: Humans are interdependent, and if they seek personal good, there will be social problems (Kitayama & Markus, 1999; Suh, 2002).

The preceding analysis offers some critical implications regarding the nature of happiness in different cultural contexts. Specifically, recent evidence suggests that there are systematic cross-cultural differences in three domains, namely: (1) cultural meanings of happiness, (2) motivations underlying happiness, and (3) correlates of happiness. (Uchida et al., 2004a,2004b).

Second, members of holistic cultures are more likely to see experiences and states to be ever-changing. If internal states are seen to be in constant flux, this implies futility in maximizing positive experiences: Positive and negative experiences will be seen as occurring (or reoccurring) in succession, so the occurrence of one implies the impending occurrence of the other. Finally, members of holistic cultures are more likely to construe the self in a way that is contextually bound and embedded within a social network. To the extent that one has a sense of self that is intrinsically tied to others, a maximalist prioritization of one's health, happiness, pleasure, and self-esteem could be construed as immature and hubristic (Hornsey et al., 2018).

One relevant factor to consider is whether individuals from Eastern cultures would even want to actively pursue their happiness. If personal happiness is not valued in their culture, then its members may not be motivated to become happier or to put in the requisite effort to do so. Indeed, culture has been found to influence people's ideal affect (i.e., how happy they would like to be) more than their actual affect (i.e., how happy they feel right now). For example, Chinese individuals value high-arousal positive emotions less (and value low-arousal positive emotions more) than do Americans (Tsai, Knutson, & Fung, 2006). If the deliberate pursuit of happiness is viewed as a chase for high-arousal positive emotions, members of Eastern cultures may not be interested in happiness-seeking strategies.

## *Latin America*

Happiness in Latin America is a complex phenomenon. Latin America is a region with particular features: less economic welfare and human development than Western Europe and America, inefficient public education and health system, weak political institutions, high corruption and violence levels, and other similar characteristics (Medici & Lewis, 2019; Ramos & Moriconi, 2018; Torres, 2019). However, Latin America is one of the happier regions in comparison with others (World Happiness Report, 2018). How can we understand this phenomenon?

This scenario can be explained by the role of family and social relations, as demonstrated by Rojas (2019) and Beytía (2016). Empirical evidence has demonstrated that the near ties with family, social capital relations, and person-based interpersonal relationships can foster happiness. Notably, Beytía (2016) found that higher density and quality of the family have a significant effect on subjective well-being in the region. Also, Latin America is a collectivistic culture (Hofstede, 1980), social networks, family, friends, and community are a vital happiness factor. In contrast, Scandinavian countries, which are characterized to be individualistic culture and wealthy countries with excellent education and public health system, their citizen is suffering higher rates of mental illness like depression, suicide, and low levels of happiness (Yamamoto, 2015).

Social relations quality and networks do not represent the unique, relevant variable to understand happiness in Latin America. Nevertheless, Graham and Felton (2006) argue and highlight the relative income difference as a critical variable in the happiness in Latin-American citizens. Authors found that it can negatively affect happiness in Latin-America countries. Economic inequality has a role: the higher income gap between citizens, less happiness they are.

We measured happiness in the sample from Peru to examine youngsters' behavior and happiness. We used two sources: (i) the five-item Satisfaction With Life Scale developed by Kahneman, Diener, and Schwarz (1999) and (ii) eight-item Flourishing Scale proposed by Diener, Wirtz, Tov, and Kim-Prieto (2010).

First, the five-item Satisfaction with Life Scale analyzes happiness as the overall sense that life is good and meaningful. It measures the cognitive judgments of satisfaction with life. This scale addresses social relationships. Items, like rewarding relationships and contributing to other's happiness, build this scale. Also, it includes the perception of having a meaningful life and purposeful. Notably, five-item Satisfaction with Life Scale can measure well-being.

Second, the eight-item Flourishing Scale measures social-psychological prosperity. Prosperity perception can imply analyzing competence, self-acceptance, and other features that capture significant aspects of personnel prosperity. It is essential to examine prosperity since this happiness dimension was not widely examined in the Latin America region. This research gap is addressed in the present chapter. Authors such as Ryff (1989), Ryff and Singer (1998), and Ryan and Deci (2000) argue that prosperity is key to personal happiness.

It was analyzed 500 cases that correspond to Peruvian youngsters between 17 and 24 years old. The non-probabilistic method was performed. Table 1 shows the five-item Satisfaction with Life Scale application in the Peruvian sample. It was used a scale from 1 to 7 where 1 totally disagrees and 7 totally agree. Results show that the studied sample widely perceives that they are competent and capable in the activities that are important to them (average of 5.77/7). Second, they feel that their social relationships are supportive and rewarding (average of 5.66/7). Third, they perceive that leading a purposeful and meaningful life (average of 5.51/7). In general, the score of the five items is higher than five that suggests that they have high levels of life satisfaction.

**Table 1** Five-item satisfaction with life scale application in a Peruvian sample

| Item | Average | Standard deviation |
|---|---|---|
| I lead a purposeful and meaningful life | 5.51 | 1.57 |
| My social relationships are supportive and rewarding | 5.66 | 1.44 |
| I am engaged and interested in my daily activities | 5.49 | 1.44 |
| I actively contribute to the happiness and well-being of others | 5.36 | 1.47 |
| I am competent and capable in the activities that are important to me | 5.77 | 1.37 |
| General average | 5.56 | 1.46 |

Self-elaboration (2020)

**Table 2** Eight-item flourishing scale application in a Peruvian sample

| Item | Average | Deviation standard |
|---|---|---|
| I am a good person and live a good life | 5.86 | 1.42 |
| People respect me | 5.91 | 1.32 |
| I am optimistic about my future | 5.76 | 1.51 |
| In most ways my life is close to my ideal | 5.32 | 1.49 |
| The conditions of my life are excellent | 5.49 | 1.43 |
| I am satisfied with my life | 5.39 | 1.56 |
| So far I have gotten the important things I want in life | 5.23 | 1.56 |
| If I could live my life over, I would change almost nothing | 5.08 | 1.73 |
| General average | 5.50 | 1.50 |

Self-elaboration (2020)

Table 2 shows the eight-item Flourishing Scale application in the Peruvian sample. It shows that the studied sample widely perceives that others respect their (average of 5.91/7). Similarly, they feel that they are a right person and live a good life (average of 5.86/7). Third, Peruvian youngster feels optimistic about their future (average of 5.76/7). All eight items have a score higher than five that suggest high levels of subjective well-being.

Results show high levels of life satisfaction and subjective well-being. It suggests that, on average, studied Peruvian youngsters are happy. This empirical evidence corroborates that Latin-American citizen has high levels of happiness (World Happiness Report, 2018). Also, this finding suggests that examined Peruvian youngsters have high levels of life satisfaction and subjective well-being, and key drivers are the rewarding and supportive social relations and the feeling that they are competent persons and have a good life.

## Conclusions

The happiness concept has evolved through ancient times. Greek philosophers were the pioneers to define happiness widely. Remarkably, Aristotle indicated that happiness is a well-lived life, and primary depends on ourselves, Kant's happiness perception was based on the idea of personal satisfaction. Notably, contemporary research evidence supports this finding since satisfaction is a happiness dimension. Modern philosophers like Nietzche defined happiness in terms of power over the world, while Aquino as God's relationship. Scholars and psychologists like Bentham (1789), Schachtel (1954), Kitayama et al. (1995, 2000), and Kesebir and Diener (2008) have found that happiness definition is built by several dimensions such as pleasure, subjective well-being, live satisfaction, attitude, social relations, and other complementary variables.

Also, it was found four widely accepted factors that can encourage happiness in human experience. First was physical activity since Rasciute and Downward (2010), Wang et al. (2012), Min et al. (2017), Souto (2014) and Khazaee-pool et al. (2014) demonstrated that the more physical activity, the higher levels of happiness. The second factor identified was the social experience. Authors like Aaker et al. (2011), Shin et al. (2017), Holder and Coleman (2009), and others found that social relations can foster happiness, human well-being, and other health benefits. The third factor explained was tourism since it can promote live satisfaction as indicated by Sheldon and Schkade (2005), Dann (1991), and others. Even more, this relation was demonstrated in the long term, as found by Nawijn (2010). The last factor identified was altruism. Notably, several scholars associated altruism activities with higher wellbeing such as Midlarsky and Kahana (1994), Omato and Snyder (1995), Aknin et al. (2012), and others.

Also, this chapter discussed how happiness perception and experience could change according to culture. It can be concluded that the European and North American region based their perception of happiness in terms of self: self-esteem, autonomy, personal achievement, and competencies. In contrast, the East Asian region has a happiness perception that refers to family, social relations quality, social validation, and interdependence. Latin America's happiness perception has similarities in the East Asian region since social relation is a factor that can promote happiness (Rojas, 2019; Beytía, 2016). However, the relative income difference in Latin America matters than other regions since it can negatively affect happiness more than other regions (Graham & Felton, 2006).

Our experiment supported the idea that Latin-American citizens have high levels of happiness. Notably, the examined sample was Peruvian youngsters who are characterized to have high levels of happiness in terms of life satisfaction and well-being. One of the main findings is that they perceive rewarding and supportive of their social relations and feeling that they are competent persons and have a good life.

## References

Aaker, J., Rudd, M., & Mogilner, C. (2011). If money does not make you happy, consider time. *Journal of Consumer Psychology, 21*(2), 126–130.

Aknin, L., Kiley, J., & Dunn, E. (2012). *Giving Leads to Happiness in Young Children., 7*(6), 1–4.

Appau, S., & Awaworyi, S. (2018). Charity, Volunteering type and subjective wellbeing. *VOLUNTAS: International Journal of Voluntary and Nonprofit Organizations, 30*, 1118–1132.

Bartolini, S., Bilancini, E., & Pugno, M. (2013). Did the decline in social connections depress Americans' happiness? *Journal: Social Indicators, 110*(3), 1033–1059.

Baruth, M., Wilcox, S., Wegley, S., Buchner, D. M., Ory, M. G., Phillips, A., Schwamberger, K., & Bazzarre, T. L. (2011) Changes in physical functioning in the active living every day program of the active for life initiative. *International Journal of Behavioral Medicine, 18*(3), 199–208.

Bentham, J. (1789). *Introduction to the principles of morals and legislation.* London: Oxford University Press.

Bett, R. (2005). Nietzsche, the Greeks, and Happiness (with special reference to Aristotle and Epicurus). *Philosophical Topics, 33*(2), 45–70.

Beytía, P. (2016). The singularity of Latin American patterns of happiness. *Handbook of Happiness Research in Latin America*, 17–29.

Bimonte, S., & Faralla, V. (2014). Happiness and nature-based vacations. *Annals of Tourism Research, 46*, 176–178.

Broadhead, W. E., Kaplan, B. H., James, S. A., Wenger, E. H., Schoenbach, V. J., Grimson, R., Heyden, S., Tibblin, G., & Gehlback, S. H. (1983). The epidemiologic evidence for a relationship between social support and health. *American Journal of Epidemiology, 117*, 521–537.

Campodonico, A. (2011). Las valoraciones del deseo: felicidad, ley natural y virtudes en Tomás de Aquino. *Tópicos (México), 40*, 51–62.

Corvo, P. (2011). The pursuit of happiness and the globalized tourist. *Social Indicators Research, 102*, 93–97.

de Souto, P. (2014). Direct and indirect relationships between physical activity and happiness levels among older adults: A cross-sectional study. *Aging & Mental Health, 18*(7), 861–868.

Delle, A., Brdar, I., Freire, T., & Wissing, M. (2010). The eudaimonic and hedonic components of happiness: qualitative and quantitative findings. *Social Indicators Research, 100*(2), 185–207.

Demır, M., & Weitekamp, L. (2007). I am so happy' cause today I found my friend: friendship and personality as predictors of happiness. *Journal of Happiness Studies, 8*(2), 181–211.

Diener, E., Wirtz, D., Tov, W., & Kim-Prieto, C. (2010). New well-being measures: Short scales to assess flourishing and positive and negative feelings. *Social Indicators Research, 97*(2), 143–156.

Dunn, E., Aknin, L., & Norton, M. (2008). Spending Money on Others Promotes Happiness. *Science, 319*(5870), 1687–1688.

Easterlin, R. (1973). Easterlin Does Money Buy Happiness? *The Public Interest, 30*, 3–10.

Fanning, A., & O'Neill, D. (2018). The wellbeing consumption paradox: Happiness, health, income, and carbon emissions in growing versus non-growing economies. *Journal of Cleaner Production, 212*, 810–821.

Feldman, F. (2010). *What is this thing called HAPPINESS?* New York: Oxford University Press.

Fujita, F., & Diener, E. (2005). Life satisfaction set point: Stability and change. *Journal of Personality and Social Psychology, 88*(1), 158–164

Gallup World Poll. (2017). World Happiness Report.

Gilbert, D. (2006). Stumbling on happiness. Alfred A. Knopf.

Goulart de Camargos, M., Ribeiro, B., Leite de Almeida, C., & Paiva, C. (2019). What is missing for you to be happy? Comparison of the pursuit of happiness among cancer patients, informal caregivers, and healthy individuals. *Journal of Pain and Symptom Management*, 1–14.

Graham, C. (2008). *Measuring quality of life in Latin America: What happiness research can (and cannot) contribute.* New York: Inter-American Development Bank.

Graham, C., & Felton, A. (2006). Inequality and happiness: Insights from Latin America. *The Journal of Economic Inequality, 4*, 107–122.

Guyer, P. (2000). *Kant on freedom, law, and happiness.* Cambridge University Press.
Hamermesh, D., & Abrevaya, J. (2013). Beauty is the promise of happiness? *European Economic Review, 64,* 351–368.
Headey, B., Muffels, R., & Wooden, M. (2008). Money does not buy happiness: or does it? A reassessment based on the combined. *Social Indicators Research, 87*(1), 65–82.
Hills, P., & Argyle, M. (1998). Musical and religious experiences and their relationship to happiness. *Personality and Individual Differences, 25*(1), 91–102.
Holder, M., & Coleman, B. (2009). The contribution of social relationships to children's happiness. *Journal of Happiness Studies, 10,* 329–349.
Hornsey, M., Bain, P., Harris, E., Lebedeva, N., Kashima, E., Guan, Y., Blumen, S. (2018). How much is enough in a perfect world? Cultural variation in ideal levels of happiness, pleasure, freedom, health, self-esteem, longevity, and intelligence. *Psychological Science,* 1–12.
Kahneman, D., Diener, E., & Schwarz, N. (1999). *Foundations of hedonic psychology: Scientific perspectives on enjoyment and suffering.* New York: Russell Sage Foundation.
Kesebir, P., & Diener, E. (2008). In pursuit of happiness: empirical answers to philosophical questions. *Perspectives on Psychological Science, 3,* 117–125.
Khazaee-pool, M., Sadeghi, R., Majlessi, F., & Rahimi Foroushani, A. (2014). Effects of physical exercise programme on happiness among older people. *Journal of Psychiatric and Mental Health Nursing, 22*(1), 47–57.
Kitayama, S., Markus, H., & Matsumoto, H. (1995). Culture, self, and emotion: A cultural perspective on "self-conscious" emotions. En J. Tangney, & K. Fischer, Self-conscious
Kitayama, S., Markus, H., & Kurokawa, M. (2000). Culture, emotion, and well-being: Good feelings in Japan and the United States. *Cognition & Emotion, 14,* 93–124.
Kraut, R. R. (1979). Two conceptions of happiness. *The Philosophical Review, 88*(2), 167–197.
Kristin, L., & Lyubomirsky, S. (2014). The how, why, what, when and who of happiness. Mechanism underlying the success of positive activity interventions. In *Positive emotion: Integrating the light sides and dark sides* (p. Chapter 25). Oxford Scholarship Online.
Kye, B. K., Kim, H. S., Kil, H. J., & Jeon, J. (2013). A study on the influence of smart education on the cognitive, affective and psychomotor domains of learners. Seoul, Korea Education and Research Information Service
Layard, R. (2006). Happiness and public policy: A challenge to the profession. *The Economic Journal, 116,* 24–33.
Layous, K., & Lyubomirsky, S. (2014). The how, why, what, when, and who of happiness.
Liu, K. (2013). Happiness and tourism. *International Journal of Business and Social Science, 4*(15), 1–4.
Lok, I., & Dunn, E. (2020). Under what conditions does prosocial spending promote happiness? *Collabra: Psychology, 6*(5), 1–14.
Luhmann, M., Hofmann, W., Eid, M., & Lucas, R. E. (2012). Subjective well-being and adaptation to life events: A meta-analysis. *Journal of Personality and Social Psychology, 102*(3), 592–615.
Lykken, D., & Tellegen, A. (1996). Happiness is a stochastic phenomenon. *Psychological Science, 7*(3), 186–189.
Lyubomirsky, S. (2008). *The how of happiness: A new approach to getting the life you want.* New York: The Penguin Press.
Lyubomirsky, S., Sheldon, K.M., & Schkade, D. (2005). Pursuing happiness: The architecture of sustainable change. *Review of General Psychology, 9,* 111–131.
McCabe, S., & Johnson, S. (2013). The happiness factor in tourism: Subjective well-being and social tourism. *Annals of Tourism Research, 41,* 42–65.
Medici, A., & Lewis, M. (2019). *Health policy and finance challenges in Latin America and the Caribbean: An economic perspective.* Oxford Research Encyclopedia of Economics and Finance.
Michalos, A. (2008). Education, happiness and wellbeing. *Social Indicators Research, 87*(3), 347–366.
Milman, A. (1998). The impact of tourism and travel experience on senior travelers' psychological well-being. *Journal of Travel Research, 37*(2), 166–170.

Mogilner, C., Kamvar, S., & Aaker, J. (2010). The shifting meaning of happiness. *Social Psychological and Personality Science, 2*(4), 395–402

Mookerjee, R., & Beron, K. (2005). Gender, religion and happiness. *The Journal of Socio-Economics, 34*(5), 674–685.

Nawijn, J. (2010). Happiness through vacationing: just a temporary boost or long-term benefits? *Journal of Happiness Studies, 12,* 651–655.

Nawijn, J. (2012). *Leisure travel and happiness: An empirical study into the effect of holiday trips on individuals' subjective wellbeing.* Nieuwegein: NRIT Media.

Nelson, M. E., Rejeski, W. J., Blair, S. N., Duncan, P. W., Judge, J. O., King, A. C. (2007). Physical activity and public health in older adults: Recommendation from the American College of Sports Medicine and the American Heart Association. *Medicine and Science in Sports and Exercise, 39,* 1435–1445.

Oswald, A. (1997). Happiness and economic performance. *Economic Journal, 107,* 1815–1831.

Pérez-Asenjo, E. (2011). If happiness is relative, against whom do we compare ourselves? Implications for labour supply. *Population Economics, 24,* 1411–1442.

Philipp, S. (2000). Race and the pursuit of happiness. *Journal of Leisure Research, 32*(1), 121–124.

Phillips, D. (1967). Mental health status, social participation and happiness. *Journal of Health and Social Behavior, 8*(4), 285–291.

Plomin, R., & Spinath, F. (2004). Intelligence: genetics, genes, and genomics. *Journal of Personality and Social Psychology, 86*(1), 112–129.

Post, S. (2005). Altruism, happiness, and health: It's good to be good. *International Journal of Behavioral Medicine, 12,* 66–77.

Ramos, M., & Moriconi, M. (2018). Corruption in Latin America: Stereotypes of politicians and their implications for affect and perceived justice. *Social Psychological and Personality Science, 9*(2), 111–122.

Rasciute, S., & Downward, P. (2010). Health or happiness? What is the impact of physical activity on the individual? *KYKLOS, 63*(2), 256–270.

Rátz, T., & Michalkó, G. (2011). The contribution of tourism to well-being and welfare: The case of Hungary. *Int. J. Sustainable Development, 14*(3/4), 332–346.

Richards, J., Jiang, X., Kelly, P., Chau, J., Bauman, A., & Ding, D. (2015). Don't worry, be happy: Cross-sectional associations between physical activity and happiness in 15 European countries. *BMC Public Health, 15*(53), 1–8.

Roddy, Zhang, W., Doherty, M. (2005). Aerobic walking or strengthening exercise for osteoarthritis of the knee? A systematic review. *Annals of the rheumatic diseases, 64*(4), 544–8.

Rojas, M. (2019). Latin America and well-being. In R. Estes, & J. Sirgy (Eds.), *Human well-being research and policy making* (pp. 1–9).

Røysamb, E., Tambs, K., Reichborn-Kjennerud, T., Neale, M., & Harris, J. (2003). Happiness and health: Environmental and genetic contributions to the relationship between subjective well-being, perceived health, and somatic illness. *Journal of Personality and Social Psychology, 85*(6), 1136–1146.

Schachtel, H. J. (1954). *The real enjoyment of living.* Dutton: Edición: 1st.

Seetanah, B., & Fauzel, S. (2018). Investigating the impact of climate change on the tourism sector: Evidence from a sample of island economies. *Tourism Review, 74*(2), 194–203.

Shin, J.-E., Suh, E., Eom, K., & Kim, H. (2017). What does "Happiness" prompt in your mind? Culture, word choice, and experienced happiness. *Journal of Happiness Studies, 19*(3), 649–662.

Skogstad, M., & Laumann, K. (2013). The role of exercise during adolescence on adult happiness and mood. *Leisure Studies, 33*(4), 341–356.

Stuck, A. E., Walthert, J.M., Nikolaus, T., Bula, C. J., Hohmann, C., Beck, J. C. (1999). Risk factors for functional status decline in community-living elderly people: A systematic literature review. *Social Science and Medicine, 48,* 445–469.

Thirdi, M., & Jowkar, B. (2011). El efecto del ejercicio y la actividad física sobre la felicidad de las mujeres posmenopáusicas. *Iranian Journal of Ageing, 6*(2).

Torres, C. (2019). *Latin American education: Comparative perspectives.* Routledge.

Uchida, Y., Norasakkunkit, V., & Kitayama, S. (2004a). Cultural constructions of happiness: Theory and empirical evidence. *Journal of Happiness Studies, 25,* 223–239.

Uchida, Y., Norasakkunkit, V., & Kitayama, S. (2004b). Cultural constructions of happiness: Theory and empirical evidence. *Journal of Happiness Studies, 5,* 223–239.

Urstad, K. (2007). *Freedom and happiness in socrates and callicles.*

Utz, S., & Beukeboom, C. (2011). The role of social network sites in romantic relationships: Effects on jealousy and relationship happiness. *Journal of Computer-Mediated Communication, 16*(4), 511–527.

Wang, F., Orpana, H., Morrison, H., de Groh, M., Dai, S., & Luo, W. (2012). Long-term association between leisure-time physical activity and changes in happiness: Analysis of the prospective national population health survey. *American Journal of Epidemiology, 176*(12), 1095–1100.

Warburton, D. E., Nicol, C., Bredin, S. S. (2006). Health benefits of physical activity: The evidence. *CMAJ, 7,* 801–809.

WHO. (2017). Depression and other common mental disorders. *Global Health Estimates.* Retrieved 02 12, 2020, from https://apps.who.int/iris/bitstream/handle/10665/254610/WHO-MSD-MER-2017.2-eng.pdf?sequence=1.

World Happiness Report. (2018). *Search of happiness levels in Latin-America.* Retrieved 03 03, 2020, from https://worldhappiness.report/ed/2018/.

World Health Organization. (2018). *Fact sheets. Physical activity.* Retrieved 02 26, 2019, from https://www.who.int/news-room/fact-sheets/detail/physical-activity

Yamamoto, J. (2015). The social psychology of Latin American. In M. Rojas (Ed.), *Handbook of happiness research in Latin America.*

# The Cultural Perspective: Are Some Societies Happier Than Others?

**Vivian Eternod**

**Abstract** The second chapter explores happiness from a cross-cultural perspective. Worldwide measures indicate huge differences in happiness between societies. For instance, the 2019 World Happiness Report (Helliwell, Layard, & Sachs, 2019) ranks Finland, Sweden, and Norway to be the happier countries, while other countries like South Sudan and Afghanistan are at the bottom of the list. On the other hand, the 2019 Global Emotions Report shows that Latin American countries lead in the positive experiences ranking. The chapter examines a handful of perspectives that explain cultural variation in happiness. One perspective quantifies key variables like GDP per capita, social support, healthy life expectancy, freedom to make life choices, perceptions of corruption and generosity. Another perspective is grounded on the fulfillment of psychological needs like learning, autonomy, using one's skills, respect, and count with others. Nevertheless, and despite the efforts to measure worldwide happiness, there is an unexplained happiness factor that is recognized by researchers and makes a difference in perception of happiness among countries with similar socioeconomic conditions. Hofstede's cultural dimensions provide a broader cross-cultural frame to attend the unexplained factor of happiness. Members of collective society define themselves through the group they belong, in contrast, within the individualistic society the person is oriented around the self, as an independent being. This might explain why a factor like the freedom to life choices might gave a greater impact on happiness in an individualistic society and social support is more important in a collective society. Finally, the chapter explains the importance of addressing cultural variations on happiness perception in business decisions.

**Keywords** Worldwide happiness measurements · Consumer values · Marketing strategies

---

V. Eternod (✉)
Business School, Tecnologico de Monterrey, San Pedro Garza García, Mexico
e-mail: vivian.eternod@tec.mx

© The Author(s), under exclusive license to Springer Nature Singapore Pte Ltd. 2021
T. Dutta and M. K. Mandal (eds.), *Consumer Happiness: Multiple Perspectives*, Studies in Rhythm Engineering,
https://doi.org/10.1007/978-981-33-6374-8_1

## Introduction: Are Some Societies Happier Than Others?

A topic of great concern among governments, scholars, and international organizations worldwide is to measure the level of happiness between countries and to identify the main factors that may explain such differences. Several publications measure happiness and well-being thoroughly and present global comparisons for public analysis. For example, the World Values Survey (WVS) focuses on changes in cultural values and their impact on social and political life. The WVS first report was issued in 1981 and limited to developed countries; nevertheless, it rapidly raised attention from researches worldwide, and nowadays, the WVS report includes societal well-being among other 13 topics related to cultural values, and it is conducted by specialists in more than 90 countries. According to WVS researchers Inglehart and Welzel, the world could be culturally explained by using two cultural dimensions: survival versus self-expressions values and traditional versus secular-rational values. Basically, according to Inglehart & Welzel (2010), as people have a stronger feeling of having free choice, they are happier. In this context, happiness is directly related to self-expression which is enabled by positive socioeconomic factors like democratization, economic development, and social tolerance. On the other hand, when people are on a survival mode, concerned with their economical or personal security, democracy, and social tolerance are not the priority and the distrust that the insecurity diminishes their sense of happiness. As a result, the countries with a higher rating on self-expression values are happier (Sweden, Denmark, Norway, Canada, and Iceland).

Since 2005, the Gallup company has continually surveyed adults from 160 countries and included more than 100 questions referring to several topics like law and order, food and shelter, institutions and infrastructure, good jobs, well-being, and brain gain. The Gallup company released a special report in 2006 called *Gallup Global Emotions* that showed the worldwide ranking on both positive and negative experiences. The Gallup Global Emotions Report does not consider socioeconomic factors like gross national product (GNP), rather the report measures the intangibles within a society's life. On the one hand, the positive experiences related to people's perceptions about their living standards, personal freedoms, and the presence of social networks. On the other hand, negative experiences reflect people's health problems, sadness, and stress (Gallup, 2020).

The Gallup company issues the Gallup Global Emotions report yearly, and Latin American countries have been consistently high on the Positive Experience Index; Latin America also ranks up in the list of the most emotional nations. These results led the authors of the report to suggest that "at least partly" they reflect the cultural tendency in the region to focus on the bright side of things, because even when Latin Americans does not rate their lives as high as the Nordic countries and are not the wealthiest either they tend to focus on the positives of life, they laugh, smile and experience enjoyment like no other countries. On the other side, at the top of the negative experiences are ranked Africa and the Middle East which are experiencing severe conflicts.

The United Nations Organization also recognized the importance of measuring happiness worldwide and gathered an independent group of experts that crafted the World Happiness Report (Helliwell, Layard, & Sachs, 2019), which is probably the best-known publication on this topic since its first issue in 2012. The WHR includes 156 countries and presents an annual country ranking formed by six key variables: GDP per capita, social support, healthy life expectancy, freedom, generosity, and absence of corruption. Each variable is measured quantitatively and added to an overall grade. The top countries at the WHR ranking tend to have higher values on the six key variables mentioned before. Denmark, Switzerland, Norway, and Finland are frequently found at the top of the list.

Certainly, a country's economic and political conditions have a straightforward impact on the happiness of their people; it should not be a surprise that wealthy and developed countries are at the top of the WHR. Nevertheless, there are other perspectives to worldwide happiness that show completely different results than what is suggested in the WHR, like the Gallup Global Emotion Report suggests that the happier people in the world live in Latin America. We may assume that there are other factors, in addition to economic and public policies influencing a social perception of their happiness.

Cross-cultural researchers have analyzed and measured the impact of culture on happiness through different viewpoints, and they have brought to the public's attention a great amount of evidence that explains the impact that cultural values have on the happiness and well-being of societies, independently of their wealth.

## The Influence of Income on Happiness Worldwide

Worldwide measurements of happiness from the Organization for Economic Co-operation and Development (OECD), the World Happiness Report and the Gallup Global Emotions Report, among others, present enough support to identify a few critical factors that have a strong influence on the level of happiness on a given society: income level (Bulmahn, 2000; Kahneman & Deaton, 2010), social support (Mariano Rojas, 2012), political instability (Coupe & Obrizan, 2016) and inequality (Graafland & Lous, 2018; Graham & Felton, 2006). The impact of these factors is shaped by cultural values, and appropriate analysis is presented in the next section. Nevertheless, as the relation money-happiness has been a heated topic of debates, the next section is dedicated to explaining the viewpoints of some of the most recognized researchers on the matter.

Even though income and happiness are positively related, such relation is not constant neither equal for all societies. Kahneman & Deaton (2010) propose that happiness has two components: cognitive and emotional. The cognitive component indicates how people evaluate their life standard when they think about it; it is also known as life satisfaction. The emotional component expresses people content with their life, whether life is good or enjoyable, and the frequency and intensity of their emotions like joy, affection, sadness, or anger. According to Kahneman and Deaton's research study on approximately 45,000 people from the USA, income impacts in a different manner the cognitive and the emotional component of happiness. Life evaluation, which is the cognitive component, has a positive and steady relation with income. The emotional component, on the other side, does increase with higher income but only until the income reaches the level of ~$75,000, and after that point, there is no further increase in happiness. Low income does intensify the emotional pain from life misfortunes such as suffering from a serious illness, a divorce or loss of a beloved person, and being alone. These results led the researchers to conclude that money can buy life satisfaction and also softens life adversities but does not buy happiness.

Bulmahn (2000) and Lakshmanasamy (2010) agree with the results of the study made by Kahneman and Deaton and explain that when a higher income improves material living conditions; it also increases pretensions and expectations. As a result, after a certain point of saturation, higher income and increased consumption do not have an impact on happiness because it is neutralized by changes in pretensions and needs. Lakshmanasamy explains that in some countries, individuals care a lot about their income level in comparison to others, which is called relative income. So, a general or absolute increase in income will not increase happiness, because people develop aspirations for status positions.

Graham & Felton (2006) summarize their research results on a cross-cultural comparison by explaining that economic growth leads to higher average happiness at low levels of per capita incomes but not at higher ones. Basic living conditions from people with low and very low income can be improved with even a small increase in their income. But at a higher income level, people's needs tend to become toward personal growth, cognitive and aesthetic need, self-actualization and transcendence and the effect related to income increase becomes much weaker (Drakopoulos & Grimani, 2013). Suh and Oishi (2002) conclude that wealth is a predictor of well-being among poor nations but not across affluent ones.

From the cross-cultural perspective, there are some societies where money is given more importance, and a lot of emphasis is on being material minded. In materialistic societies, possessions are seen as a sign of success and consumers tend to display their possessions because they believe that material goods help them make a positive impression on others as creating happiness, and as being central to their lives. While in many societies, matters like comfort, leisure, and relationships get precedence over being materialistic (Ger & Belk, 1996; Kilbourne, Grünhagen, & Foley, 2005; Richins & Dawson, 1992). A materialistic society would show a stronger relation between money and happiness, whereas a non-materialistic society the link between money and happiness will be mostly determined by the living conditions and the nature of the needs that are yet to be fulfilled (Minkov, 2009).

## Cultural Values and Their Effect on Happiness

Cross-cultural comparison brings into the analysis of the consumer a thought-provoking framework that reveals the particularities of the own culture and uncovers straightforward differences among societies. The discussion gets fascinating particularly on the happiness debate due to contradictory positions from the specialists on the topic.

The World Happiness Report and the Gallup World Poll show that in general countries with higher economic development and political stability tend to show higher rankings on life evaluations, but there are some cases where happiness does not correspond with objective measurements such as gross domestic product (GDP). For instance, according to the analysis done with OECD endorsement, a group of Latin American countries (Costa Rica, Mexico, Panama, and Brazil) report higher life evaluations than might be expected, given their per capita income levels, as do Denmark and Finland. Contrariwise, some central and Eastern European countries (Hungary, Bulgaria, Latvia, and Georgia), as well as Botswana and Hong Kong, report lower life evaluations than might be expected based on income alone (Exton, Smith, & Vandendriessche, 2015).

Social and anthropological scientist have developed extensive research to identify and understand which are the other factors in stake—beyond objective measurements of well-being—that makes a society happy, and culture has been found as an element of mayor significance (Arrindell et al., 1997; De Mooij & Hofstede, 2002; Diener, Ng, Harter, & Arora, 2010; Diener, Oishi, & Lucas, 2003; Hofstede & Hofstede, 2001; Tov & Diener, 2009; Triandis, Kitayama, & Markus, 1994). Cultural values can shape the concept of happiness within a society and redefine how is it experienced in terms of the amount, extent, or degree of happiness (Diener et al., 2003). The meaning of happiness may change according to the culture's values, for example, the Chinese conception of happiness prioritize harmony of interpersonal relationships, achievement at work, and contentment with life (Lu & Shih, 1997), while happiness in the USA can be defined as a form of personal accomplishment (Uchida & Kitayama, 2009), which is related with important elements of the American culture like the American Dream and the frontier spirit (Hochschild, 1996).

### *Individualistic–collective Characteristics*

From the cultural viewpoint, the individualism–collectivism dimension relates to how one values the individual relative to the group (Hofstede & Hofstede, 2001). On the one hand, individualism emphasizes individual freedom, individual achievement, and the pursuit of individual positive feelings. People are expected to care basically about themselves and their core families. Beyond their family boundaries, altruism, selflessness, and even self-sacrifice may be practiced, but based on an individual, deliberate, personal decision. The culture of individualism prevails in Western countries in Europe and America. On the other hand, members of collective societies identify themselves about the groups they are members of, people put relatively more emphasis on human relationships, including families, colleagues, and neighbors. They seek and obtain fulfillment by maintaining and enhancing the harmony of the group, even at the expense of subordinating their goals and aspirations. Collectivist culture zones include Japan, China, Korea, Taiwan, Venezuela, Guatemala, Indonesia, Ecuador, Argentina, Brazil, and India.

## Individualistic Internal Focus and Happiness

The study of the individualistic/collective dimension and happiness has been largely studied; nevertheless, the results are not conclusive. A high proportion of the studies on the individualism/collectivism and happiness indicates that individualistic societies tend to be happier than collective ones (Diener, Suh, Smith, & Shao, 1995; Lu, Gilmour, & Kao, 2001; Myers & Diener, 1995). Ye, Ng, & Lian (2015) explain that the main characteristics of individualism like self-esteem, individual achievement, and autonomy are linked more directly to happiness and life satisfaction than collective characteristics such as a higher emphasis on human relationships where happiness feelings are affected relatively more by the evaluation of others. In collective societies, the relationships between happiness and individual effort and achievement are not clear. This may make their happiness levels lower than in the individualist countries.

The motivation source is also a critical factor that contributes to higher happiness results in individualistic vs. collective societies (Ahuvia, 2002; Kasser & Ryan, 1996; Ryan, Sheldon, Kasser, & Deci, 1996; Sheldon & Kasser, 1998). Supposedly, people in individualistic societies are free of the opinions of others, and they can emphasize internal goals like personal growth, close personal relationships, making a social contribution, and maintaining one's health which is associated with higher levels of happiness. On the other side, as collective societies care a great deal about the opinion of others, external goals for financial success, social recognition, and having an appealing appearance are more important in collective societies and did not produce similar positive results because those goals do not necessarily come from within the person (Kasser & Ryan, 1996; Sheldon & Kasser, 1998). Moreover, pursuing goals out of anxiety, guilt, or a desire to please others is associated with lower levels of self-esteem regardless of what those goals are (Carver & Baird, 1998). Therefore, it is not enough to pursue internal goals; one must do so out of internal motivation.

## Collective Social Support, Health, and Happiness

Evidence shows that individualistic societies are not always happier than collective ones. Carballeira, González, & Marrero, (2015) compared Mexican (collective) and Spaniards (individualistic) happiness and life satisfaction. Their results show that the collective sample had higher levels of all the happiness/subjective well-being indicators.

Rojas (2018) explains that in general Latin American countries have a human-relations orientation, and the culture in Latin America cherishes warm and close relationships with family, nuclear and well as extended, and friends and allows the experience and manifestation of emotions, and these affective experiences are highly relevant in Latin American's happiness. Schmuck & Sheldon (2001) suggests that when people are less concern on immediate material gratification and more on close interpersonal relations will perceive higher well-being and happiness and these factors seem to appear more often within collective societies.

Triandis et al. (1988) argued that unpleasant life events and the availability of close, interpersonal support are critical factors that might influence one's susceptibility to disease. Cultural collectivism might function as a prophylactic against disease as it promotes harmony within a narrow, supportive ingroup; this cohesion and availability of caring others would buffer against stress, thereby decreasing the incidence of disease. This theoretical argument may apply to the area of mental health as well (Arrindell et al., 1997).

Additionally, people with a higher perception of social support feel more satisfied with their lives, show higher satisfaction with their couple is better psychologically adjusted, and have less negative emotions and more positive (Marrero Quevedo & Carballeira Abella, 2010). Social research has proven the importance of social networks on life quality and physical and mental health. A social network that is sensitive, stable, functional, and active generates better satisfaction conditions that impact positively the quality of life and therefore the physical and psychological health of the individual.

Social support has strong linkages to happiness through its effects on physical and mental health. It provides protection and allows for higher satisfaction due to the benefits of mutual contact and assistance (Montero, 2003). Especially for people living in poor conditions, which account for almost half of the world population according to the World Bank (Bank, 2018), social networking enables individuals with disadvantages to obtaining affective, moral, economic, and social support and besides allows the configuration of survival mechanisms that help in solving day-to-day problematic situations and cover the needs arising from the government absence or inefficiency (Madariaga, Abello, & Sierra, 2003; Ávila-Toscano, 2009).

Social networks impact positively a wide set of factors that result in a higher life satisfaction such as a decrease in feelings associated with isolation and exclusion, an increase of the resources and options related to well-being, and promoted by net interchanges, emotional support, and promotion of health. (Huenchuan, Guzmán, & Montes de Oca Zavala, 2003). For instance, Mexico's happiness level and ranking in the world do not correspond to their income ranking and their economic problems, and the relevance of interpersonal relationships and the predominance of non-materialistic values are key factors accountable for the happiness level in this country (Mariano Rojas, 2012).

Additionally, social networks have an important role in the offering of support and cooperation that is based on the interchange of feeling and affective expressions like a coping mechanism for difficulties, and so, it is a coping mechanism for poverty and social disadvantage. A study with elderly Chinese found that income is more significant than gender and education in determining happiness, but it is less important than personal network size and particularly perceived social support (Chen, Lu, Gupta, & Xiaolin, 2014).

Cross-cultural research by Goodwin & Hernandez Plaza (2000) shows that collectivism predicted family and friends support and perceptions of available support which positively impacts self-esteem and life satisfaction. Moreover, global perceived support and support from friends after an event where significant correlates of self-esteem, which along with global support and support from family members, was a significant correlate of life satisfaction.

**Institutional Collectivism Versus In-Group Collectivism**

The GLOBE project (House, Hanges, Javidan, Dorfman, & Gupta, 2004) investigates cultural values on 62 countries and proposes two independent collective dimensions: institutional collectivism that refers to the extent to which societal institutions encourage the collective distribution of resources and collective action and in-group collectivism that reflects the degree to which individuals take pride in their membership in small groups such as their families and circles of close friends (House et al. 2002).

Institutional collectivism is a little different from in-group collectivism in that it gives more emphasis to social support. Ye et al., (2015) found that institutional collectivism is positively related to happiness, whereas in-group collectivism was negatively related. This led us to conclude that within the characteristics of a collective society is social support what helps buffer stress and disease leading to higher happiness while other treats like caring a great deal about others opinion and subordinating goals and aspirations to the group interests diminish individual happiness.

## *Power Distance*

Power distance refers to the extent to which the less powerful persons in a society accept inequality in power and consider it as normal (Hofstede, 1984). Even though every society has a certain degree of inequality, there are some societies in which inequality is part of the culture (Shepelak, 1987; Zacharias & Vakulabharanam, 2011). In societies with high power distance rankings, inequalities are accepted as part of the social system, where every person has a place, high or low, and it is extremely difficult to change positions. Powerful members look down on their subordinates but also take on certain responsibility toward the less privileged ones. On the other side, in societies with low-power distance rankings, people are equal, and they consider social inequalities obsolete, and every member of the society has equal rights and coexist in harmony.

## Inequality's Psychological Effect on Happiness

The psychological aspect of inequality has a strong impact on happiness (Clark & d'Ambrosio, 2015;
 Graafland & Lous, 2018; Graham & Felton, 2006; Ye et al., 2015). Hofstede (1980) points out that higher levels of power distance might go hand in hand with inequality in a vast range of areas like employment, social status, income, civil, and health rights (Bottero, 2004; Kogevinas, Pearce, Susser, & Boffetta, 1997). These conditions may induce unprivileged people to feel powerless and at the mercy of external forces that they cannot control. These feelings of powerlessness are connected with sadness and fear (TenHouten, 2016) and diminish mental well-being (Arrindell et al., 1997). While in societies with low-power distance, people tend to feel more powerful and in control with their lives.

Income inequality is found to have a negative effect on life satisfaction (Graafland & Lous, 2018). However, relative income inequality has an even larger effect in happiness and well-being according to several studies in Asia (Hao, 2010; Oshio, Nozaki, & Kobayashi, 2011; Smyth & Qian, 2008) and Latin America (Graham & Felton, 2006). Even though huge social and economic inequalities can be found in these regions, for example, Latin America is the region with the higher inequalities' rankings worldwide; it is not level of income per se that decreases happiness, but the inequality relative to other people. One of the major insights of happiness research is that individuals have a strong tendency to compare their situation to that of other people (Frey, Gallus, & Steiner, 2014). Relative disadvantage within the reference group would report lower levels of happiness (Oshio et al., 2011).

## Survival Versus Self-expression

The survival vs self-expression cultural dimension shows the improvement societies have made moving up the ladder on basic human needs. According to Inglehart & Welzel (2010), survival values refer to the sense of being in danger, tend to be found in places characterized by low material wealth and limited physical security; so, there are lower levels of trust and tolerance, self-expression values are meanwhile found among those who can afford to take survival for granted, and include tolerance for diversity, higher levels of trust, demand for participation in decision-making, and prioritizing issues like environmental protection.

Positive socioeconomic factors like economic development, democratization, and social tolerance enable self-expression which is directly related to happiness. In contrast, when people are in a survival mode, concerned with their economical or personal security, democracy, and social tolerance are not the main concern and the distrust that the insecurity produces weakness their sense of happiness. As a result, the countries with a higher rating on self-expression values are happier (Sweden, Denmark, Norway, Canada, and Iceland) (Inglehart & Welzel, 2010).

It is very probable that negative experiences, including experiences of conflict, explain the high levels of negative affect in several Middle Eastern countries like

Iraq, Iran, Syria, Israel, Armenia, and Serbia rather than simply to a more negative outlook on life (Exton et al., 2015).

## *Masculinity–Femininity*

The masculinity–femininity dimension refers to the cultural expectations for men and women. Masculine cultures have a specific concept of how men and women are expected to behave and to fulfill their lives (Hofstede, 1984; House et al., 2004). For instance, men must be assertive, ambitious, competitive, and seek material success, while women should serve and care for the non-material quality of life, for children and the weak. Feminine cultures, on the other hand, have a more fluid role definition allowing the overlapping of social roles for the sexes. Men do not need to be ambitious or competitive but may go for a different quality of life rather than material success and both genders. The more feminine societies would offer both sexes, especially women, greater opportunities for the fulfillment of multiple social roles (professional, marriage, parenthood).

Masculine cultures relate well-being and happiness to material success and assertiveness (Japan, Austria, Venezuela), while feminine societies (Sweden, Poland, and Costa Rica) emphasize other aspects like quality of life, interpersonal relationships, and concern for the weak. Additionally, feminine countries give women more space to control their life and enhance the average well-being levels in the whole society (Arrindell et al., 1997). These conditions are associated with higher self-assessed health levels, such as lower rates of sickness and lower usage of drugs, hence contributing to higher happiness (Barnett, Biener, & Baruch, 1987). On the other hand, masculine societies are characterized by higher job stress and lower overall satisfaction at work (Arrindell et al., 1997). Consequently, job stress and job dissatisfaction being important correlates and determinants of satisfaction with life as a whole.

Hofstede et al., (1998) found that femininity is a positive predictor of well-being only across rich nations, but not so in poor countries. Moreover, not all countries have the conditions to nurture a feminine mentality. Wealthier countries have the financial resources to develop a welfare society capable to offer both men and women equal opportunities and a greater sense of equity in society (Hofstede, 1984). In feminine countries like Sweden and the Netherlands, it is considered important to establish a welfare society characterized by a minimum quality of life for every person and by a welfare system that makes sure that the financial means to achieving such a society should be collected from those who have the means (Arrindell et al., 1997). Undoubtedly, it is extremely difficult for the poorer countries to develop a feminine society like Denmark, Switzerland, Norway, or Finland, which have higher rankings in happiness worldwide (Diener et al., 1995),

## *Indulgence Versus Restraint*

Various methods for quantifying culture include tendencies toward being emotional versus neutral, or tendencies toward pleasure-seeking and hedonism (Exton et al., 2015). Indulgence vs restraint cultural dimension emphasized freedom from control, and the importance of leisure and enjoying life as specific cultural phenomena. It reflects the degree to which it is culturally acceptable to indulge in leisurely and fun-oriented activities, either with family and friends or alone, and spend one's money, at one's discretion (Hofstede & Hofstede, 2001). All this predicts relatively high happiness (Minkov, 2009). Other societies impose more severe restrictions on the enjoyment of life—in terms of indulgence in leisure, fun, and spending. Individuals have a perception that their actions are restraint by various social norms and prohibitions and a feeling that enjoyment of leisurely activities, spending, and other similar types of indulgence is somewhat wrong. It would appear justifiable that societies, where people do not feel much freedom to act as they please, to spend their disposable income, and to indulge in leisure, have lower percentages of happy people (Minkov, 2009).

An OECD study (Exton et al., 2015) reveals that the world's top ten highest-scoring countries and territories on this dimension (out of the 93 studied) include six in Latin America and the Caribbean (Venezuela, Mexico, Puerto Rico, El Salvador, Colombia and Trinidad, and Tobago), two in Africa (Nigeria and Ghana), one in Europe (Sweden) and one English-speaking country (New Zealand). Those characterized by low scores, and thus a high degree of restraint, include one South Asian country (Pakistan), two Middle Eastern countries (Egypt and Iraq) and Latvia, Ukraine, Albania, Belarus, Lithuania, Bulgaria, and Estonia. From this perspective, cultural values might determine both the extent to which positive experiences are likely to be sought out and experienced and the extent to which they are likely to be impacting overall happiness (Exton et al., 2015).

Compared with other cultural dimensions, the relation between indulgence versus restraint and happiness and has less empirical evidence in part because this dimension was added lastly to Hofstede's cultural dimension's model and the results obtained so far are not conclusive. Schinzel (2013) proves the relation of indulgence and happiness in Luxemburg, while Delle Fave et al. (2016) did not find empirical evidence of such relationships in a study made in 12 countries. Nevertheless, high levels of indulgence are found consistently in several Latin American countries (Delle Fave et al., 2016; Exton et al., 2015; Hofstede & Hofstede, 2001) which have a happiness ranking that does not correspond to their conditions (Helliwell et al., 2019; Gallup, 2020).

## The Impact of Culture in Marketing Strategies

Several studies demonstrate that consumer behavior is to some extent shaped by culture (De Mooij & Hofstede, 2002; Leo, Bennett, & Härtel, 2005; Murphy, Gordon, & Anderson, 2004). Consumers' needs and desires are molded through cultural aspirations and norms, which specify an acceptable range of responses to specific situations. Therefore, marketing strategies are adapted to address buyers' attitudes and preferences within a different cultural context.

It is indispensable to analyze cross-culture differences conscientiously before expanding operations to other countries with different cultural values than one's own. Lack of sensitivity to cultural differences can lead to serious losses (De Mooij & Hofstede, 2002).

### *Impact of Culture on Marketing Communication*

In terms of marketing communication strategies, there are clear differences to consider between collective and individualistic societies. For instance, in collective societies, people´s image within the community is of utmost importance; thus, the focus of advertising is upon status, symbolism, prestige, and also on family or in-group benefits (De Mooij & Hofstede, 2002). In contrast, in individualist societies, the focus is on elements such as design and performance (Soares, Farhangmehr, & Shoham, 2007). Individualist consumers are more interested in knowing about the capabilities of the product before they purchase, as opposed to collectivists who rely a great deal on other factors for decision-making such as status and symbolism (Dhar, 2007).

Regarding communications styles, in collective societies, it is advisable the use-friendly and respectful messages. Nonverbal communications and symbols related to status, prestige, and hierarchy are effective (Shavitt, Johnson, & Zhang, 2011) as well as creating a professional and trust gaining relationship (Nayeem, 2012). Whereas, individualistic culture values direct and straight forward information on the product's features and may find interesting the use of comparisons vs competitive products.

Another important issue to consider is that buyers' collective cultures tend to rely a great deal on external information sources, such as family and friends, while individualist consumers base decisions upon information gathered through personal experiences (Doran, 2002). Thus, the communication strategy has to reach an increased number of influencers on the collective society.

## *Cultural Implications on Consumer Information Search*

As explained before in collectivist cultures, it is expected that one will involve family and friends in the information search process (Doran, 2002). Especially, if it is a high involvement purchase, they are likely to first consult with their colleagues, friends, and families, and to consider their advice or opinions before collecting information. They may also consider advice from friends and family in terms of which information sources are likely to be reliable such as magazines, word of mouth, spending time with dealers, and test driving, to complete the information search process. As a whole, consumers from collectivist cultures may look for social approval from others (Nayeem, 2012).

In contrast, since individualism is mainly reflected in being independent of others and being in control of one's surrounding environment, consumers from individualist cultures may be less likely to rely on others for example family members, peers, and social groups in their purchase decisions. Individualist consumers rely on internal knowledge based on their personal experience and seek out new experiences to expand upon that knowledge (Doran, 2002). They are less likely to rely on other people's opinions. For example, in terms of automobile purchases, they may spend a large amount of time looking at websites or speak to several dealers and test-drive several cars to extend their internal knowledge through their own experience. They are likely to utilize a much greater variety of patterns of information sources rather than relying on friends and family. Friends and family might make suggestions, but decisions are made individually. To be more specific, the final decision might involve consultation with friends or family, but the purchaser usually feels that the decision is their own (Green, Deschamps, & Paez, 2005). These variations in consumer information preferences might be utilized in a marketing communication strategy to create a more effective message when targeting consumers from these two different cultural backgrounds.

A study between Asian-born (collective) and Australia-born (individualistic) consumers in Australia revealed that collective consumers rely heavily on friends and family as the most important source of information. Other sources of information Asian-born consumers use include television advertisements, newspapers, billboards, and magazines (Doran, 2002). Therefore, marketers need to be aware of this situation and provide similar, or even the same, types of information/messages by using these above-mentioned sources to communicate with Asian-born consumers, so that they do not become confused. The information/messages could include less information on the mechanical and innovative features of the automobiles. It is not that Asian-born consumers do not want to know about these features, but the first impression they are likely to prefer is that which emphasizes the prestige of the automobile. (Nayeem, 2012).

The research showed that Australian-born consumers use the Internet as the most preferred information source. Therefore, in terms of communicating with Australian-born consumers, marketers might utilize the internet as a promotion or communication technique for distributing automobile information and to emphasize the cutting edge, innovative aspects of their products (Nayeem, 2012).

## *Cultural Influences on Brand Image Function*

Images and symbolic meanings attached to brands or companies are shaped by culture. For instance, collective societies tend to express high social status by using certain brands that supposedly convey a status message to other society members. The literature shows that collectivism exposes psychological benefits from selecting products and services with high brand credibility, confidence benefits, and a good reputation. Brand image is much more important than in the individualistic cultures (Frank et al, 2015) where brand image is used to reduce time spent on decision-making as the attributes associated with the brands are already familiar.

For example, in individualist cultures, consumers may prefer particular brands or products because those products provide expected functional benefits. Individualist consumers are not worried about social status or prestige when they make purchasing decisions; they are more likely to focus on the performance of the product. By comparison, in collectivist cultures, consumers may prefer particular brands/products because those products can be used for symbolic purposes which are important within their culture. Collectivist consumers may purchase products that represent status, or that reassert their similarity to members of their reference group, while individualist consumers may purchase products that differentiate them from referent (Aaker & Maheswaran, 1997).

A study on automobile purchases by Nayeem (2012) shows that individualist consumers are likely to choose motor power and better performance over prestige or brand, whereas collectivist consumers may choose high-priced automobiles due to the associated status and prestige element (Wong & Ahuvia, 1998). Besides, consumers may even draw on cultural values when searching/collecting information concerning purchases that require high involvement from the consumer.

As for collective societies, the company's image reflects status and conveys messages. For individualistic cultures, image is a source of information and helps to reduce time and effort in the purchase decision process (Eternod, Martínez, & Moliner, 2015).

## *Culture and Repurchase Intent Precedents*

Repurchase intent is the driver for increased profitability, and it is important to understand how cultural differences may impact repurchase intend and the strategies to follow. Empirical research done by (Frank, Enkawa, & Schvaneveldt, 2015) on the antecedents on repurchase intent shows that for collective customers brand image is crucial and firms are advised to shift their limited resources toward building a good public brand image in countries with a more collectivist culture.

In countries with a more individualistic culture, (Frank et al., 2015) recommend that firms put greater weight on satisfying customers. This conclusion for customer satisfaction appears to be valid for services but only for some products. Manufacturers of products thus are advised to conduct additional market research to assess the strength of this effect in their specific industry.

Moreover, building personal relationships with customers appears to be more effective in more collectivist cultures. This country difference emerges only in some service industries where it has a substantial effect size and thus makes a substantial strategic difference (Frank et al., 2015).

## Conclusion

Formal reports on happiness worldwide rankings like the World Happiness Reports, the Gallup, or the World Values Survey are of utmost importance to understand and compare the happiness level around the world. Even when these reports have a vast amount of data and well-established criteria of analysis, still there are questions left unanswered. For instance, why some wealthy and developed countries have a lower happiness level than some poorer and underdeveloped ones.

Cultural values impact societies' happiness beyond wealth and political stability. Individualism vs collectivism appears to be the cultural dimension with more research to it. A very high percentage of the evidence shows that individualism nurtures happiness, but collective societies have taken advantage of their networking quality to cope with scarceness and maintain mental and physical health. Even though the cultural dimension of power distance has less amount of research than individualism vs collectivism, it has a greater effect on happiness (Clark & d'Ambrosio, 2015; Graafland & Lous, 2018; Graham & Felton, 2006; Ye et al., 2015). Societies with a high level of power distance tend to accept human inequalities as normal and this diminishes people's self-esteem leading to the sadness and despair of the unprivileged ones. But even when people struggle with all type of circumstances, is their willingness to enjoy life, to focus on non-material values and to search for harmony, what could help them raise a few steps in the happiness ladder.

## References

Aaker, J. L., & Maheswaran, D. (1997). The effect of cultural orientation on persuasion. *Journal of Consumer Research, 24*(3), 315–328. https://doi.og/10.1086/209513

Ahuvia, A. C. (2002). Individualism/collectivism and cultures of happiness: A theoretical conjecture on the relationship between consumption, culture and subjective well-being at the national level. *Journal of Happiness Studies, 3*(1), 23–36. https://doi.org/10.1023/a:1015682121103.

Arrindell, W. A., Hatzichristou, C., Wensink, J., Rosenberg, E., van Twillert, B., Stedema, J., & Meijer, D. (1997). Dimensions of national culture as predictors of cross-national differences in subjective well-being. *Personality and Individual Differences, 23*(1), 37–53. https://doi.org/10.1016/S0191-8869(97)00023-8.

Ávila-Toscano, J. H. (2009). Redes sociales, generación de apoyo social ante la pobreza y calidad de vida. *Revista Iberoamericana De Psicología, 2*(2), 65–74.

Bank, T. W. (2018). Nearly half the world lives on less than $5.50 a day [Press release]. Retrieved from https://www.worldbank.org/en/news/press-release/2018/10/17/nearly-half-the-worldlives-on-less-than-550-a-day.

Barnett, R. C., Biener, L. E., & Baruch, G. K. (1987). *Gender and stress*. Free Press.

Bottero, W. (2004). *Stratification: Social division and inequality*: Routledge.

Bulmahn, T. (2000). Modernity and happiness—The case of Germany. *Journal of Happiness Studies, 1*(3), 375–399.

Carballeira, M., González, J. -Á., & Marrero, R. J. (2015). Diferencias transculturales en bienestar subjetivo: México y España. *Anales De Psicología, 31*, 199–206.

Carver, C. S., & Baird, E. (1998). The American dream revisited: Is it what you want or why you want it that matters? *Psychological Science, 9*(4), 289–292.

Chen, A., Lu, Y., Gupta, S., & Xiaolin, Q. (2014). Can customer satisfaction and dissatisfaction coexist? An issue of telecommunication service in China. *Journal of Information Technology, 29*(3), 237–252. https://doi.org/10.1057/jit.2013.26.

Clark, A. E., & d'Ambrosio, C. (2015). Attitudes to income inequality: Experimental and survey evidence. In *Handbook of income distribution* (Vol. 2, pp. 1147–1208). Elsevier.

Coupe, T., & Obrizan, M. (2016). The impact of war on happiness: The case of Ukraine. *Journal of Economic Behavior & Organization, 132*, 228–242. https://doi.org/10.1016/j.jebo.2016.09.017.

De Mooij, M., & Hofstede, G. (2002). Convergence and divergence in consumer behavior: Implications for international retailing. *Journal of Retailing, 78*(1), 61–69.

Delle Fave, A., Brdar, I., Wissing, M. P., Araujo, U., Castro Solano, A., Freire, T., . . . Soosai-Nathan, L. (2016). Lay definitions of happiness across nations: The primacy of inner harmony and relational connectedness. *Frontiers in Psychology, 7*. https://doi.org/10.3389/fpsyg.2016.00030.

Dhar, M. (2007). *Brand Management 101: 101 lessons from real-world marketing:* John Wiley & Sons.

Diener, E., Ng, W., Harter, J., & Arora, R. (2010). Wealth and happiness across the world: Material prosperity predicts life evaluation, whereas psychosocial prosperity predicts positive feeling. *Journal of Personality and Social Psychology, 99*(1), 52–61. https://doi.org/10.1037/a0018066.

Diener, E., Oishi, S., & Lucas, R. (2003). Personality, culture, and subjective well-being: Emotional and cognitive evaluations of life. *Annual Review of Psychology, 54*, 403–425. https://doi.org/10.1146/annurev.psych.54.101601.145056.

Diener, E., Suh, E. M., Smith, H., & Shao, L. (1995). National differences in reported subjective wellbeing: Why do they occur? *Social Indicators Research, 34*(1), 7–32.

Doran, K. (2002). Lessons learned in cross-cultural research of Chinese and North American consumers. *Journal of Business Research, 55*(10), 823–829.

Drakopoulos, S. A., & Grimani, K. (2013). *Maslow's needs hierarchy and the effect of income on happiness levels*.

Eternod, V., Martínez, M. d. C., & Moliner, B. (2015). A cross cultural perspective of the image satisfaction-loyalty relation on services. *International Journal of Arts and Commerce, 4*(9), 20.

Exton, C., Smith, C., & Vandendriessche, D. (2015). Comparing happiness across the world: Does culture matter? *OECD Statistics Working Papers, 2015*(4), 0–1.
Frank, B., Enkawa, T., & Schvaneveldt, S. J. (2015). The role of individualism vs. collectivism in the formation of repurchase intent: A cross-industry comparison of the effects of cultural and personal values. *Journal of Economic Psychology, 51,* 261–278. https://doi.org/10.1016/j.joep.2015.08.008.
Frey, B. S., Gallus, J., & Steiner, L. (2014). Open issues in happiness research. *International Review of Economics, 61*(2), 115–125. https://doi.org/10.1007/s12232-014-0203-y.
Gallup, I. (2020). *Gallup 2019 Global Emotions Report.* Retrieved from https://www.gallup.com/analytics/248906/gallup-global-emotions-report-2019.aspx.
Ger, G., & Belk, R. W. (1996). Cross cultural differences in materialism. *Journal of Economic Psychology, 17*(1), 55–77.
Goodwin, R., & Hernandez Plaza, S. (2000). Perceived and received social support in two cultures: Collectivism and support among British and Spanish students. *Journal of Social and Personal Relationships, 17*(2), 282–29. 1https://doi.org/10.1177/0265407500172007.
Graafland, J., & Lous, B. (2018). Economic freedom, income inequality and life satisfaction in OECD countries. *Journal of Happiness Studies, 19*(7), 2071–2093. https://doi.org/10.1007/s10902-0179905-7.
Graham, C., & Felton, A. (2006). Inequality and happiness: Insights from Latin America. *The Journal of Economic Inequality, 4*(1), 107–122. https://doi.org/10.1007/s10888-005-9009-1.
Green, E. G. T., Deschamps, J.-C., & Paez, D. (2005). Variation of individualism and collectivism within and between 20 countries: A typological analysis. *Journal of cross-cultural psychology, 36*(3), 321–339.
Hao, G. (2010). The impact of income on happiness: Absolute and relative measures. *Nankai Economic Studies, 5.*
Helliwell, J., Layard, R., & Sachs, J. (2019). *World Happiness Report 2019.* Retrieved from New York: https://worldhappiness.report/.
Hochschild, J. L. (1996). *Facing up to the American dream: Race, class, and the soul of the nation* (Vol. 51). Princeton University Press.
Hofstede, G. (1980). Culture and organizations. *International Studies of Management & Organization, 10*(4), 15–41.
Hofstede, G. (1984). *Culture's consequences: International differences in work-related values* (Vol. 5). Sage.
Hofstede, G. H., & Hofstede, G. (2001). *Culture's consequences: Comparing values, behaviors, institutions and organizations across nations.* Sage.
Hofstede, G., Hofstede, G. H., Arrindell, W. A., & Hofstede, G. H. (1998). *Masculinity and Femininity: The Taboo Dimension of National Cultures:* SAGE Publications.
House, R., Javidan, M., Hanges, P., & Dorfman, P. (2002). Understanding cultures and implicit leadership theories across the globe: An introduction to project GLOBE. *Journal of world business, 37*(1), 3–10.
House, R. J., Hanges, P. J., Javidan, M., Dorfman, P. W., & Gupta, V. (2004). *Culture, leadership, and organizations: The GLOBE study of 62 societies.* Sage publications.
Huenchuan, S., Guzmán, J. M., & Montes de Oca Zavala, V. (2003). Redes de apoyo social de las personas mayores: Marco conceptual. *Notas de población.*
Inglehart, R., & Welzel, C. (2010). The WVS cultural map of the world. *World Values Survey.*
Kahneman, D., & Deaton, A. (2010). High income improves evaluation of life but not emotional wellbeing. *Proceedings of the National Academy of Sciences, 107*(38), 16489–16493.
Kasser, T., & Ryan, R. M. (1996). Further examining the American dream: Differential correlates of intrinsic and extrinsic goals. *Personality and Social Psychology Bulletin, 22*(3), 280–287.
Kilbourne, W., Grünhagen, M., & Foley, J. (2005). A cross-cultural examination of the relationship between materialism and individual values. *Journal of Economic Psychology, 26*(5), 624641.
Kogevinas, M., Pearce, N., Susser, M., & Boffetta, P. (1997). Social inequalities and cancer.

Lakshmanasamy, T. (2010). Are you satisfied with your income? The economics of happiness in India. *Journal of Quantitative Economics, 8*(2), 115–141.
Leo, C., Bennett, R., & Härtel, C. E. J. (2005). Cross-cultural differences in consumer decision-making styles. *Cross Cultural Management, 12*(3), 32–62. https://doi.org/10.1108/13527600510798060.
Lu, L., Gilmour, R., & Kao, S.-F. (2001). Cultural values and happiness: An East-West dialogue. *The Journal of Social Psychology, 141*(4), 477–493.
Lu, L., & Shih, J. B. (1997). Sources of happiness: A qualitative approach. *The Journal of Social Psychology, 137*(2), 181–187.
Madariaga, C., Abello, R., & Sierra, O. (2003). Redes sociales, infancia, familia y sociedad. *Universidad del Norte. Barranquilla: Ediciones Uninorte.*
Marrero Quevedo, R. J., & Carballeira Abella, M. (2010). El papel del optimismo y del apoyo social en el bienestar subjetivo. *Salud Mental, 33*(1), 39–46.
Minkov, M. (2009). Predictors of differences in subjective well-being across 97 nations. *CrossCultural Research, 43*(2), 152–179.
Montero, M. (2003). *Teoría y práctica de la psicología comunitaria* (Vol. 5). Buenos Aires: Paidós.
Murphy, E. F., Jr., Gordon, J. D., & Anderson, T. L. (2004). Cross-cultural, cross-cultural age and cross-cultural generational differences in values between the United States and Japan. *Journal of Applied Management and Entrepreneurship, 9*(1), 21–48.
Myers, D. G., & Diener, E. (1995). Who is happy? *Psychological Science, 6*(1), 10–19.
Nayeem, T. (2012). Cultural influences on consumer behaviour. *International Journal of Business and Management, 7*(21), p78.
Oshio, T., Nozaki, K., & Kobayashi, M. (2011). Relative income and happiness in Asia: Evidence from nationwide surveys in China, Japan, and Korea. *Social Indicators Research, 104*(3), 351–367. https://doi.org/10.1007/s11205-010-9754-9.
Richins, M. L., & Dawson, S. (1992). A consumer values orientation for materialism and its measurement: Scale development and validation. *Journal of Consumer Research, 19*(3), 303316.
Rojas, M. (2012). Happiness in Mexico: The importance of human relations. In H. Selin & G. Davey (Eds.), *Happiness across cultures: Views of happiness and quality of life in non-western cultures* (pp. 241–251). Dordrecht: Springer Netherlands.
Rojas, M. (2018). Happiness in Latin America has social foundations. *World Happiness Report*, 115145.
Ryan, R. M., Sheldon, K. M., Kasser, T., & Deci, E. L. (1996). *All goals are not created equal: An organismic perspective on the nature of goals and their regulation.*
Schinzel, U. (2013). Why are people in Luxembourg happy? An exploratory study of happiness and culture measured by the dimension of a language as identifier in the Grand Duchy. *Journal of Customer Behaviour, 12*(4), 315–340.
Schmuck, P. E., & Sheldon, K. M. (2001). *Life goals and well-being: Towards a positive psychology of human striving.* Hogrefe & Huber Publishers.
Shavitt, S., Johnson, T., & Zhang, J. (2011). Horizontal and vertical cultural differences in the content of advertising appeals. *Journal of International Consumer Marketing, 23*, 297–310. https://doi.org/10.1080/08961530.2011.578064.
Sheldon, K. M., & Kasser, T. (1998). Pursuing personal goals: Skills enable progress, but not all progress is beneficial. *Personality and Social Psychology Bulletin, 24*(12), 1319–1331.
Shepelak, N. J. (1987). The role of self-explanations and self-evaluations in legitimating inequality. *American Sociological Review*, 495–503.
Smyth, R., & Qian, X. (2008). Inequality and happiness in urban China. *Economics Bulletin, 4*(23), 1–10.
Soares, A. M., Farhangmehr, M., & Shoham, A. (2007). Hofstede's dimensions of culture in international marketing studies. *Journal of Business Research, 60*(3), 277–284. https://doi.org/10.1016/j.jbusres.2006.10.018.
Suh, E., & Oishi, S. (2002). Subjective well-being across cultures. *Online Readings in Psychology and Culture, 10*. https://doi.org/10.9707/2307-0919.1076.

TenHouten, W. D. (2016). The emotions of powerlessness. *Journal of Political Power, 9*(1), 83–121. https://doi.org/10.1080/2158379X.2016.1149308.

Tov, W., & Diener, E. (2009). Culture and subjective well-being. In E. Diener (Ed.), *Culture and well-being: The collected works of Ed Diener* (pp. 9-41). Dordrecht: Springer Netherlands.

Triandis, H. (1988). Collectivism v. Individualism: A reconceptualisation of a basic concept in cross-cultural social psychology. In G. K. Verma & C. Bagley (Eds.), *Cross-Cultural Studies of Personality, Attitudes and Cognition* (pp. 60–95). London: Palgrave Macmillan UK.

Triandis, H. C., Kitayama, S. E., & Markus, H. R. E. (1994). *Emotion and culture: Empirical studies of mutual influence.* American Psychological Association.

Uchida, Y., & Kitayama, S. (2009). Happiness and unhappiness in east and west: Themes and variations. *Emotion, 9*(4), 441.

Veenhoven, R. (2012). Does happiness differ across cultures? In H. Selin & G. Davey (Eds.), *Happiness across cultures: Views of happiness and quality of life in non-western cultures* (pp. 451–472). Dordrecht: Springer Netherlands.

Wong, N., & Ahuvia, A. (1998). Personal taste and family face: Luxury consumption in Confucian and Western societies. *Psychology and Marketing, 15*, 423–441. https://doi.org/10.1002/(SICI)1520-6793(199808)15:5<423::AID-MAR2>3.0.CO;2-9

Ye, D., Ng, Y.-K., & Lian, Y. (2015). Culture and happiness. *Social Indicators Research, 123*(2), 519–547. https://doi.org/10.1007/s11205-014-0747-y

Zacharias, A., & Vakulabharanam, V. (2011). Caste stratification and wealth inequality in India. *World Development, 39*(10), 1820–1833.

# Happiness Digital Technology and Social Networks

**Francisco Mochón**

**Abstract** This chapter focuses on the analysis of the impact of new technologies on happiness. Traditionally, studies on the impact of technology on subjective well-being have been carried out by philosophers or thinkers and have generally been relatively negative. In this chapter, a pragmatic approach is adopted. It starts from accepting that technology offers tools that can have positive or negative effects on the happiness of individuals, depending largely on how the technology is used. As regards the new digital technology, the analysis takes as a central point the study of the impact on happiness of social networks and the Internet. In this sense, it is considered that although social networks can have a negative impact on the subjective well-being of individuals, in general their impact is positive due in large part to the possibilities that connectivity opens up. It is also argued that precisely the positive effect of social networks on happiness is one of the factors that contribute to explaining the impressive success of social networks. Social networks offer users something they strongly want, communicate and stay in touch with family and friends. This work closes with the analysis of the impact of a specific social network, Facebook, on the subjective well-being of individuals.

**Keywords** Happiness · Subjective well-being · Technology · Digital technology · Internet · Social networks

## Introduction

Happiness is important because it constitutes a final goal for human beings. Happiness is something to which one aspires and its search motivates human action. An example of this is that a resolution of the United Nations of 2012 states that "the pursuit of happiness is a fundamental human objective" (Rojas & Martínez, 2012).

For centuries, the study of happiness has been dominated by non-scientific traditions that are based on the idea that it is up to experts to judge the happiness of

F. Mochón (✉)
UNED, Madrid, Spain
e-mail: franciscomochon@gmail.com

© The Author(s), under exclusive license to Springer Nature Singapore Pte Ltd. 2021
T. Dutta and M. K. Mandal (eds.), *Consumer Happiness: Multiple Perspectives*, Studies in Rhythm Engineering,
https://doi.org/10.1007/978-981-33-6374-8_2

human beings (Rojas, 2014). Thus, the tradition of imputation is inspired by the work of philosophers, social thinkers and academics, and is based on the fact that it is a third party who defines what is the good life (McMahon, 2006; Tatarkiewicz, 1976). Experts propose criteria for making a judgment and make a list of observable attributes. Based on these observable attributes, the expert imputed the happiness—or well-being—of the people (Rojas, 2017).

The tradition of presumption recognizes that happiness is something that people experience. However, instead of inquiring directly and asking people about their welfare state, the tradition uses theories about nature and human behavior. In this way, lists of factors that are presumed to be closely related to a satisfactory life experience are obtained. Within this tradition, happiness is associated with achieving a set of factors that are believed to be relevant to achieve happiness.

Both the tradition of imputation and that of presumption are based on measuring the well-being of people based on a judgment made by a third person based on variables or attributes that are observable. This has led to the conception of well-being as a list of attributes (possessions, deficiencies and actions) and not as an experience of the people (Rojas, 2006,2007).

## The Scientific Study of Happiness

In the second half of the twentieth century, what has become the scientific study of happiness is born. Pioneering works arise from different disciplines: sociology (Veenhoven, 1984; Campbell, 1976; Campbell, Converse, & Rodgers, 1976), economics (Easterlin, 1973,1974; Praag, 1968,1971), psychology (Diener, 1984) (Andrews & Withey, 1976) and political science (Lane, 1991). In these works, it is evident that it is possible to study happiness scientifically based on its direct measurement.

The scientific study of happiness is based on a conception of happiness as a human experience and on the measurement of happiness by asking directly those who experience it. Under the new approach, the information that people provide about their well-being experience is valuable both for knowing their welfare situation and for studying the importance that different personal and social environment factors have for their happiness (Layard, 2006).

Happiness is a human experience so it cannot be conceived in the absence of human beings who experience it. Happiness is neither an academic creation nor an invention of philosophers, but an experience that happens to human beings. Consequently, the work of academics should consist in investigating it in order to understand what its explanatory factors are. The starting point is that happiness refers to people's experience of well-being, and that each subject is the one who can best report this experience because it is who experience it. Logically, happiness is inherently subjective, since it is an experience of the subject and this experience cannot exist without the person (Ferrer-i-Carbonell, 2002).

Happiness research requires high-level techniques to dealing with large information sets in order to extract that information which is relevant. In the study of happiness, there are many observations—as many as persons in the world—there are many variables, and there are many interrelations and synergies to take account of. In consequence, happiness research benefits from sophisticated models that allow for a better understanding of people's happiness. Without losing contact with what real human beings experience, it is important to use techniques that allow researchers to process all the information reaching valuable conclusions (Mochón & Rojas, 2014).

## Global Synthesis of Life: Satisfaction with Life and Happiness

People have wellness experiences and can make a global synthesis of them. This synthesis constitutes a global appreciation of how happy they are. This global synthesis of life, as well as the essential experiences of well-being, constitutes the object of the scientific study of happiness. The overall synthesis of life is usually made with phrases such as I am happy, I am satisfied with my life, my life is going well and I feel good about myself. The term happiness is used as a concept that refers to the overall synthesis of satisfaction with life. In other words, happiness refers to how the individual evaluates the overall quality of his life (Diener, Emmons, Larsem, & Griffin, 1985). As such, the happiness of individuals will depend entirely on an individual perception and it will be linked to concepts of quality of life and well-being.

The happiness of human beings depends on many factors, some of personal nature and others related to the conditions of their physical and social environment. In this sense, happiness is conditioned by a wide group of variables, among which the following stand out, social relations (family, friends and colleagues), the nature of work activities, parenting conditions, personality traits, availability and use of free time, the place where one lives, safety, the existence of children and their ages, the couple's relationship, household income, the macroeconomic environment, economic occupation, unemployment, health, values they have, expectations and the possibility of participation in political decisions (Frey, 2001; Ahn & Mochón, 2010; Praag & Ferrer-i-Carbonell, 2004; Rojas, 2016; De Juan, Mochón, & Rojas, 2014; Ahn, Mochón, & De Juan, 2012; Argyle, 2002).

## The Measurement of Happiness: The Use of Social Networks as a Possibility

Ed Diener and his collaborators presented a method to measure happiness based on the idea that individuals can consistently identify their level of satisfaction with life on scale, and as such, what must be done to ask people questions (Diener et al., 1985; Crooker & y Near, 1998; Grinde, 2002). This way of measuring happiness is the one that justifies conducting surveys like the World Values Survey, and it is the most widely used method. A numerical response scale is usually used (e.g., in the range of 0–10), where 0 represents the lowest satisfaction (lowest happiness) and 10 the highest satisfaction (greatest happiness). In addition to the question about the global synthesis of life, one can also ask the person about their satisfaction in different domains of life (Cummins, 1996).

As a result of the growth of social networks, a new possibility has emerged to measure happiness. This new approach consists of inferring the feelings of social network users on the basis of a semantic analysis of the words used in their communications and messages. Likewise, a study done by the Vermont Complex Systems Center uses information from Twitter to infer how happy or unhappy people in different states of the USA feel. Specifically, the researchers Dodds and Danforth have developed a method that, by incorporating the direct human evaluation of words, allows us to quantify levels of happiness on a continuous scale from a diverse collection of texts (Dodds & Danforth, 2009; Frank, Mitchell, Dodds, & Danforth, 2013). In the study carried out by Dodds and Danforth, on the basis of ten million "tweets," a code for determining to what extent each analyzed message can be catalogued as happy or sad was developed. The study focused on certain keywords that were deemed to be indicative. Following this approach in the article by Mochón y Sanjuan, the happiness of a large group of Latin American countries is measured through the use of social networks (Mochón & Sanjuán, 2014). Specifically, the social network used is Twitter. The paper shows that it is possible to calculate, via objective and empirical means, factors that allow us to measure happiness through the use of social networks.

## Happiness and Social Relations

The idea that the welfare of individuals is conditioned by their social relationships and social context is something generally accepted. It is argued that there is an important interaction between the social context and the attitude of individuals to their environment, which has a notable impact on the subjective satisfaction of people (Diener, Seligman, Choi, & Oishi, 2018; Mochón & Juan, 2015; , 2017). It has been pointed out that relationships that people cultivate in their lives are some of the most valuable treasures a person can own. Given that several researchers have dealt with the impact of technology on social relationships and consequently on happiness, we should highlight the social component of subjective well-being.

In the literature on happiness have been identified a group of socioeconomic determinants of the subjective well-being of individuals, among which is the network of social relationships; this is social capital and relational goods (Mochón & Juan, 2017; Frey & Stutzer, 2000, 2002; Layard, 2003; Veenhoven, 1993, 2000, 2001; Iglesias, Pena, & Sánchez, 2013; Gui & Sugden, 2005).

The network of social relationships is the result of situations as varied as family and marriage, relationships with friends and neighbors, relationships in the workplace or the use of new technologies related to the Internet, email, social networks, SMS,

WhatsApp. The key is that this type of relationships affects happiness and also promotes integrity and trust in others (Nussbaum & Sen, 1993).

New technologies have introduced a new way of relating among friends, family and co-workers. Social networks allow you to interact daily with all your friends by sending messages, photographs and videos, which makes it easy to share experiences and keep the relationship alive. In this sense, it is worth highlighting the growing importance of WhatsApp as a communication tool. Thus, among the young WhatsApp is used not only to exchange messages but as a tool to discuss the doubts of class and solve problems that may arise from the duties. In this sense, it can be affirmed that social networks have contributed to create new links between people. In any case, its impact on happiness is a controversial issue that requires careful analysis, as we will see in the following epigraphs.

## Technology and Happiness

In a historical perspective, the relationship between technology and happiness has been a constant object of study by economists and social scientists since the advent of the Industrial Revolution. Generally, attention has focused on the relationship between material prosperity and well-being. In this sense, Gregg Easterbrook in his book The Progress Paradox said that although thanks to advances in technology almost all aspects of Western life have vastly improved in the past century—and in the present, surprisingly most men and women feel less happy than in previous generations (Easterbrook 2004).

In any case, the key work to analyze the relationship between prosperity (caused by technological progress) and well-being was made by Professor Richard Easterlin (Easterlin 1974,2017; Easterlin & Angelescu, 2009; Easterlin et al. 2010) who showed that in developed countries there was no real correlation between a nation's income level and its citizens' happiness. The results of Easterlin's work, known as *The Easterlin's Paradox*, state that at a point in time happiness varies directly with income both among and within nations, but over time happiness does not trend upward as income continues to grow. The original conclusion for the USA was based on data from 1946 (which is when formal surveys of happiness started) to 1970; later evidence through 2014 confirmed the initial finding. The trend in USA happiness has been flat or even slightly negative over a roughly seven decades stretch in which real incomes more than tripled.

From Easterlin's research work, it is inferred therefore that you could give people more income and consequently more choice possibilities and not have much of an impact on their sense of well-being. In other words, it seems as if from a certain level of income, people get used to high levels of income and value less and less the increases in income.

We find this same idea if we refocus the analysis on the relationship between happiness and technology and leave income aside. It seems as if people adapt very easily to the advantages that technology brings and no longer make them happy.

So, let us imagine that at the end of the nineteenth century we asked anyone if they would be happier if they could have a vehicle that would allow them to travel in a day hundreds of kilometers, or if they could cross the Atlantic in a few hours or talk to a person who it is located thousands of kilometers away. It is very likely that we would say yes. However, few people today associate their happiness by having cars, traveling by plane or talking by phone with a relative who lives in another continent. The usefulness of advances in technology is recognized, but we quickly become accustomed to these advantages. But not only that, it is even considered that these advances can be a source of stress and frustration. Therefore, it is not clear that advances in technology make us happier (Surowiecki, 2005).

This facility to adapt to the advantages of new technology coincides with one of the conclusions obtained by research in happiness that people adapt very quickly to the good news. Thus, for example, it has been shown that if a person touches the lottery, at that time he will feel euphoric and very happy, but after a reasonable time he will return to his habitual levels of happiness (Rojas, 2009).

The fact that we adapt very quickly to advances in technology does not mean that technology does not have positive or negative effects on our quality of life and consequently on happiness. The relevant thing is that its net impact is not always easy to determine. We will start with the positive effects of technology on happiness, and later we will comment on the negative effects.

## Positive Effects of Technology on Happiness

The theory of economic growth has shown clearly that the main driver of growth and improvement of living conditions has been technological progress. In this sense, it would seem logical to think that new technologies not only make people live better but that they should also be happier.

In this sense, new technologies, as consumers, have a positive effect at least during certain periods of time by providing us with a wide range of new products, such as cars or household appliances and by improving the quality of them.

Technology can also be used to communicate with one another. Thus, for example, the Internet or a mobile phone is communication tool that can be used to enrich social relationships. As a communication tool, technology can be used as a means to connect, to share knowledge or to empower people. In this sense, its impact on happiness is positive. But the relationship between happiness and technology, when it is used as a tool to communicate, is, as we will see, quite complex.

Technology has also radically changed the nature of work for most workers. This matters because the workplace is very important to people sense of well-being. With the Industrial Revolution, mechanization allowed workers to escape from agriculture. Although they were often thrown initially into hard industrial jobs, over time, and thanks to the significant increases in productivity, very substantial improvements in working conditions and wages have taken place. More recently, the appearance of the digital society, and the advent of knowledge-based businesses, means that workplaces have become less formal and more open, often creating a really nice work environment (Surowiecki. 2005; Hochschild, 1997). Thanks to technology, the individuals have become global. We live in a world where borders have been erased, overcoming the limitations of place and space. Some people work in offices, while others do it from their homes or even in a coffee shop. We move fluidly in and out of the hazy world of the Internet-based "cloud" with part of our belongings in the physical world and other part in the virtual world.

In any case, where technology has had a more significant impact on the well-being of people is in the field of health. An example of this is the considerable increase in life expectancy that has taken place in the vast majority of countries in the last hundred years. The highlight is the majority of people are happy to be alive, and if they live longer they will feel happier.

## Negative Effects of Technology on Happiness

The origin of criticism of technology has focused on what Heidegger's terminology is known as the question of technology—that is, the impact of technology on our humanity (Heidegger, 1977). In this sense, it has been questioned peoples' ability to use technology to their own ends. Heidegger highlights the role of technology in bringing about the decline of human beings by constricting our experience of things as they are. He argues that we increasingly view human beings, only technologically—that is, we view people only as raw material for technical operations. We treat even human capabilities as though they were only means for technological procedures. People are mere human resources to be arranged, rearranged and disposed of (Blitz, 2014). We tend to believe that technology is a means to our ends and a human activity under our control. But in truth we now conceive of means, ends and ourselves as fungible and manipulable. For these reasons, Heidegger denounces technology harmful effects and the view that technology is a neutral tool to be wielded for either good or evil.

Following the contribution of Heidegger, the two main criticisms of technology for their impact on happiness have a somewhat contradictory meaning. On the one hand, it is pointed that technological progress is leading to an ever more rigid, controlled, soulless society, in which it is easier for people to be manipulated and monitored. In this sense, Jacques Ellul shows his concern for the emergence of a technological tyranny over humanity (Ellul, 1964). On the other hand, it has been criticized, referring especially to the role of television, how the most popular media of a time in history shapes the discourse of the world (Postman, 1985). From a different perspective, Putman has pointed out that technology is contributing to the reduction in all the forms of in-person social intercourse. The consequence of this is a fragmented society, in which traditional relationships are harder to sustain and a reduction of the social capital (Putnam, 1995).

From this pioneering contributions, the idea that technology disrupts social relationships and fractures the community has gained followers and as will be seen later has become central to the critique of the Internet. From this perspective, technology, and more specifically the Internet, supposedly isolates people from what critics always call the real world. One of the first times this criticism was pointed out was in a famous study conducted among the residents of the city of Pittsburgh (USA) and published in September 1998 (Kraut et al., 1998). In this article, it is pointed out that the Internet, being a communication tool, instead of allowing people to connect with a much wider set of potential friends and exposing them to information they might otherwise never have come across, made people more depressed and lonely than they would otherwise have been. According to the authors of this work, the Internet could change the lives of average citizens as much as did the telephone in the early part of the twentieth century and television in the 1950s and 1960s. For this reason, it is interesting to try to find out whether the Internet is improving or harming participation in community life and social relationships. According to the results of this research work, the Internet was used extensively for communication. Nonetheless, greater use of the Internet was associated with declines in participants' communication with family members in the household, declines in the size of their social circle and increases in their depression and loneliness. The authors described this result as a paradox, since the Internet, as a communication tool, should improve the subjective well-being of individuals.

Although this research work had a great impact, its statistical support is not very solid, only 169 people from 73 households were interviewed. In fact, a few years later, some of the authors re-analyzed the issue and found that negative effects of Internet dissipated (Kraut et al. 2001). In the new research work, the authors report that the people investigated generally experienced positive effects of using the Internet on communication, social involvement and well-being. However, using the Internet predicted better outcomes for extraverts and those with more social support but worse outcomes for introverts and those with less support.

The criticism of technology, and particularly the Internet, for its impact on social relations is especially relevant from the perspective of happiness and deserves special attention. Keep in mind that one of the main conclusions of the scientific study of society is the existence of a high correlation between happiness and social relations. Logically a tool as broad and ubiquitous as the Internet will have a multitude of effects, some may be negative but others not. In addition, in essence, the Internet is a communications technology, one that, like the telephone, allows people to expand their affective and informational networks and this is something that people value positively. Obviously, the Internet is not the ideal place to establish all kinds of communications, but in any case it is a public communication area that works openly and without gatekeepers. Therefore, criticizing technology and, in particular, the Internet in a generalized manner due to its alleged negative effects on subjective well-being may be excessive (Surowiecki, 2005).

A less controversial way in which technology can negatively affect people's happiness is in its relentless generation of newness (Surowiecki, 2005). One of the implications of studies on happiness is that people have a hard time being happy with what they have when they know that others have more or have better things (Buss, 2000; Praag et al., 2003). Nowadays, technological change takes place so quickly that if we buy any technological product (a mobile phone, a computer, a television, etc.), we know that in a few months there is going to be a better, faster version of the products. We will be left with obsolete products, while other people will have new and more technologically advanced products, which may negatively affect our well-being. This feeling there is no way to avoid it and it is in the heart of the modern consumer.

And then, there are the worries about artificial intelligence [AI] and the technological displacement of labor. Simply by focusing on robotics, it has the potential to transform lives and work practices. Its impact will be increasing, as the interactions between robots and people multiply. Although there is no consensus on the effects that this will have on employment, what is indisputable is that its impact will be very important and difficult doubts arise. How should the benefits of robotics be distributed? The universal basic income will no longer be a possibility and will become an obligation, and given the important effect that employment has on subjective well-being, how will all this affect the happiness of the individual? (Mercader, 2017).

## Digital Technology, Social Networks and Happiness

In a recent research carried out in Spain (Mochón & Juan, 2017), the incidence of social networks on happiness is analyzed. It is observed that individuals, regardless of their age, who use social networks have, on average, a greater life satisfaction than those who do not use them. The results of the survey show that, in addition, those with more than 65 years of age who use social networks feel more satisfied even than those of mature age. It seems that social networks can be a good way to combat loneliness. The feeling of being communicated at any time of the day with your friends and family and being able to share images, videos, etc., with them, it makes individuals more satisfied.

In some other research work, it has also been found that virtual relationships can be as intimate as in-person relationships (Brannan & Mohr, 2018). In fact, Bargh and colleagues found that online relationships are sometimes more intimate (Bargh, McKenna, & Fitsimons, 2002). This can be especially true for those individuals who are more socially anxious and lonely—such individuals who are more likely to turn to the Internet to find new and meaningful relationships (McKenna, Green, & Gleason, 2002, McKenna 2008). In other words, these research works suggest that for people who have a hard time meeting and maintaining relationships, due to shyness, anxiety or lack of face-to-face social skills, the Internet can offer a safe, nonthreatening place to develop and maintain relationships. Likewise, some researchers have shown that young people are using digital technology and online social media within their everyday lives to enrich their social relationships (Hynan et al. 2014).

In any case, the effects are not always positive; depending on how the Internet is used and in particular the social networks, these can be beneficial or harmful (Pénard, Poussing, & Suire, 2013; Holsten, 2018; Zhan & Zhou, 2018). In this sense, one reason why Internet technology can have negative effects on happiness is due to the corporate and governmental power to surveil users (attendant loss of privacy and security). To this, we must add the effect of the addictive technologies that have captured the attention and mind space of the youngest generation.

Thus, although until recently social networks were presented as an instrument of socialization, in recent dates doubts have grown. Especially since 2017, criticism of the social networks has proliferated, largely due to the scandals related to Facebook (Carmody, 2018). It has been argued that the platforms are designed to hook the user and get us to spend as much time as possible in them creating addiction. It has also been criticized that social networks tighten the debate as they filter the information showing only a view of the facts and contaminate it with false information, and that even they can be a tool to manipulate democratic electoral processes.

## The expert's Opinion

To try to offer a global vision of the impact of digital technology on happiness, we will analyze the results of a research that adopts a similar approach as that used by researchers to measure happiness: Ask the interested parties (Anderson & Rainie, 2018). In this sense, Pew Research Center and Elon University's Imagining the Internet Center decided to query 1,150 technology experts, scholars and health specialists on the following question:

- Over the next decade, how will changes in digital life impact people's overall well-being physically and mentally?

The conclusions of this investigation can be summarized by saying that 47% of those queried predict that individuals' well-being will be more helped than harmed by digital life in the next decade, while 32% say people's well-being will be more harmed than helped and the remaining 21% predict there will not be much change in people's well-being compared to now (Anderson & Rainie, 2018).

As a general comment, it can be said that many of those who argue that human well-being will be harmed also acknowledge that digital tools will continue to enhance various aspects of life. They also note there is no turning back in the sense that new technologies are here to stay. At the same time, hundreds of them suggested that digital tools could mitigate the problems and emphasize the benefits. Moreover, many of the hopeful respondents also agree that some harm will arise in the future, especially to those who are vulnerable.

To analyze the answers of the interviewees in a systematic way, these can be classified into three categories: (1) the positive effects of digital technology, (2) the negative effects of digital technology and (3) remedies to mitigate the possible negative effects.

1. *The positive effects of digital technology.* The benefits of digital life on happiness are analyzed in terms of the following four factors [73]:

- *Connection.* Digital life links people to people, contributing to spread the knowledge, facilitating education and supplying entertainment anywhere globally at any time in an affordable manner. People need to be connected, and the Internet is a communication tool par excellence. In subjects specific to society, science, education or politics, the Internet connects people by facilitating rewarding information and relationships.
- *Commerce, government and society.* Digital life revolutionizes civic, business, consumer and personal logistics, opening up a world of opportunity and options.

  To show the advantages of a hyperconnected society, let us think about the massive benefits to life from access to finance, to online shopping, to limitless free research opportunities, to keeping in touch with loved ones in faraway places.
- *Crucial intelligence.* Digital life is essential to tapping into an ever-widening array of health, safety and science resources, tools and services in real time. Advances in computer science have meant that information is increasingly distributed globally and openly. For example, the relatively recent trends toward openness in scientific publication, scientific data and educational resources are likely to make people across the world better off by expanding individuals' access to a broad set of useful information, by decreasing barriers to education and by enhancing scientific progress.
- *Contentment.* Digital life empowers people to improve, advance or reinvent their lives, allowing them to self-actualize and meet soul mates. The Internet helps to break down barriers and supports people in their ambitions and objectives. Internet helps people achieve their desire to improve their education to communicate with others, to share their experiences, to create networks of enterprise, commerce, culture, sports. All these are supported by digital technologies.
- *Continuation toward quality.* Emerging tools will continue to expand the quality and focus of digital life; the big picture results will continue to be a plus overall for humanity. The future artificial intelligence (AI) will enhance human well-being. Throughout history, it has been shown that human beings need tools and want improvements, and AI is facilitating them and will continue to do so. And the saying goes "First we make our tools, then our tools form us."

2. *The negative effects of digital technology.* The negative impact of digital technology on happiness is analyzed in terms of the following five factors [73]:

    - *Digital deficits.* People's cognitive capabilities will be challenged in multiple ways, including their capacity for analytical thinking, memory, focus, creativity, reflection and mental resilience. The digital society is characterized by an intrusive connectivity that has harmful cognitive and emotional consequences.
    - *Digital addiction.* Internet businesses are organized around dopamine-dosing tools designed to hook the public. The current generation of tools for consuming attention is very effective and can cause addictive effects. Network

effects and economies of scale have placed control of these tools in a very small number of very powerful companies.
- *Digital distrust/divisiveness.* Personal agency will be reduced, and emotions such as shock, fear, indignation and outrage will be enhanced. Although technologies are created with a sincere desire to advance understanding of mood, cognition, etc., or with the pretension to facilitate the control of our response, the actual implementation of these techniques and devices is likely to be quite different. It is possible that they may finally be used to reduce well-being because a population in a state of fear and anxiety is a far more malleable and profitable.
- *Digital duress.* Information overload + declines in trust and face-to-face skills + poor interface design = rises in stress, anxiety, depression, inactivity and sleeplessness. There are organizations that are actively vying people's attention, distracting them with smartphone notifications, highly personalized news, addictive games, Buzzfeed-style headlines and fake news.
- *Digital dangers.* The structure of the Internet and pace of digital change invite ever-evolving threats to human interaction, security, democracy, jobs, privacy, etc. In addition, many people are unable to adapt to the behaviors and needs that digital technology requires.

3. *Remedies to mitigate the possible negative effects.* Five possible lines of action are presented to combat the possible problems that digital technologies may cause [73]:

- *Reimagine systems.* Societies can revise both tech arrangements and the structure of human institutions—including their composition, design, goals and processes. The challenge to be overcome is neither more nor less than simply learning to call what we have created what it really is and then regulate and manage it accordingly.
- *Reinvent tech.* Things can change by reconfiguring hardware and software to improve (AI), virtual reality (VR), augmented reality (AR) and mixed reality (MR). We can resort to human-centered technology design to improve our experiences and outcomes, to better serve us.
- *Regulate.* Governments and/or industries should create reforms through agreement on standards, guidelines, codes of conduct and passage of laws and rules. Security and privacy cause great concern for what is necessary to come to some kind of detente.
- *Redesign media literacy.* Formally educate people of all ages about the impacts of digital life on well-being and the way tech systems function, as well as encourage appropriate, healthy uses. The primary change needs to come is in education. From a very early age, people need to understand how to interact with networked, digital technologies.
- *Recalibrate expectations.* People must gradually evolve and adjust to digital changes. People must learn how to reign in the pitfalls, threats, bad guys and ill-meaning uses.

## Happiness and Social Networks: Special Consideration of the Facebook Case

The approach that has been followed so far to analyze the relationship between social networks and subjective well-being has been somewhat speculative and generic. To finish this work, it may be interesting to adopt a more concrete approach. Many of the doubts regarding the positive impact of social networks have arisen as a result of criticisms of the behavior of the best known social network, Facebook. Therefore, it makes sense that we give special consideration to the relationship between happiness and Facebook.

To follow a systematic approach, we will carry out the analysis following the different missions that according to the CEO of the company, Martin Zuckerberg, Facebook has had throughout its history. At first, Facebook was created to allow users to know each other and communicate (Zuckerberg, 2019a). Subsequently, the platform evolved so that it also served as a sounding board for "flashy" news and gossip (Sherman, 2018a). According to Zuckerberg, Facebook's next mission was to promote the creation of meaningful online communities that make us grow as people (Zuckerberg, 2017). The last and current mission of Facebook is to allow users that in their relationships through Internet can have privacy (Zuckerberg, 2019b).

We will analyze in what sense the different missions that Zuckerberg has entrusted to Facebook positively or negatively affect the subjective well-being of users. This analysis will be carried out regardless of Facebook's commercial intentions, its strategies to optimize the value of advertising and its frequent violations of data privacy (Sherman 2018b,c; Kessler 2018). The hypothesis that is defended in this work is that the positive impact of Facebook on the happiness of its users is one of the explanatory reasons for the success of the social network.

As Zuckerberg has repeatedly stated, Facebook's initial mission was to create a platform that would help create a more open and connected world (Zuckerberg, 2019a). To do this, Facebook has been careful to offer users something that people want, be connected and the ability to learn from each other. The fact that stands out is that Facebook efficiently allows its users to be connected and this has a positive impact on their happiness. The impact on the happiness of the users takes place regardless of whether the true objective of the social network is to optimize advertising revenue. In other words, one thing is the facilities and services that Facebook provides to its users and another very different is how the company manages the information it obtains from its users.

## User Data, Advertising Revenue, Marketing and Happiness

From an economic perspective, the key to Facebook's success is its ability to precisely segment users according to their tastes and preferences, which maximizes the effectiveness of advertising (Sherman, 2018a). On the other hand, Facebook knows that the more restrictive users are in relation to the type of information they provide or the lower the amount of information that users share on the social network, the lower the "rent" that Facebook can charge for targeted ad placements. This explains why Facebook has no interest in limiting the information that users disseminate through their social network.

Consider, for example, that an advertiser wishes to promote a high-quality television set, which has certain technological innovations. The advertiser would be willing to pay a relatively high price per thousand impressions if he knows that his ad will only be seen by customers with the greatest purchase potential for this type of television (defined by gender, income, location, purchase history, interest and willingness to pay for a smart television of superior quality, etc.). For these reasons, Facebook can be a good option to promote the television, as it offers advertisers the option to adapt their advertising spending to match their strategic objective: pay per impression, per click or similar. Since Facebook has a very complete knowledge of consumers, it offers advertisers extraordinary precision to reach the right audience.

These types of issues determine the amount of revenue that Facebook can get from online advertising. Logically, Facebook pays the utmost attention to all the details that can condition the generation of income from user data. Facebook tracks all the movements of its users on and off the platform. He knows what they like, what they share and what they do not. Facebook introduces this knowledge accumulated in algorithms to achieve two objectives: Maximize the participation (engagement) of its users on the platform, and persuade them to buy products and services from their advertisers. Above all, Facebook is an attention and persuasion optimization platform. Connecting friends and family is a laudable goal but not the engine of an ad-based strategy.

The economic value of user data to generate advertising revenue can explain why Facebook has violated the privacy of the information. The fact that stands out is that most Facebook users do not seem to care much about data privacy. Initially, the Cambridge Analytica case triggered a sharp drop in the confidence of Facebook users in the company (Sherman, 2018a). The distrust remained after Zuckerberg's apologies during his testimony in the Congress and his promises that he would try to correct the rulings. Despite Zuckerberg's promises, less than 30% of Facebook users in the USA really believe that Facebook is committed to protecting their private data. However, only between 1 and 3% said they would cancel their Facebook account due to these facts. Moreover, the number of daily active Facebook users grew by 13% in March 2018 despite the widespread dissemination of the Cambridge Analytica case and the audience at the Congress. These data suggest that users do not believe that Facebook is really committed to protecting their privacy, but they continue to use the platform. It seems as if the vast majority of Facebook users were not upset about what the company can do with their data.

In any case, from the point of view of the analysis that is being carried out in this work, the important thing is not the advertising revenue that Facebook obtains. The relevant thing is how users feel about the possibilities of interacting with their friends and family that Facebook offers them. Users know how Facebook has behaved and are aware that it has repeatedly violated the privacy of the data it uses. However, this has not led them to leave the platform. It seems as if users think that one thing is the moral principles that guide Zuckerberg and another is what Facebook gives them. Probably, the most coherent explanation of why Facebook users do not leave the platform, although some of their practices are not recommended, is because of the value they give to the advantages of being connected. In other words, users recognize the advantages that the social network gives them by being able to connect with their friends and family.

Facebook users continue on the platform, although they know that it has some negative things because the positive aspects it provides compensate for the negatives. The platform gives them relationships and allows them to interact for free with the people they want. Many of Facebook users have made the platform an integral part of their daily lives to capture memories, stay connected, entertain, share information and avoid fear of missing out (FOMO). No one wants to feel disconnected. Therefore, despite all the controversies and scandals, the number of users continues to grow.

This suggests that users value the services provided by Facebook very much because they have a positive impact on their subjective well-being. In other words, the most logical explanation that users continue to prefer being on Facebook is because it positively affects their subjective well-being.

But in addition to the possibility of connecting with friends and family and establishing relationships, what other things does Facebook offer to people who attract and engage them?

## Facebook Offers the Type of Information that Users Want

The evolution followed by Facebook has turned it not only into a "site" in which it connects with family and friends but also a source of information. Do not forget that Facebook seeks to maximize user engagement and this determines the type of information provided. Facebook has become a main source of news for a large segment of its user community, and the type of content they are seeing on the platform is illustrative of Facebook's objective. Facebook's News Feed has given priority to information worthy of "Like." That information attracts attention and encourages content creators to be increasingly sensational to increase engagement. The consequence of this is that the sources of news and publications strongly biased to reaffirm partisan beliefs, with news of celebrities and funny videos, are those that have prospered on Facebook, at the expense of serious journalism.

In addition, it must be remembered that Facebook offers the type of information that its users want, and many of them want novel and fun news. On the other hand, some of the characteristics that have fun and novel news also have fake news and gossip. Therefore, it should not surprise us that fake news and gossip are frequently found on Facebook.

From a psychological perspective, it may be that gossip and invented stories simply entertain us more than the real ones. Likewise, fake news and gossip are generally exciting and sensational and on many occasions they appear a villain who we can blame for what we do not like. Fake news can be silly, but it is often fun and helps satisfy the desire to ease the burdens of our real life. In short, there are several reasons that justify the false news having a great impact and spread more than the real ones (Rita Watson, 2018; Vosoughi, Roy, & Aral, 2018). Likewise, behavioral science confirms that we are prepared to be attracted to rumors and gossip that provide entertainment or reaffirm our political or social beliefs, regardless of the reliability of the source.

In any case, Zuckerberg has explicitly rejected on several occasions Facebook's responsibility to disseminate erroneous information or unproven gossip. He has argued that Facebook is a technology company and not a media company. Facebook does not have people dedicated to producing content or editing content. Although this is true, Facebook understands and encourages people's intrinsic attraction to fake news and gossip. Users like fake news and gossip, and Facebook knows it. That is why it is easy to understand that Facebook encourages the spread of false news and gossip in a concerted manner. Moreover, Facebook has the collaboration of its users to disseminate this type of information. In any case, Facebook encourages false news and gossip, not only because users value them positively and favorably influence their happiness, but to increase engagement.

The importance of gossip and gossip is not a thing of our time, encouraged by social networks or something that cannot be taken lightly. In fact, some authors highlight the capital role that gossip had for the evolution of language and for Homo sapiens to achieve supremacy over other humans (Dumbar, 1998). Therefore, it cannot surprise us that being aware of the gossip of the people is one of the reasons why people are attracted to social networks. In fact, a determining factor in the success of Facebook in its origins was the interest of the students of Harvard University to be aware of the gossip of their peers, and this is something that Zuckerberg has always been fully aware of. Before creating Facebook, in 2003 Zuckerberg launched Facemash, a Web site that offered information about students, which had been obtained in violation of all kinds of information and privacy ownership rules (Rebecca O'Brien, 2017).

In summary, Facebook offers the type of information that users want and thus contributes positively to their happiness. In addition, by facilitating the type of information desired, it manages to engage users and this is a key variable for the success of a social network. These types of facts lead us to think that to explain the success of Facebook it may be interesting to include an additional variable in the analysis, the happiness or subjective well-being of the users.

## A New Mission: Create Communities

In this review, we are doing to see what Facebook offers that people like the next fact to analyze is to what extent Facebook encourages the creation of communities.

In a long open letter published by Zuckerberg in 2017, he stated that from its creation until June 22, 2017, Facebook's mission had been to "make the world more open and connected" (Zuckerberg, 2017). But as of the aforementioned date, "Our complete mission statement is: to empower people to build communities and make the world more united." According to Zuckerberg, Facebook needs people to have the opportunity to express themselves to obtain a diversity of opinions. But communities must be propitiated because they bring together enough points in common so that we can all move forward together. In Zuckerberg's opinion, we need to be connected with the people we already know and care about, but we also need to meet new people with new perspectives.

This change of mission is justified by Zuckerberg for the desire to try to combat something pointed out by researchers specialized in the field of social capital and happiness, the decline in the number of affiliations to communities observed worldwide (Helliwel 2006). Although it seems surprising, the fact is that during the last decades, belonging to all types of groups and communities has decreased. That implies that a large number of people now need to find a sense of purpose and support somewhere else. Communities allow us to feel that we are part of something bigger than ourselves, that we are not alone and that we have something better ahead of us to work for. Zuckerberg points out that studies have shown that the more connected we are, the happier we feel and the healthier we will be (Zuckerberg 2017).

For all, the communities to which we belong are relevant and have meaning. Whether it is a religious, sporting or cultural community, communities give us the strength to expand our horizons and worry about broader issues than strictly personal ones. But not all communities are equal. A community to exchange photographs of toy poodles is not the same as a community to help children with Down syndrome. The objective of Zuckerberg is to promote the creation of significant communities. These are those to which by joining, they quickly become the most important part of our experience in social networks and an important part of our support structure in the online world or social support in the real world. Significant online communities, such as those that aim to end poverty, cure diseases, change the culture of domestic violence or stop climate change, strengthen physical communities by helping people to join online and in physical life, including over great distances.

As Zuckerberg acknowledged, making the new mission a reality is something that Facebook cannot achieve on its own. "This is not going to happen from top to bottom. There is no one in the world who can snap your fingers and make this happen. People have to love it." The change begins locally, when a sufficient number of people have a sense of purpose and can begin to worry about broader issues. A successful mission change can only be achieved by empowering people and helping them build communities and bring people together in shared projects. In order for there to be significant communities that Facebook can support, there is a need for leaders to launch projects. And that is why community leaders are so important for the new mission.

Facebook's own experience shows that every great community has great leaders. Leaders establish culture, inspire others, offer a safety net and take care of community members. Therefore, to realize its new mission, Facebook is giving more leaders the power to build communities. Historically, this is not how groups have been created on Facebook. The groups have been quite flat where everyone is equal, and that makes sense when talking about a family group or friends who do not really have a leader. But it makes no sense when trying to lead a group with thousands of people. In this sense, Facebook has begun to implement new tools to facilitate the creation of communities and to facilitate the task of leaders. Facebook has designed a clear roadmap and provided new tools that facilitate information on who the members are and how they get involved: information on participation requests and the time of day in which their members are most active. Tools have also been created to manage the requests of new members. The leader can sort and filter requests by location and gender, and group them so that he can accept or reject them all at once. The new tools help to quickly eliminate bad actors and their content to maintain a positive and safe environment. In any case, the fact that stands out is that the key element of any community is its leader. The artificial intelligence facilities that Facebook puts at the service of the leaders, although very important, are mere tools, and the important thing is the leader. In this sense, Zuckerberg affirms that we all have the power to be leaders. And if enough of us work to build a community and bring people together, we could change the world (Zuckerberg, 2017).

Zuckerberg to impose this mission change has set very specific objectives "We want to help a billion people to participate in meaningful communities." This is how Facebook will achieve the greatest opportunities and build the world the company wants for generations to come. "If we can do this, not only will the decline the number of people belonging to communities that have been observed over the past decades change, but it will also begin to strengthen our social fabric and the world will be more united. We have to build a world where everyone has a sense of purpose and community. This is how we will unite the world. We have to build a world in which we care about a person in India or China or Nigeria or Mexico as much as a person in our neighborhood" (Zuckerberg, 2017).

In addition to these altruistic goals about social capital and happiness, when Zuckerberg decided to change Facebook's mission, he also took into account other facts. According to the information available, users who belong to relevant communities feel more committed to the social network, tend to be connected longer and more regularly and are more loyal. So, to carry out the mission change and promote communities there are also business reasons. This change is also intended to boost growth and engagement. In any case, the subjective well-being of individuals will be also positively affected by this Facebook mission change. Promote the creation of communities is something that increases the social capital of users and consequently their happiness.

## From Creating Communities to Enhancing Privacy

Zuckerberg's ability to reinvent the mission that Facebook must carry out, it never ceases to amaze us. On March 6, 2019, Zuckerberg published another extensive note in which he announced the construction of a social media and message platform focused on privacy (Zuckerberg, 2019b). That is something that may seem surprising because it involves a profound change of orientation in the way Facebook acts. According to Zuckerberg, in the last 15 years, Facebook has helped people connect with friends, communities and interests in the open, but increasingly, people also want to connect privately. "When thinking about the future of the Internet, I believe that a communications platform focused on privacy will be even more important than today's open platforms. Privacy gives people the freedom to be themselves and connect more naturally" (Zuckerberg, 2019b).

The real reasons for launching a private social network are pointed out by Zuckerberg; "Today we see that private messages, ephemeral stories and small groups are by far the fastest growing areas of online communication. There are a number of reasons for this. Many people prefer the privacy of communicating one by one or only with a few friends. People are cautious of having a permanent record of what they have shared. And we all hope to be able to do things like payments privately and securely."

In Zuckerberg's opinion, the future of communication will increasingly change to private and encrypted services where people can trust that what they say is kept safe and their messages and content will not stay forever.

Public social networks will continue to be very important in people's lives, to connect with all their acquaintances, discover new people, ideas and content, and give them a broader voice. People find this valuable every day, and there are still many useful services to build on them. But now, with all the ways that people also want to interact privately, there is also the opportunity to create a simpler platform that focuses first on privacy.

Zuckerberg himself points out that many people may not believe that Facebook can or wants to build a platform focused on privacy, "because, frankly, we currently do not have a solid reputation for creating privacy protection services, and we have historically focused on tools to share in open. But we have repeatedly shown that we can evolve to build the services that people really want, even in private messages and stories." In this sense, he gives an example of what the group has developed in WhatsApp. The plan is to focus on the most fundamental and private use case, messaging, make it as safe as possible and then build more ways for people to interact in addition to calls through video chats, groups, stories, businesses, payments, commerce and, ultimately, a platform for many other types of private services.

It turns out, therefore, that according to Zuckerberg there is a growing demand for privacy. And since Facebook gives users what they want, Facebook's new approach is privacy. This sudden concern for privacy is not only because, as Zuckerberg says, "privacy gives people the freedom to be themselves and connect more naturally." As always, in addition to trying to meet the wishes of users there are commercial reasons in the decisions that Zuckerberg makes. As Zuckerberg also points out private messages, end-to-end encrypted communications, ephemeral stories and small groups are by far the fastest growing areas of online communication. But, in addition to these reasons explicitly stated by Zuckerberg there are other reasons not clearly stated. Most likely the real reason for the new orientation change is to start laying the groundwork for "business, payments and commerce" in order to prepare the ground for the successful launch of the cryptocurrency, the libra.

But again, from the perspective of user happiness, the fact that stands out is that the possibility of maintaining private relationships through social networks is something that is valued positively. Therefore, this new orientation in the way Facebook acts positively affects the happiness of users.

## The Different Missions of Facebook and Happiness

From the review that has been presented of the different missions that Facebook has set, a clear relationship between Facebook and the subjective Bienestar of individuals is inferred. In each of the missions that have been commented, "keeping people communicated," "promoting the creation of communities" and "promoting privacy" there are two types of reasons that justify them. An altruist is publicly declared by Zuckerberg and another commercial. From the perspective followed in this work, the fact to note is that each of the three missions has been carried out because they were something desired by users. In this sense, Facebook has managed to give users what they wanted and in this way has contributed positively to their subjective well-being. Zuckerberg's ability is that in each case there were also, or first, clear business reasons to carry them out. In this sense, it can be said that between Facebook and the happiness of users there is a mutually beneficial interaction. Facebook offers users what they want (connection, information, communities and privacy), which has a positive impact on the subjective well-being of users, and they show their satisfaction by connecting to Facebook and boosting their growth.

## Subjective Well-Being and Social Networks: A Resume

Once the analysis of the relationship between subjective well-being and Facebook, the best known social networks, has been presented, we will resume the analysis from a general perspective. As noted in previous paragraphs, the effects on happiness of social networks are not always positive, since these depend on how the Internet is used and the objectives of social networks themselves. From the perspective of the science of happiness, the dilemma between the positive and negative effects of social networks on happiness is solved by the individuals themselves. Simply ask people who use social networks what is the state of their subjective well-being, and compare it with those of those who do not use them. As noted, various research papers indicate that in general people who use social networks are happier than those who do not use them. A justification for these results can be formulated by saying that if there is something that the empirical analysis of happiness has made clear is the importance of social and family relationships in the happiness of individuals. Those who have good and strong relationships with their family, friends, acquaintances and groups that have common interests tend to be happier. In this sense, social networks are a tool that allows to intensify the relationships of individuals and to be aware of what happens to their relatives and people who are interested in them and for this reason they usually have a positive impact on happiness. Facebook is a good example.

## Final Thoughts

Does technology make us less happy or more happy? This is the question we have tried to answer throughout this article. From the analysis made in previous pages, it is inferred that the most objective analysis is not the one made by social thinkers or philosophers. The studies of this type of authors are interesting to become aware of trends and anticipate possible future issues. However, they are not usually the most appropriate way to obtain the specific response to an issue, as in our case, to know the incidence of technology on the subjective well-being of individuals.

The most reliable results are obtained when research work is carried out on the impact of technology in specific cases. From them, it is evident that technology tends to have a positive impact on the subjective well-being of individuals but that it can also generate negative effects. On the other hand, it should not be forgotten that technology in many cases what it offers is tools; the impact of these on happiness to a large extent will depend on how we use them. Technology can be a very important source of well-being, although it is essential to learn to ration its use. Again, what we have learned from Facebook example is very illustrative. You have to know how to discriminate between its possible uses and discern those platforms that are worth getting involved with and in which we should not enter.

A similar conclusion is reached when analyzing the opinion of the technology experts, scholars and health specialists. They affirm that technology will continue to improve many aspects of our life but that in certain aspects it may harm the subjective well-being of the individuals, especially those who are vulnerable.

In any case, and given that progress and technological innovation are essential for the advancement of society, it is necessary to pay special attention to regulation. Only through proper regulation can we mitigate the possible damages derived from technology and emphasize its benefits.

## References

Ahn, N., & Mochón, F. (2010). La felicidad de los españoles: Factores explicativos. *Revista De Economía Aplicada, XVIII, 54*, 5–31.
Ahn, N., Mochón, F., & De Juan, R. (2012). La felicidad de los jóvenes. *Papers Revista De Sociología, 97*(2), 407–430.
Anderson, J. and Rainie, L. (2018). *The future of well-being in a tech-saturated world*. Pew Research Center. Internet & Technology. April 17. https://www.pewinternet.org/2018/04/17/the-future-of-well-being-in-a-tech-saturated-world/.
Andrews, F. M., & Withey, S. B. (1976). *Social Indicators of Well-Being*. New York: Plenum Press.
Argyle, M. (2002) *The psychology of happiness*. Routledge.

Bargh, J. A., McKenna, K. Y. A., & Fitsimons, G. G. (2002). Can you see the real me? Activation and expression of the true self on the Internet. *Journal of Social Issues, 58,* 33–48.

Blitz, M. (2014) Understanding heidegger on technology. *The New Atlantis,* Number 41, Winter 2014 (pp. 63–80). https://www.thenewatlantis.com/publications/understanding-heidegger-on-technology.

Brannan, D., & Mohr, C. D. (2018). Love, friendship, and social support. In R. Biswas-Diener & E. Diener (Eds), *Noba textbook series: Psychology.* Champaign, IL: DEF publishers.

Buss, D. (2000). The evolution of happiness. *American Psychologist, 55*(1), 15–23.

Campbell, A. (1976). Subjective measures of well-being. *American Psychologist, 31,* 117–124.

Campbell, A., Converse, P.E., & Rodgers, W.L. (1976). *The quality of American life. Perceptions, evaluations and satisfactions.* Russel Sage Foundation

Carmody, T. (2018) Facebook is in a trust crisis. Adweek. January 23. https://www.adweek.com/digital/facebook-is-in-a-trust-crisis/

Crooker, K., & Near, J. (1998). Happiness and satisfaction: Measures of affect and cognition? *Social Indicators Research, 44,* 195–224.

Cummins, R. A. (1996). The domains of life satisfaction: An attempt to order chaos. *Social Indicators Research, 38,* 303–332.

De Juan, R., Mochón, F., & Rojas, M. (2014). Expectations and happiness: evidence from Spain. *Journal of Social Research & Policy, 5*(2), 89–102.

Diener, E. (1984). Subjective well-being. *Psychological Bulletin, 95,* 542–575.

Diener, E., Emmons, R. A., Larsem, R. J., & Griffin, S. A. (1985). The satisfaction with life scale. *Journal of Personality Assessment, 49*(1), 71–75.

Diener, E., Seligman, M., Choi, H., & Oishi, S. (2018). Happiest people revisited. *Perspectives on Psychological Science,* March 29. https://doi.org/10.1177/1745691617697077.

Dodds, P., & Danforth, C. (2009). Measuring the happiness of LargeScale written expression: songs, blogs, and presidents. *Journal of Happiness Studies, 11*(4), 441–456.

Dumbar, R. (1998). *Grooming, gossip, and the evolution of language.* Cambridge, Mass., Harvard University Press.

Easterbrook, G. (2004) *The progress paradox: How life gets better while people feel worse.* Random House.

Easterlin, R. (1973). Does money buy happiness? *The Public Interest, 30,* 3–10.

Easterlin, R. (1974) Does economic growth improve the human lot? Some empirical evidence. In: P.A. David & M.W. Reder (Eds.), *Nations and households in economic growth* (pp. 89–125). Academic Press.

Easterlin, R. (2017). Paradox lost? *Review of Behavioral Economics, 4*(4), 311–339.

Easterlin, R. A., & Angelescu, L. (2009). Happiness and growth the world over: time series evidence on the happiness-income paradox. IZA Discussion Paper, num. 4.060.

Easterlin, R. A., Angelescu, L., Switek, M., Sawangfa, O, & Zweig, J.S. (2010). The happiness-income paradox revisited. In *Proceedings of the National Academy of Sciences of the United States of America (PNAS).*

Ellul, J. (1964). *The technological society.* John Wilkinson.

Ferrer-i-Carbonell, A. (2002). *Subjective questions to measure welfare and well-being.* Discussion paper TI 2002–020/3. Tinbergen Institute.

Frank, M., Mitchell, L., Dodds, P., & Danforth, C. (2013). Happiness and the patterns of life: A study of geolocated Tweets. *Scientific Reports, 3*(2625).

Frey, B., & Stutzer. (2001). *Happiness and economics: How the economy and institutions affect human well-being.* Princeton University Press.

Frey, B., & Stutzer, A. (2000). Happiness, economy and institutions. *Economic Journal, 110,* 918–938.

Frey, B., & Stutzer, A. (2002). What can economists learn from happiness research? *Journal of Economic Literature, 40,* 402–435.

Grinde, B. (2002). Happiness in the perspective of evolutionary psychology. *Journal of Happiness Studies, 3,* 331–354.

Gui y Sugden, 2005). Gui, B., & Sugden, R. (2005). Why interpersonal relations matter for economics. In B. Gui, R. Sugden (Eds.) *Economics and social interaction accounting for interpersonal relations* (pp. 1–22). Cambridge (Mass.): Cambridge University Press.

Heidegger, M. (1977). *The question concerning technology, and other essays.* Garland Publishing,

Helliwel, J.F. (2006). Well-being, social capital and public policy: what's new? *The Economic Journal*, **116**(510), C34–C45. March.

Hochschild, R. A. (1997). The time bind: When work becomes home and home becomes work. Metropolitan Books.

Holsten, H. (2018) How does internet use affect well-being? February 26. https://phys.org/news/2018-02-internet-affect-well-being.html.

Hynan, A. Murray, J. and Goldbart, J. (2014) Happy and excited: Perceptions of using digital technology and social media by young people who use augmentative and alternative communication. *Child Language Teaching and Therapy.* January 27. https://doi.org/10.1177/0265659013519258.

Iglesias, E., Pena, A., & Sánchez, J.M. (2013). Bienestar subjetivo, renta y bienes relacionales. Los determinantes de la felicidad en España. *Revista Internacional de Sociología, 71*(3), 567–592.

Kessler, L. (2018). *Why does fake news spread faster than real news?* It's all about pleasure. *PsychologyToday.* April 10. https://www.psychologytoday.com/us/blog/psychoanalysis-unplugged/201804/why-does-fake-news-spread-faster-real-news.

Kraut, R., Kiesler, S., Boneva, B., Cummings, J., Helgeson, V., & Crawford, A. (2001). Internet paradox revisited. *Journal of Social Issues*, October 12. Version 16.2. https://kraut.hciresearch.org/sites/kraut.hciresearch.org/files/articles/kraut02-paradox-revisited-16-20-2.pdf.

Kraut, R., Patterson, M., Lundmark, V., Kiesler, S., Mukopadhyay, T., & Scherlis, W. (1998). Internet paradox. A social technology that reduces social involvement and psychological well-being? *American Psychologist, 53*(9), 1017–1031

Lane, R. (1991) *The market experience.* Cambridge University Press.

Layard, R. (2003) Happiness: Has social science a clue? Three lectures. First lecture: What is happiness? Second lecture: Income and happiness: Rethinking economic policy; Third lecture: What would make a Happier society. *Lionel Robbins Memorial Lectures.* Londres: London school of economic.

Layard, R. (2006) Happiness: Lessons from a New Science. Penguin.

McKenna, K. A. (2008). MySpace or your place: Relationship initiation and development in the wired and wireless world. In S. Sprecher, A. Wenzel, & J. Harvey (Eds.), *Handbook of relationship initiation* (pp. 235–247). New York, NY: Psychology Press.

McKenna, K. Y. A., Green, A. S., & Gleason, M. E. J. (2002). Relationship formation on the Internet: What's the big attraction? *Journal of Social Issues, 58,* 9–31.

McMahon, D. (2006). *Una Historia de la Felicidad*, Taurus.

Mercader, J. R. (2017). Robotización mecanización pérdida de empleo. *Trabajo y Derecho, nº 27,* 13 a 24.

Mochon, F., & De Juan, R. (2015). *Happiness and social capital: Evidence from Latin American countries. Handbook of happiness research in Latin America.* Editorial Springer

Mochón, F., & de Juan, R. (2017). Capital social y bienes relacionales. In Iglesias, J., & De Juan, R. (coords.) *La felicidad de los españoles.* Tecnos.

Mochón, F., & Rojas, M. (2014). Editor's note. *International Journal of Interactive Multimedia and Artificial Intelligence* (Special Issue on AI Techniques to Evaluate Economics and Happines), 2(5).

Mochón, F., & Sanjuán, O. (2014). A first approach to the implicit measurement of happiness in latin america through the use of social networks. *International Journal of Artificial Intelligence and Interactive Multimedia, 2*(5), 17–23.

Nussbaum, M. C., & Sen, A. K. (1993). *The quality of life.* Oxford: Clarendon Press.

Pénard, T., Poussing, N., & Suire, R. (2013). Does the Internet make people happier? *Journal of Socio-Economics, 46,* 105–116.

Postman, N. (1985) *Amusing ourselves to death: Public discourse in the age of show business.* Penguin.

Putnam, R. D. (1995). Bowling alone: America's declining social capital. *Journal of Democracy*, 6(1), 65–78.
Rebecca O'Brien, R. (2017). *Mark Zuckerberg at Harvard: The truth behind 'the social network'*. July 14. https://www.thedailybeast.com/mark-zuckerberg-at-harvard-the-truth-behind-the-social-network.
Rita Watson, R. (2018). Research reveals why fake news is so powerful. *PsychologyToday*. July 27. https://www.psychologytoday.com/intl/blog/social-dilemmas/201802/the-fake-news-game.
Rojas, M. (Ed.). (2016). *Handbook of happiness research in Latin America*. Springer.
Rojas, M. (2006). Life satisfaction and satisfaction in domains of life: is it a simple relationship? *Journal of Happiness Studies*, 7(4), 467–497.
Rojas, M. (2007). The complexity of well-being: A life-satisfaction conception and a domains-of-life approach. In I. Gough & A. McGregor (Eds.). *Researching well-being in developing countries*. Cambridge University Press.
Rojas, M. (2009) Consideraciones sobre el Concepto de Progreso. In M. Rojas (coord.). *Midiendo el progreso de las sociedades: reflexiones desde México* (pp. 71–78). México, Foro Consultivo Científico y Tecnológico.
Rojas, M. (2014). *El Estudio Científico de la Felicidad, Fondo de Cultura Económica*.
Rojas, M. (2017) El estudio de la felicidad. In J. Iglesias & R. De Juan (coords.) *La felicidad de los españoles*. Tecnos.
Rojas, M., & Martínez, I. (coords.). (2012). *La Medición, Investigación e Incorporación en Política Pública del Bienestar Subjetivo: América Latina*. Reporte de la Comisión para el Estudio y Promoción del Bienestar en América Latina, Foro Consultivo Científico y Tecnológico, México.
Sherman, L. (2018) Why Facebook will never change its business model. *Forbes*. April 16. https://www.forbes.com/sites/lensherman/2018/04/16/why-facebook-will-never-change-its-business-model/#3207f46064a7.
Sherman, L. (2018) Zuckerberg's broken promises show facebook is not your friend. *Forbes*, May 23. https://www.forbes.com/sites/lensherman/2018/05/23/zuckerbergs-broken-promises-show-facebook-is-not-your-friend/#2ffc21a77b0a.
Sherman, L. (2018). Zuckerberg's promises won't fix facebook, but you can. *Forbes*. May 23. https://www.forbes.com/sites/lensherman/2018/05/23/zuckerbergs-promises-wont-fix-facebook-but-you-can/#6e1fe4f71e38.
Surowiecki, J. (2005). Technology and happiness. *MIT Technology Review*. January 1. https://www.technologyreview.com/s/403558/technology-and-happiness/.
Tatarkiewicz, W. (1976). *Analysis of Happiness*, Martinus Nijhoff.
van Praag, B.M.S. (1968). *Individual welfare functions and consumer behavior*, North Holland.
van Praag, B. M. S. (1971). The welfare function of income in Belgium: An empirical investigation. *European Economic Review, 2*, 337–369.
van Praag, B.M.S., Ferrer-i-Carbonell, A. (2004). *Happiness quantified: A satisfaction calculus approach*. Oxford University Press.
van Praag, B. M. S., Frijters, P., & Ferrer-i-Carbonell, A. (2003). The anatomy of subjective well-being. *Journal of Economic Behavior and Organization, 51*, 29–49.
Veenhoven, R. (1984). *Conditions of happiness*. Kluwer Academic.
Veenhoven, R. (1993). *Happiness in nations: Appreciation of life in 56 nations*. Rotterdam: Erasmus.
Veenhoven, R. (2000). The four qualities of life. Ordering concepts and measures of the good life. *Journal of Happiness Studies, 1*, 1–39.
Veenhoven, R. (2001) *What we do know about happiness?* Working Paper. Rotterdam: Erasmus University.
Vosoughi, S., Roy, D., & Aral, S. (2018). The spread of true and false news online. *Science*. 9 de marzo. Vol. 359, Issue 6380, pp. 1146–1151. https://science.sciencemag.org/content/359/6380/1146.
Zhan, G., & Zhou, Z. (2018). Mobile internet and consumer happiness: The role of risk. *Internet Research, 28*(3), 785–803. https://doi.org/10.1108/IntR-11-2016-0340.

Zuckerberg, M. (2017) *Bringing the world closer together*. facebook.com. June 22. https://www.facebook.com/notes/mark-zuckerberg/bringing-the-world-closer-together/10154944663901634/.

Zuckerberg, M. (2019). Understanding Facebook's business model. January 24. https://newsroom.fb.com/news/2019/01/understanding-facebooks-business-model/.

Zuckerberg, M. (2019) A privacy-focused vision for social networking. March 6. https://www.facebook.com/notes/mark-zuckerberg/a-privacy-focused-vision-for-social-networking/10156700570096634/.

# Socioeconomic Status and Consumer Happiness

## Lucia Savadori and Austeja Kazemekaityte

**Abstract** When it comes to happiness and satisfaction, consumer socioeconomic status (SES) has an important moderating role. In this chapter, we outline in which way SES intervenes in shaping consumer preferences and consumer happiness. When considering consumer preferences, low socioeconomic status has been shown to impact dietary patterns, such as consumption of fruits and vegetables, high caloric food, sugar-sweetened beverages, as well as consumption of alcohol and tobacco. Studies also show that low SES consumers tend to engage in purchases of various status-signaling goods. Socioeconomic status has also been shown to intervene in delineating happiness for experiential and material goods, consumer loyalty behavior, and consumer happiness with food consumption. We discuss the factors responsible for these relationships.

**Keywords** Consumer happiness · Consumer satisfaction · Low socioeconomic status

## Socioeconomic Status and Consumer Happiness

Consumer happiness depends on several individual and contextual factors. Among these, consumer's socioeconomic status (SES) has an important moderating role. Socioeconomic status defines a relative position of an individual or a household within a society and is usually related to their income, education, and occupation. Low SES has been associated with diverse outcomes in different domains, such as health (Williams, 1990) or educational attainment (Paterson, 1991), as well as differences in cognitive performance (Mani et al., 2013), impulsivity (Haisley et al., 2008), and self-control (Mullainathan, 2012) among others.

L. Savadori (✉)
Department of Economics and Management, University of Trento, Trento, Italy
e-mail: lucia.savadori@unitn.it

A. Kazemekaityte
University of Trento, Trento, Italy
e-mail: a.kazemekaityte@unitn.it

© The Author(s), under exclusive license to Springer Nature Singapore Pte Ltd. 2021
T. Dutta and M. K. Mandal (eds.), *Consumer Happiness: Multiple Perspectives*,
Studies in Rhythm Engineering,
https://doi.org/10.1007/978-981-33-6374-8_3

Socioeconomic differences also translate into consumer behavior and consumer happiness. It is important to address consumer happiness of both high- and low-SES consumers, especially because the factors that increase happiness among one type of consumer might not have the same effect on the other type of consumer, or they might even have a counterproductive effect.

There are different channels through which low SES can influence consumer decision making. Hamilton et al. (2019) discuss financial deprivation and four perspectives through which it can affect consumer behavior. First, low SES consumers face financial constraints (scarce resources) which limits the possibility to acquire products and services they need or want. Second, consumers may have fewer options to choose from (choice restriction), such as when shopping in a small supermarket offering limited number per category of product or lower number of categories on the total as compared with bigger supermarkets. Third, consumers tend to engage in social comparison, usually, an upward one, which means that being in a relatively deprived setting may result in fewer opportunities to make a favorable social comparison. Fourth, consumers living in financial deprivation usually are uncertain about their future, especially, future income, therefore this results in a less predictable future. All these perspectives influence the way consumers make decisions and can alter patterns, distinguishing them from what is observed among medium and high SES consumers.

## Status Goods

One type of consumption decisions widely observed among people living in poverty is status consumption (Banerjee & Duflo, 2011; Guillen-Royo, 2011; Jin et al., 2011; Van Kempen, 2004). It characterizes purchases through which consumers seek to improve their relative position within a society, provided that status signal of such goods coincides for both the individual and the society. The term is sometimes used interchangeably with *conspicuous consumption*, which has been defined as consumption with a goal to display a high social status or prestige (Page, 1992; Veblen, 1899).

Such display of status among low SES consumers can begin at a young age. Adolescent need to fit in among their peers is a widely observed behavior that is not constrained to particular social groups. Such a young age is generally marked by the presence of self-doubt, self-consciousness, and personal insecurities (Jiang et al., 2015; Kara & Chan, 2013). However, children from less affluent families see possession of well-known brands as a gateway to social circles that include their more well-off peers. Parents experience strong pressure from their children to obtain such branded products; otherwise, they refuse to wear cheaper, non-branded clothing items or shoes (K. Hamilton, 2012). Research shows that adolescent self-esteem is strongly influenced by material possessions and money which are the main signals for social inclusion among this age group (Isaksen & Roper, 2012). A child or teenager tends to first be judged by appearance: brands work as a material norm symbol (Whelan & Hingston, 2018) and can determine a decision of peers to initiate a conversation with the owner of branded goods (Elliott & Leonard, 2004). Children from poorer families put more importance on these status goods as compared with children from more affluent families (Isaksen & Roper, 2012). This does not come as a surprise as belonging to a higher social circle can seem like a way to escape the dire everyday environment. Failure to comply with these norms can result not only in no communication with the peers but also teasing or bullying (Elliott & Leonard, 2004). To avoid their children experiencing social stigma, parents can get driven into taking loans. In this way, a state of low welfare might persist even further. Moreover, such experiences of status signaling in the young age can leave residual behaviors in the adulthood: as shown by studies, the adolescent experience of deprivation and attempts to fit in among the peers by acquiring and using branded goods can make the adult consider simple and cheaper everyday brands threatening to their self-esteem, this way fostering the further consumption of branded products (Whelan & Hingston, 2018).

Status signaling among adults of low SES is a particularly common behavior. Certain countries even share public government-backed messages discouraging conspicuous consumption seeing it as an activity that propagates deprivation further (Chipp et al., 2011; Danzer et al., 2014). This is a paradox of status signaling: consumers are (subjectively or objectively) deprived, they spend a part of their already scarce income on status goods to signal that they belong to a higher social circle, which push them into more deprivation. Moreover, these are usually not a one-time purchase; once you publicly signal your status, you seek to maintain it, which asks for continuous investment in status goods (Chipp et al., 2011).

For a financially deprived consumer, their marginal utility of income is much higher than for their more affluent peer. If you are struggling with basic needs in your everyday life, to spend additional resources on goods that are not commodities of basic necessity is too costly or even impossible. Nevertheless, in some cases, consumers choose to exchange part of their essential need budget to obtain goods that signal status. In a study by Colson-Shira and Bellet, 2018 consumers, living in conditions close to or below the poverty line in India, indicated that they sacrifice around 13% of their daily caloric food intake to get what they call *aspirational goods*, that is, goods whose demand grows with an increase in deprivation. Even under threat of malnourishment, consumers spent a considerable amount of money on goods that do not serve a (high-) nutritional value. Examples would be cold drinks, branded clothing, packaged products, dairy and meat (which are an expensive source of caloric intake and could be more affordably replaced with cereals, fruits, and vegetables).

Although upward comparisons within a society are highly common, the biggest impact of them is concentrated for social circles close to ours. Status signaling tends to be directed not toward those at the top of the income pyramid, but rather the ones closer to you, whether it is people who are slightly more affluent or consumers physically close to you, such as neighbors (Colson-Sihra & Bellet, 2018; Hamilton, 2012). For this reason, status signaling plays a big role when an individual moves into a new environment. If you migrate to a new area, status signal can be a way for you to establish your position within the social network of more affluent residents, this way getting access to social circles and more resources (Danzer et al., 2014). Status signaling also can have a different end-goal: a person might seek to be distinct from others around them or, vice versa, seek to integrate to a certain social group. Integration can also happen to fictional social groups. For example, poorer household members might wish to identify with middle-class characters from popular TV shows, thus engaging in the consumption of branded goods seen on the screen as a way to escape the daily routine (Tufte, 2000).

However, status consumption can increase consumer happiness, especially among the most deprived ones (Jaikumar et al., 2018). Purchases of seemingly unnecessary goods for households that are already deprived of financial resources might not seem like an optimal decision, but it can carry the benefits associated with a subjective understanding of one's well-being, one of them being the positioning of oneself in a higher social circle. A purchase of counterfeit branded goods is a widely observed practice among low SES consumers (Van Kempen, 2003). Nevertheless, although consumers are willing to pay a premium to acquire them, in the end, those goods serve the same functionality and have an added benefit of signaling higher status, provided that consumer disappointment with the strength of the signal is not significant (Van Kempen, 2004). Therefore, such purchases can have a welfare-enhancing effect as long as well serves its practical and status signaling purpose.

There are also other examples of signaling behaviors that serve a different purpose from an improvement in subjective social standing. As shown in a study by Hill et al. (2012), examples from periods of economic difficulty have shown an increase—or at least no change—in consumption of female beauty products, which do not belong to the category of basic need commodities. However, the authors show that there is a practical reason for these observations. Results of the study suggest that females are willing to sacrifice a part of their already scarcer income on appearance enhancing products to increase their attractiveness and, in turn, boost their chances of attracting a more affluent mate who would give them more security and contribute to their living with additional resources. Netchaeva & Rees (2016) expand this argument by including another reason: increase in consumption of beauty products during difficult economic periods can also help in securing better job propositions which are in greater need during recession times. In their study, job-seeking dominates mate-seeking as the main reason behind the use of appearance enhancing products.

## Nutritional Choices and Unhealthy Commodities

Socioeconomic disparities also reflect in dietary choices. For example, it is a commonly observed trend that low SES populations have higher rates of obesity both in developed and developing countries (James, 2004), their average diet is less diverse (Mayén et al., 2014) and of lower quality (Erber et al., 2010). These patterns are also a big problem for children and adolescent (Kim, 2001; Olivares et al., 2007; Shahar et al., 2005). Moreover, the experience of deprivation in childhood forms dietary behaviors for adulthood: if consumers grew up in low SES environment, they tend to consume unnecessary calories (surpassing their actual energy needs) in the adulthoods as well (Netchaeva & Rees, 2016). The cause of poor nutritional state is usually a low intake of fruits and vegetables (Höglund et al., 1998; Lindström et al., 2001). Low SES consumers or households tend to prioritize price versus health in terms of food products (Pechey & Monsivais, 2016). Products that are rich in fat and sugar and more energy-dense are usually cheaper than the healthier alternatives (Pechey & Monsivais, 2016). Price levels are an important—yet, not a single—reason behind the choice of food products that have a lower nutritional value. Other potential reasons include, among others, compensatory eating, which is a tendency to compensate for a potential scarcity of energy in the future with extra caloric intake in the presence (Sterling, 2015) as well as stress and anxiety, which are the states when consumers tend to prefer to consume foods with a higher caloric value that usually induce more pleasure and comfort (Bratanova et al., 2016; Cheon & Hong, 2017; Langer et al., 2018). Food can also work as a status signaling good: in addition to previously mentioned example by Colson-Sihra & Bellet (2018), it has also been observed that consumers with low subjective SES tend to consider meat as a type of food that signals status, which can interfere with medical and environmental advice on consuming less meat (Chan & Zlatevska, 2019). Overall, food is an important constituent of general well-being, nutritionally and psychologically, and can affect happiness via different channels. Dietary choices can serve the purpose of counteracting negative experiences in daily life and satisfying emotional needs. On the other hand, external reasons might push consumers to make such decisions (such as income level and price of products), but this leads to poorer dietary habits that directly impact well-being.

In addition to processed foods high in fat and sugar, other unhealthy commodities—that are tobacco, alcohol, and sugar-sweetened beverages—show an increasing trend of consumption, especially for low SES populations (Stuckler et al., 2012). Sugar-sweetened beverages are immensely popular among teenagers and young adults (Singh et al., 2015), but the average consumption falls with increase in household income (Bolt-Evensen et al., 2018; Fontes et al., 2020; Han & Powell, 2013). For example, in the USA, the sugar-sweetened beverages constitute around 117 cal per day for the consumer in a household with income higher than $75,000 per year, while the daily amount of calories from such beverages almost doubles (200 cal) for consumers in the lower part of the income distribution, that is households with income less than $25,000 per year (average data for a period of 2009–2016) (Allcott et al., 2019). The reasons for this vary. The poorer nutritional knowledge and cost are among the most cited (Bolt-Evensen et al., 2018; Fontes et al., 2020). In addition to sugar-sweetened beverages, alcohol and tobacco are two commodities that are widely consumed in low SES populations (Bhan et al., 2012). As noted in previous examples, relative deprivation is sufficient to increase the consumption of certain unhealthy commodities. Mulia and Karriker-Jaffe (2012) find that low SES consumers are at higher risk of having alcohol-related problems if they live in more affluent neighborhoods. However, some studies suggest that alcohol consumption is distributed more equally among the whole income distribution (Kell et al., 2015). In terms of smoking, the rates of tobacco use are highest in the most disadvantaged neighborhoods, especially among people experiencing long-term unemployment, single parents, homeless, people with mental illnesses, prisoners, certain groups of new immigrants and ethnic minorities (Hiscock et al., 2012).

## Consumer Happiness in Poor and Affluent Societies

The history of humanity is studded with examples of deprivation and poverty that have faded from the earliest times to the present day. While our ancestors suffered from hunger and scarcity, in recent times, the wealth of nations has undergone a substantial increase that seems to want to grow more and more. But, does being wealthier consumers mean being also happier consumers? Does living in a bigger home, owning a more luxurious car, having a more varied diet, experiencing a more luxurious holiday, also translate into being a happier consumer?

The difference between rich and poor consumers can come in two forms: at the macrolevel and at the microlevel. At the macrolevel, we distinguish between poorer and richer countries: in lower-income countries, consumers are generally poorer; in wealthier countries, consumers are generally richer. At the microlevel, the difference is between richer and poorer consumers within the same society. The two-level distinction is important because the social comparison process is especially active at the microlevel, while at the macrolevel the comparison process occurs to a lesser extent. For example, looking at a neighbor who can afford a luxury car while we cannot, certainly generates a feeling of uneasiness and dissatisfaction with our utilitarian car. If instead, a person who lives in a particularly rich state has a luxury car, this does not generate in us the same extent of dissatisfaction with our utilitarian car. Comparison processes are relevant when we deal with the distinction between absolute happiness (i.e., independent of other people's happiness or on our previous happiness) and relative happiness (i.e., which depends on other people's happiness or our previous happiness).

The topic linking socioeconomic status to happiness at the macrolevel has received considerable attention from the literature, but the conclusions are not always straightforward. At the macrolevel, researchers try to answer the question of whether consumers in richer countries are happier than consumers in poorer countries. Stated in other terms, this is the age-old problem of whether money buys happiness. According to some research, wealthier people are also happier (Argyle, 2001; Hagerty & Veenhoven, 2003) but other data indicate an absence of a relationship between income, wealth, and happiness (Clark et al., 2008; Diener et al., 2010; Diener & Biswas-Diener, 2002; Easterlin, 1995; Headey et al., 2008). Some authors have also argued that the inconsistent results are determined by an unprecise measurement of well-being. According to the psychologist, Daniel Kahneman, Nobel Laureate for economics in 2002, there are at least two different ways of measuring happiness. The first, defined as *emotional well-being,* refers to the emotional quality of experiences, for example, the frequency with which we feel joy or sadness during the day. The second, called *life evaluation,* refers to the evaluation that people make of their lives when they look back and must say, in general, how happy they are with the life they have lived up to that moment. The two forms of happiness are influenced by different factors. Income and education influence life evaluation, while health, care for others, loneliness, and smoking affect the emotional well-being. Having more money and being better educated (higher SES), therefore, increases happiness for life in general, measured by questions such as "In general, how satisfied are you with your life overall?". But when we turn to emotional well-being, having more money and more education increases happiness only up to an annual income of around $75,000 (Kahneman & Deaton, 2010). A higher SES therefore buys satisfaction for life but not happiness; on the contrary, not having money makes consumers both less satisfied with life and less happy (Kahneman & Deaton, 2010).

A related issue is a diverse capacity by rich and poor societies of extracting satisfaction from climbing the happiness ladder. Consumers living in poorer societies are still climbing the happiness ladder. For this reason, every step forward on this hypothetical scale will produce greater happiness for consumers in developing countries, compared to consumers in already developed countries, who are already in the regions of high value and every step forward will be neutralized from the lack of meaningful social comparison. Social comparison is especially important at the microlevel when consumers can make comparisons with their neighbors and it is especially strong for those goods that are termed *learned preferences* as opposed to *innate preferences* (Tu & Hsee, 2016). While innate preferences are those that satisfy the needs of survival, such as, for example, the preference for hot rather than freezing water, learned preferences, instead, are the result of years of interactions between human beings with each other and have the value of signaling the distinction between belonging to one social status or another. For example, a Gucci watch is an asset that makes consumers happy not for its intrinsic value but for its status-symbol value. According to some authors, goods like these also have a shorter duration in the degree to which they can make a person happy: happiness lasts less because it adapts more quickly (Tu & Hsee, 2016). Indeed, happiness derived from status-symbol goods needs a social comparison process to survive, whereas happiness derived from essential goods does not. Furthermore, goods that have a status-symbol value do not have a stable comparison scale from which to derive happiness, because they are based on comparison with others, which is a changing element of the context. For example, we could derive a lot of happiness from buying a luxury watch, if the others do not own it but if the others become like us, our advantage to have a luxury item disappears. In other words, the value of a good that derives its' value from a social comparison process undergoes a form of hedonic adaptation (Tu & Hsee, 2016).

The problem, therefore, becomes that of understanding how consumers in the richest and most developed countries can still increase their happiness, given the limited room for maneuver. One recommendation is to identify needs that have not yet been met. For example, even in the most advanced societies, many people suffer from boredom, depression, or lack of free time. Not to mention more serious health problems such as real mental illness. One suggestion is to try to satisfy this type of needs. By doing so a substantial increase in happiness will be obtained. Being able to have free time (Hsee et al., 2010), find the true meaning of life and have a satisfactory network of social relationships (Baumeister & Leary, 1995), or being able to help others (Aknin et al., 2013; Dunn et al., 2008), are all types of goods that can increase happiness in richer countries where social comparison and hedonic adaptation reduce the possibility of a further increase in happiness.

## Socioeconomic Status and Happiness With Material and Experiential Purchases

Buying an experiential good offers a greater feeling of happiness than buying a material good (Millar & Thomas, 2009; Thomas & Millar, 2013; Van Boven & Gilovich, 2003). The difference between the two types of goods, and the happiness they produce, has been widely examined by several researchers. The first type of goods indicates those products that make us feel an emotional experience deriving from the senses, such as a holiday, a film, a day at the spa, a cooking seminar, a perfume, a dinner at a restaurant, a trip to the museum, a horseback ride, an experience of river rafting or a relaxing massage, to name a few. Tangible goods, on the other hand, are concrete purchases, which offer an experience of pleasure linked to the fact of owning them, such as a car, a watch, a smartphone, a T-shirt, a necklace, a house, a boat, and so on.

The reasons why experiential goods make us happier than material ones are still being evaluated, but it seems that two factors contribute to this difference. The first is the centrality of the type of experience for the definition of our identity: an experiential good contributes to a more positive definition of our identity and achieves the need for self-fulfillment and personal growth. In other words, after having lived an experience, a consumer will feel *internally* richer, while after purchasing a material good a consumer will feel *externally* richer. The two would have a different weight on happiness: the first would contribute more to make consumers happy than the second one (Kasser & Ahuvia, 2002; Kasser & Ryan, 1996; Van Boven, 2005). The second factor is social interaction. Experiential goods usually go hand in hand with greater social interaction, compared to material goods (Van Boven, 2005). Going to dinner in a restaurant determines a greater possibility of weaving relationships with others than simply owning a new dress. And social relationships are known to make us happy (Diener, 2009).

Socioeconomic status determines a different happiness with the consumption of the two types of goods. People with lower SES status are happier after consuming material goods, while people with higher SES status seems to be happier with experiential goods (Kasser & Ahuvia, 2002). For example, Thomas and Miller found confirmation for the fact that consumers are happier for experiential purchases than for material ones; and they found also confirmation for the explanation that the reason is that experiential goods enrich the person and the identity more in terms of personal growth. However, not all consumers behaved in this way. Consumers with low SES were happier when buying an object that they obtained and kept in their possession (a watch, a piece of jewelry, a car, a smartphone) than they were when buying a life experience. On the contrary, high SES consumers were equally happy with both types of purchases (Thomas & Millar, 2013). Similarly, it was found that the utilitarian value of a purchase (i.e., finding it useful) was a predictor of social confidence (i.e., how much I think I impress people with the purchase I made) for low-income consumers, but not for high-income consumers; on the contrary, the hedonic orientation (i.e., enjoy the shopping trip for itself) was a predictor of social confidence for both high- and low-income consumers (Paridon et al., 2006).

According to Kasser and colleagues, the difference between people of low and high SES is also determined by the fact that material goods relieve the poorest from their constant concern with the scarcity of resources that they constantly face and this would increase happiness (Kasser & Ahuvia, 2002). And in fact, the poor buy material goods to feel "less poor", that is, to feel that they belong to a higher social class (Sangkhawasi & Johri, 2007).

## Socioeconomic Status and Happiness with Specific Product Characteristics

Socioeconomic status not only affects the different happiness resulting from the consumption of material goods rather than experiential ones. But within each of these types of goods, socioeconomic status also determines a different appreciation for some aspects rather than others. For example, researchers involved in studying what are the aspects that have the greatest impact in determining a satisfying experience with a meat steak have identified some fundamental characteristics: tenderness, juiciness, and flavor. Among these, the flavor is the one that contributes most to the overall satisfaction with the product with 49.4%, followed by tenderness (43.4%) and juiciness (7.4%) (Felderhoff et al., 2020; O'Quinn et al., 2018). But these results cannot be generalized to the various socioeconomic strata of society. It has been discovered that high-SES consumers appreciate tenderness more, while low-SES consumers appreciate juiciness and flavor more (Felderhoff et al., 2020).

Where beef food safety is concerned, certification strategies (traceability or quality labels) are used to increase perceived safety. High-income consumers are more likely to pay a premium for certified beef than low-income consumers, showing that they give to safety a priority role (Angulo & Gil, 2007). Higher-income consumers are also different in their food preferences. For example, they were found to be more likely to choose fish and other seafood products frequently (Myrland et al., 2000; Thong & Solgaard, 2017) probably because low-income consumers are highly sensitive to price (Steptoe et al., 1995). However, income does not predict a different quality perception and a different taste perception for Pringles and Coca Cola, when these products were offered in English packaging or adapted in Urdu packaging showing no income effects on satisfaction with food as a function of the type of packaging (Khan & Lee, 2020). On the same vein, higher-income consumers pay more attention to informational aspects of food consumption (label use, information use, healthy eating, food safety, consumer rights, and consumer responsibility) (Nam, 2019). Therefore, lower-income consumers do indeed show different preferences and satisfaction with some food characteristics, but only up to a certain point.

## Socioeconomic Status and Satisfaction With Health Care

Socioeconomic status also influences how satisfied we are with medical care. Lower-income consumers are generally more satisfied with the received medical care than higher-income consumers (Fox & Storms, 1981). This is probably because people are different concerning their preferences for certain types of care and what they expect to receive. The less wealthy consumers, having lower expectations and less detailed preference models, are probably happier with the care received. For example, if a person who has a sore throat thinks she/he has tonsillitis and goes to the doctor for a cure, she will be satisfied if she receives an antibiotic and a certificate of disease to stay home a week from work. In other words, receiving the diagnosis that is expected makes you satisfied (Ong et al., 2007). The lower-income consumers may not have a clear idea of the diagnosis, might not have a clear expectation regarding the cure and therefore would experience less dissatisfaction with the lack of congruence between expectation and results.

This ties in with paradoxical data that shows that patients who receive worse medical care because they live in poorer countries have greater satisfaction with the care received (Kruk et al., 2018). Just to quote some data, out of eight low-income countries surveyed, 79% of the consumers said they were very satisfied with the care received, even though they received less than half of the essential care (Kruk et al., 2018). The explanation seems linked to the factors that contribute to determining satisfaction with the medical treatments. Satisfaction with the treatment depends both on the treatment received but also on the accessibility, the costs, the state of health, the expectations, the immediate results and the gratitude. Among these factors, low expectation of medical care seems to explain the anomalous data found among the poorest consumers of medical care. In other words, poorer people would have less knowledge of what should be expected and therefore less chance of properly evaluating a medical treatment. For example, a vignette in which a doctor changed the medication of a hypertensive patient without measuring his pressure, was rated as good to excellent by 53% of the 17,966 interviewed, all living in low-income and middle-income countries (LMICs).

On a different but parallel vein, in a meta-analysis of the literature, it was found that the oldest, the least educated, those with a higher social status (occupational status) and the married were the most satisfied with the medical care (Hall & Dornan, 1990). This result is noteworthy because it is very strange to observe that social status goes in the opposite direction of education. According to the authors, apart from the need for further studies that can shed light and possibly disconfirm this relationship, a possible explanation is that being very educated but having a low-status job creates a great frustration and dissatisfaction with the doctor who has a high-status occupation. This resentment would result in dissatisfaction with the service. The relationship between high social status and greater satisfaction, however, seems easier to understand. Those who have a good job usually turn to the best doctors and also get better treatment.

The link between socioeconomic status and satisfaction with health care is still a matter of debate. More recent studies have tried to determine in which way the relationship stands, but without any success. In some studies, higher socioeconomic consumers are happier with health care, in other studies, the opposite was found with no clear explanation for the incongruent findings (Batbaatar et al., 2016).

## Socioeconomic Status, Customer Satisfaction, and Loyalty

The greater the competitive forces in a market, the greater costumers' expectation and the greater the need for the firms to hold on to existing customers (i.e., loyalty). One of the factors considered as most important in inducing a customer to repeat the purchase of the same product is the happiness of the customer with that product. Happier costumers turn into more loyal customers (Anderson & Sullivan, 1993).

However, this relationship has been questioned: there are loyal but not happy costumers and there are happy costumers who, however, do not remain faithful (Jones & Sasser, 1995). Here, we will examine the role that socioeconomic status differences play in this relationship.

Indeed, while lower-income costumers tend to satisfy this relationship, that is, those who are more satisfied are also more likely to remain faithful to the product, this is not always true for medium–high-income costumers. For the latter, being more satisfied with the product does not always imply that they will also be more loyal costumers and therefore will buy the same product again in the future (Homburg & Giering, 2001). According to the researchers, this could be explained by the fact that, while for a low-income customer making a purchase involves a very high financial risk, for a customer with more financial availability, a mistake would not be catastrophic (Kaplan et al., 1974; Murray & Schlacter, 1990). Therefore, a richer customer could repeat the same purchase even if he was not satisfied with it the previous time, giving, in fact, a second chance to the product: a risk that he can afford. The research also highlighted that only the wealthiest customers are influenced by the purchasing process (e.g., how kind the salespeople are, etc.) when they have to decide whether to buy the same product again. Probably, consumers with higher income also have a higher education which makes the relationship and the exchange of information with the seller an important factor for them. However, other data show that consumers with higher SES are also more satisfied with the three primary needs: autonomy (the need to be in control of their actions), competence (feeling capable of managing important tasks), and relational (the need to feel close and supported by significant others) (Deci & Ryan, 2000). And this greater satisfaction translates into greater loyalty, at least for performing arts type of product (White & Tong, 2019).

# References

Aknin, L. B., Barrington-Leigh, C. P., Dunn, E. W., Helliwell, J. F., Burns, J., Biswas-Diener, R., Kemeza, I., Nyende, P., Ashton-James, C. E., & Norton, M. I. (2013). Prosocial spending and well-being: Cross-cultural evidence for a psychological universal. In *Journal of Personality and Social Psychology* (Vol. 104, Issue 4, pp. 635–652). American Psychological Association. https://doi.org/10.1037/a0031578

Allcott, H., Lockwood, B. B., & Taubinsky, D. (2019). Should we tax sugar-sweetened beverages? An overview of theory and evidence. *Journal of Economic Perspectives, 33*(3), 202–227.

Anderson, E. W., & Sullivan, M. W. (1993). The Antecedents and Consequences of Customer Satisfaction for Firms. *Marketing Science, 12*(2), 125–143. https://doi.org/10.1287/mksc.12.2.125.

Angulo, A. M., & Gil, J. M. (2007). Risk perception and consumer willingness to pay for certified beef in Spain. *Food Quality and Preference, 18*(8), 1106–1117. https://doi.org/10.1016/j.foodqual.2007.05.008

Argyle, M. (2001). The psychology of happiness, 2nd ed. In *The psychology of happiness, 2nd ed.* Routledge.

Banerjee, A., & Duflo, E. (2011). *Poor Economics A Radical Rethinking of the Way to Fight Global Poverty*. PublicAffairs. https://www.amazon.com/Poor-Economics-Radical-Rethinking-Poverty/dp/1610390938.

Batbaatar, E., Dorjdagva, J., Luvsannyam, A., Savino, M. M., & Amenta, P. (2016). Determinants of patient satisfaction: A systematic review. *Perspectives in Public Health, 137*(2), 89–101. https://doi.org/10.1177/1757913916634136.

Baumeister, R. F., & Leary, M. R. (1995). The need to belong: Desire for interpersonal attachments as a fundamental human motivation. *Psychological Bulletin, 117*(3), 497–529. https://doi.org/10.1037/0033-2909.117.3.497.

Bhan, N., Srivastava, S., Agrawal, S., Subramanyam, M., Millett, C., Selvaraj, S., & Subramanian, S. V. (2012). Are socioeconomic disparities in tobacco consumption increasing in India? A repeated cross-sectional multilevel analysis. *British Medical Journal Open, 2*(5), e001348. https://doi.org/10.1136/bmjopen-2012-001348.

Bolt-Evensen, K., Vik, F. N., Stea, T. H., Klepp, K. I., & Bere, E. (2018). Consumption of sugar-sweetened beverages and artificially sweetened beverages from childhood to adulthood in relation to socioeconomic status - 15 years follow-up in Norway. *International Journal of Behavioral Nutrition and Physical Activity, 15*(1). https://doi.org/10.1186/s12966-018-0646-8

Bratanova, B., Loughnan, S., Klein, O., Claassen, A., & Wood, R. (2016). Poverty, inequality, and increased consumption of high calorie food: Experimental evidence for a causal link. *Appetite, 100*, 162–171. https://doi.org/10.1016/j.appet.2016.01.028.

Chan, E. Y., & Zlatevska, N. (2019). Jerkies, tacos, and burgers: Subjective socioeconomic status and meat preference. *Appetite, 132*, 257–266. https://doi.org/10.1016/j.appet.2018.08.027

Cheon, B. K., & Hong, Y. Y. (2017). Mere experience of low subjective socioeconomic status stimulates appetite and food intake. *Proceedings of the National Academy of Sciences of the United States of America, 114*(1), 72–77. https://doi.org/10.1073/pnas.1607330114.

Chipp, K., Kleyn, N., & Manzi, T. (2011). Catch up and keep up: relative deprivation and conspicuous consumption in an emerging market. *Journal of International Consumer Marketing, 23*(2), 117–134. https://doi.org/10.1080/08961530.2011.543053.

Clark, A. E., Frijters, P., & Shields, M. A. (2008). Relative income, happiness, and utility: an explanation for the easterlin paradox and other puzzles. *Journal of Economic Literature, 46*(1), 95–144. https://doi.org/10.1257/jel.46.1.95.

Colson-Sihra, E., & Bellet, C. (2018). The conspicuous consumption of the poor: forgoing calories for aspirational goods. *SSRN Electronic Journal*. https://doi.org/10.2139/ssrn.3270814.

Danzer, A. M., Dietz, B., Gatskova, K., & Schmillen, A. (2014). Showing off to the new neighbors? Income, socioeconomic status and consumption patterns of internal migrants. *Journal of Comparative Economics, 42*(1), 230–245. https://doi.org/10.1016/j.jce.2013.05.002.

Deci, E. L., & Ryan, R. M. (2000). The "what" and "why" of goal pursuits: Human needs and the self-determination of behavior. *Psychological Inquiry, 11*(4), 227–268. https://doi.org/10.1207/S15327965PLI1104_01.

Diener, E. (2009). *Subjective Well-Being BT - The Science of Well-Being: The Collected Works of Ed Diener* (E. Diener (ed.); pp. 11–58). Springer Netherlands. https://doi.org/10.1007/978-90-481-2350-6_2

Diener, E., & Biswas-Diener, R. (2002). Will money increase subjective well-being? A literature review and guide to needed research. *Social Indicators Research, 57*(2), 119–169. https://doi.org/10.1023/A:1014411319119.

Diener, E., Ng, W., Harter, J., & Arora, R. (2010). Wealth and happiness across the world: Material prosperity predicts life evaluation, whereas psychosocial prosperity predicts positive feeling. In *Journal of Personality and Social Psychology* (Vol. 99, Issue 1, pp. 52–61). American Psychological Association. https://doi.org/10.1037/a0018066

Dunn, E. W., Aknin, L. B., & Norton, M. I. (2008). Spending money on others promotes happiness. *Science, 319*(5870), 1687 LP–1688. https://doi.org/10.1126/science.1150952

Easterlin, R. A. (1995). Will raising the incomes of all increase the happiness of all? *Journal of Economic Behavior & Organization, 27*(1), 35–47. https://doi.org/10.1016/0167-2681(95)00003-B

Elliott, R., & Leonard, C. (2004). Peer pressure and poverty: Exploring fashion brands and consumption symbolism among children of the 'British poor'". *Journal of Consumer Behaviour, 3*(4), 347–359. https://doi.org/10.1002/cb.147.

Erber, E., Beck, L., Hopping, B. N., Sheehy, T., De Roose, E., & Sharma, S. (2010). Food patterns and socioeconomic indicators of food consumption amongst Inuvialuit in the Canadian Arctic. *Journal of Human Nutrition and Dietetics, 23*(SUPPL. 1), 59–66. https://doi.org/10.1111/j.1365-277X.2010.01097.x.

Felderhoff, C., Lyford, C., Malaga, J., Polkinghorne, R., Brooks, C., Garmyn, A., & Miller, M. (2020). Beef quality preferences: Factors driving consumer satisfaction. *Foods, 9*(3). https://doi.org/10.3390/foods9030289

Fontes, A. S., Pallottini, A. C., Vieira, D. A. D. S., Fontanelli, M. de M., Marchioni, D. M., Cesar, C. L. G., Alves, M. C. G. P., Goldbaum, M., & Fisberg, R. M. (2020). Demographic, socioeconomic and lifestyle factors associated with sugar-sweetened beverage intake: A population-based study. *Revista Brasileira de Epidemiologia, 23*. https://doi.org/10.1590/1980-549720200003.

Fox, J. G., & Storms, D. M. (1981). A different approach to sociodemographic predictors of satisfaction with health care. *Social Science & Medicine. Part A: Medical Psychology & Medical Sociology, 15*(5), 557–564. https://doi.org/10.1016/0271-7123(81)90079-1.

Guillen-Royo, M. (2011). Reference group consumption and the subjective wellbeing of the poor in Peru. *Journal of Economic Psychology, 32*(2), 259–272. https://doi.org/10.1016/j.joep.2009.12.001.

Hagerty, M. R., & Veenhoven, R. (2003). Wealth and happiness revisited—growing national income does go with greater happiness. *Social Indicators Research, 64*(1), 1–27. https://doi.org/10.1023/A:1024790530822.

Haisley, E., Mostafa, R., & Loewenstein, G. (2008). Subjective relative income and lottery ticket purchases. *Journal of Behavioral Decision Making, 21*(3), 283–295. https://doi.org/10.1002/bdm.588.

Hall, J. A., & Dornan, M. C. (1990). Patient sociodemographic characteristics as predictors of satisfaction with medical care: A meta-analysis. *Social Science and Medicine, 30*(7), 811–818. https://doi.org/10.1016/0277-9536(90)90205-7.

Hamilton, K. (2012). Low-income families and coping through brands: inclusion or stigma? *Sociology, 46*(1), 74–90. https://doi.org/10.1177/0038038511416146.

Hamilton, R., Mittal, C., Shah, A., Thompson, D. V., & Griskevicius, V. (2019). How financial constraints influence consumer behavior: an integrative framework. *Journal of Consumer Psychology, 29*(2), 285–305. https://doi.org/10.1002/jcpy.1074.

Han, E., & Powell, L. M. (2013). Consumption patterns of sugar-sweetened beverages in the United States. *Journal of the Academy of Nutrition and Dietetics, 113*(1), 43–53. https://doi.org/10.1016/j.jand.2012.09.016

Headey, B., Muffels, R., & Wooden, M. (2008). Money does not buy happiness: or does it? A reassessment based on the combined effects of wealth, income and consumption. *Social Indicators Research, 87*(1), 65–82. https://www.jstor.org/stable/27734646.

Hill, S. E., Rodeheffer, C. D., Griskevicius, V., Durante, K., & White, A. E. (2012). Boosting beauty in an economic decline: Mating, spending, and the lipstick effect. *Journal of Personality and Social Psychology, 103*(2), 275–291. https://doi.org/10.1037/a0028657.

Hiscock, R., Bauld, L., Amos, A., Fidler, J. A., & Munafò, M. (2012). Socioeconomic status and smoking: A review. In *Annals of the New York Academy of Sciences* (Vol. 1248, Issue 1, pp. 107–123). https://doi.org/10.1111/j.1749-6632.2011.06202.x.

Höglund, D., Samuelson, G., & Mark, A. (1998). Food habits in Swedish adolescents in relation to socioeconomic conditions. *European Journal of Clinical Nutrition, 52*(11), 784–789. https://doi.org/10.1038/sj.ejcn.1600644.

Homburg, C., & Giering, A. (2001). Personal characteristics as moderators of the relationship between customer satisfaction and loyalty—An empirical analysis. *Psychology and Marketing, 18*(1), 43–66. https://doi.org/10.1002/1520-6793(200101)18:1%3c43::AID-MAR3%3e3.0.CO;2-I.

Hsee, C. K., Yang, A. X., & Wang, L. (2010). Idleness Aversion and the Need for Justifiable Busyness. *Psychological Science, 21*(7), 926–930. https://doi.org/10.1177/0956797610374738.

Isaksen, K. J., & Roper, S. (2012). The Commodification of Self-Esteem: Branding and British Teenagers. *Psychology & Marketing, 29*(3), 117–135. https://doi.org/10.1002/mar.20509.

Jaikumar, S., Singh, R., & Sarin, A. (2018). I show off, so I am well off": Subjective economic well-being and conspicuous consumption in an emerging economy. *Journal of Business Research, 86*, 386–393. https://doi.org/10.1016/j.jbusres.2017.05.027.

James, P. T. (2004). Obesity: The worldwide epidemic. *Clinics in Dermatology, 22*(4 SPEC. ISS.), 276–280. https://doi.org/10.1016/j.clindermatol.2004.01.010

Jiang, J., Zhang, Y., Ke, Y., Hawk, S. T., & Qiu, H. (2015). Can't buy me friendship? Peer rejection and adolescent materialism: Implicit self-esteem as a mediator. *Journal of Experimental Social Psychology, 58*, 48–55. https://doi.org/10.1016/j.jesp.2015.01.001.

Jin, Y., Li, H., & Wu, B. (2011). Income inequality, consumption, and social-status seeking. *Journal of Comparative Economics, 39*(2), 191–204. https://doi.org/10.1016/j.jce.2010.12.004.

Jones, T. O., & Sasser, W. E. J. (1995). Why satisfied customers defect. *Harvard Business Review, 73*, 88–99.

Kahneman, D., & Deaton, A. (2010). High income improves evaluation of life but not emotional well-being. *Proceedings of the National Academy of Sciences, 107*(38), 16489 LP–16493. https://doi.org/10.1073/pnas.1011492107.

Kaplan, L. B., Szybillo, G. J., & Jacoby, J. (1974). Components of perceived risk in product purchase: A cross-validation. *Journal of Applied Psychology, 59*(3), 287–291. https://doi.org/10.1037/h0036657.

Kara, C., & Chan, K. (2013). Development of materialistic values among children and adolescents. *Young Consumers, 14*(3), 244–257. https://doi.org/10.1108/YC-01-2013-00339.

Kasser, T., & Ahuvia, A. (2002). Materialistic values and well-being in business students. *European Journal of Social Psychology, 32*(1), 137–146. https://doi.org/10.1002/ejsp.85.

Kasser, T., & Ryan, R. M. (1996). Further Examining the American Dream: Differential correlates of intrinsic and extrinsic goals. *Personality and Social Psychology Bulletin, 22*(3), 280–287. https://doi.org/10.1177/0146167296223006.

Kell, K. P., Judd, S. E., Pearson, K. E., Shikany, J. M., & Fernández, J. R. (2015). Associations between socio-economic status and dietary patterns in US black and white adults. *British Journal of Nutrition, 113*(11), 1792–1799. https://doi.org/10.1017/S0007114515000938.

Khan, H., & Lee, R. (2020). Does packaging influence taste and quality perceptions across varying consumer demographics? *Food Quality and Preference, 84.* https://doi.org/10.1016/j.foodqual.2020.103932

Kim, Y. (2001). Food and nutrient consumption patterns of korean adults by socioeconomic status. *Korean Journal of Community Nutrition, 6*(4), 645–656.

Kruk, M. E., Gage, A. D., Arsenault, C., Jordan, K., Leslie, H. H., Roder-DeWan, S., Adeyi, O., Barker, P., Daelmans, B., Doubova, S. V, English, M., Elorrio, E. G., Guanais, F., Gureje, O., Hirschhorn, L. R., Jiang, L., Kelley, E., Lemango, E. T., Liljestrand, J., … Pate, M. (2018). High-quality health systems in the Sustainable Development Goals era: time for a revolution. *The Lancet Global Health, 6*(11), e1196–e1252. https://doi.org/10.1016/S2214-109X(18)30386-3

Langer, S. L., Soltero, E. G., Beresford, S. A. A., McGregor, B. A., Albano, D. L., Patrick, D. L., & Bowen, D. J. (2018). Socioeconomic status differences in food consumption following a laboratory-induced stressor. *Health Psychology Open, 5*(2). https://doi.org/10.1177/2055102918804664

Lindström, M., Hanson, B. S., Wirfält, E., & Östergren, P.-O. (2001). Socioeconomic differences in the consumption of vegetables, fruit and fruit juices: The influence of psychosocial factors. *European Journal of Public Health, 11*(1), 51–59. https://academic.oup.com/eurpub/article/11/1/51/447298

Mayén, A.-L., Marques-Vidal, P., Paccaud, F., Bovet, P., & Stringhini, S. (2014). Socioeconomic determinants of dietary patterns in low- and middle-income countries: A systematic review. *the American Journal of Clinical Nutrition, 100*(6), 1520–1531. https://doi.org/10.3945/ajcn.114.089029.

Millar, M., & Thomas, R. (2009). Discretionary activity and happiness: The role of materialism. *Journal of Research in Personality, 43*(4), 699–702. https://doi.org/10.1016/j.jrp.2009.03.012

Mulia, N., & Karriker-Jaffe, K. J. (2012). Interactive Influences of Neighborhood and Individual Socioeconomic Status on Alcohol Consumption and Problems. *Alcohol Anf Alcoholism, 47*(2), 178–186. https://academic.oup.com/alcalc/article/47/2/178/187989

Mullainathan, S. (2012). Psychology and development economics. In *Behavioral Economics and Its Applications* (pp. 85–114). https://www.scopus.com/inward/record.uri?eid=2-s2.0-84883974426&partnerID=40&md5=a4ac8d4af852ac7f082df138fa76a6f5

Murray, K. B., & Schlacter, J. L. (1990). The impact of services versus goods on consumers' assessment of perceived risk and variability. *Journal of the Academy of Marketing Science, 18*(1), 51–65. https://doi.org/10.1007/BF02729762.

Myrland, Ø., Trondsen, T., Johnston, R. S., & Lund, E. (2000). Determinants of seafood consumption in Norway: lifestyle, revealed preferences, and barriers to consumption. *Food Quality and Preference, 11*(3), 169–188. https://doi.org/10.1016/S0950-3293(99)00034-8

Nam, S.-J. (2019). The effects of consumer empowerment on risk perception and satisfaction with food consumption. *International Journal of Consumer Studies, 43*(5), 429–436. https://doi.org/10.1111/ijcs.12521.

Netchaeva, E., & Rees, M. (2016). Strategically stunning. *Psychological Science, 27*(8), 1157–1168. https://doi.org/10.1177/0956797616654677.

O'Quinn, T. G., Legako, J. F., Brooks, J. C., & Miller, M. F. (2018). Evaluation of the contribution of tenderness, juiciness, and flavor to the overall consumer beef eating experience1. *Translational Animal Science, 2*(1), 26–36. https://doi.org/10.1093/tas/txx008.

Olivares, S. C., Bustos, N. Z., Lera, L. M., & Zelada, M. E. (2007). Estado nutricional, consumo de alimentos y actividad física en escolares mujeres de diferente nivel socioeconómico de Santiago de Chile. *Revista Medica De Chile, 135*(1), 71–78. https://doi.org/10.4067/s0034-98872007000100010.

Ong, S., Nakase, J., Moran, G. J., Karras, D. J., Kuehnert, M. J., & Talan, D. A. (2007). Antibiotic Use for Emergency Department Patients With Upper Respiratory Infections: Prescribing Practices, Patient Expectations, and Patient Satisfaction. *Annals of Emergency Medicine, 50*(3), 213–220. https://doi.org/10.1016/j.annemergmed.2007.03.026

Page, C. (1992). A History of Conspicuous Consumption. *ACR Special Volumes*, *SV-08*. https://www.acrwebsite.org/volumes/12197/volumes/sv08/SV-08/full

Paridon, T. J., Carraher, S., & Carraher, S. C. (2006). The income effect in personal shopping value, consumer self-confidence, and information sharing (word of mouth communication) research. *Academy of Marketing Studies Journal*, *10*(2), 107–124.

Paterson, L. (1991). Socio-economic status and educational attainment: a multi-dimensional and multi-level study. *Evaluation & Research in Education*, *5*(3), 97–121. https://doi.org/10.1080/09500799109533303.

Pechey, R., & Monsivais, P. (2016). Socioeconomic inequalities in the healthiness of food choices: Exploring the contributions of food expenditures. *Preventive Medicine*, *88*, 203–209. https://doi.org/10.1016/j.ypmed.2016.04.012.

Sangkhawasi, T., & Johri, L. M. (2007). Impact of status brand strategy on materialism in Thailand. *Journal of Consumer Marketing*, *24*(5), 275–282. https://doi.org/10.1108/07363760710773094.

Shahar, D., Shai, I., Vardi, H., Shahar, A., & Fraser, D. (2005). Diet and eating habits in high and low socioeconomic groups. *Nutrition*, *21*(5), 559–566. https://doi.org/10.1016/j.nut.2004.09.018.

Singh, G. M., Micha, R., Khatibzadeh, S., Shi, P., Lim, S., Andrews, K. G., Engell, R. E., Ezzati, M., Mozaffarian, D., Fahimi, S., Powles, J., Elmadfa, I., Rao, M., Wirojratana, P., Abbott, P. A., Abdollahi, M., Gilardon, E. A., Ahsan, H., Al Nsour, M. A. A., … Zajkás, G. (2015). Global, regional, and national consumption of sugar-sweetened beverages, fruit juices, and milk: A systematic assessment of beverage intake in 187 countries. *PLoS ONE*, *10*(8). https://doi.org/10.1371/journal.pone.0124845

Steptoe, A., Pollard, T. M., & Wardle, J. (1995). Development of a Measure of the Motives Underlying the Selection of Food: the Food Choice Questionnaire. *Appetite*, *25*(3), 267–284. https://doi.org/10.1006/appe.1995.0061

Sterling, P. (2015). Principles of allostasis: Optimal design, predictive regulation, pathophysiology, and rational therapeutics. In *Allostasis, Homeostasis, and the Costs of Physiological Adaptation* (pp. 17–64). https://doi.org/10.1017/CBO9781316257081.004

Stuckler, D., Mckee, M., Ebrahim, S., & Basu, S. (2012). Manufacturing Epidemics: The Role of Global Producers in Increased Consumption of Unhealthy Commodities Including Processed Foods, Alcohol, and Tobacco. *PLoS Med*, *9*(6). https://doi.org/10.1371/journal.pmed.1001235

Thomas, R., & Millar, M. (2013). The effects of material and experiential discretionary purchases on consumer happiness: moderators and mediators. *The Journal of Psychology*, *147*(4), 345–356. https://doi.org/10.1080/00223980.2012.694378.

Thong, N. T., & Solgaard, H. S. (2017). Consumer's food motives and seafood consumption. *Food Quality and Preference*, *56*, 181–188. https://doi.org/10.1016/j.foodqual.2016.10.008

Tu, Y., & Hsee, C. K. (2016). Consumer happiness derived from inherent preferences versus learned preferences. *Current Opinion in Psychology*, *10*, 83–88. https://doi.org/10.1016/j.copsyc.2015.12.013

Tufte, T. (2000). *Living with the rubbish queen: Telenovelas, culture and modernity in Brazil*. Indiana University Press. https://scholar.googleusercontent.com/scholar.bib?q=info:8FbKaDVeSKUJ:scholar.google.com/&output=citation&scisdr=CgXcXIVEEIGR-fH6nvI:AAGBfm0AAAAAXwD_hvLgCj2wxe1YSDP-a7_7bne6i2kt&scisig=AAGBfm0AAAAAXwD_hsWInKsVn4kRhH9JvMcNK_ugz7qX&scisf=4&ct=citation&cd=-1.

Van Boven, L. (2005). Experientialism, materialism, and the pursuit of happiness. *Review of General Psychology*, *9*(2), 132–142. https://doi.org/10.1037/1089-2680.9.2.132.

Van Boven, L., & Gilovich, T. (2003). To Do or to Have? That Is the Question. In *Journal of Personality and Social Psychology* (Vol. 85, Issue 6, pp. 1193–1202). American Psychological Association. https://doi.org/10.1037/0022-3514.85.6.1193

Van Kempen, L. (2003). Fooling the eye of the beholder: Deceptive status signalling among the poor in developing countries. *Journal of International Development*, *15*(2), 157–177. https://doi.org/10.1002/jid.973.

Van Kempen, L. (2004). Are the poor willing to pay a premium for designer labels? a field experiment in Bolivia. *Oxford Development Studies, 32*(2), 205–224. https://doi.org/10.1080/13600810410001699957.

Veblen, T. (1899). *The Theory of the Leisure Class.*

Whelan, J., & Hingston, S. T. (2018). Can everyday brands be threatening? Responses to brand primes depend on childhood socioeconomic status. *Journal of Consumer Psychology, 28*(3), 477–486. https://doi.org/10.1002/jcpy.1029.

White, C. J., & Tong, E. (2019). On linking socioeconomic status to consumer loyalty behaviour. *Journal of Retailing and Consumer Services, 50,* 60–65. https://doi.org/10.1016/j.jretconser.2019.05.001.

Williams, D. R. (1990). Socioeconomic differentials in health: A review and redirection. *Social Psychology Quarterly, 53*(2), 81–99. https://doi.org/10.2307/2786672.

# Consumer Happiness: Neuroscience Perspective

# Subliminal Messaging and Application in Sports: Moving Beyond the Conscious

Özge Ercan

**Abstract** The brain has a very superior system, and there are some complex structures which are still undiscovered. The subconscious is one of these unclear features. The subconscious unconsciously places the stimuli in the environment into the mind with the help of sensory organs and processes these pieces of information accumulated there. It manages and directs behavior with images and messages. Although they are felt consciously, the center of emotions is actually subconscious. Thus, the individual involuntarily realizes her or his way of life, relationships with other people and many other factors thanks to her or his subconscious. Consumption behaviors are among these involuntary behaviors. Brand preference, brand loyalty, brand awareness, purchase intention and behavior are affected by these subliminal messages. Hence, companies have endeavored to place subliminal messages that are specially designed and coded below the perception limit into the subconscious of their targeted consumers by using five senses. They bring consumers together with the subliminal messages by preparing psychological, sociological and neurological sub-grounds that meet their needs, demands and expectations. With subliminal messages, the awareness level of the consumers is exceeded, and the codes related to the brand are created subconsciously. This enables consumers to choose the brand and develop their buying behavior. However, these messages, sent to the subconscious of consumers, also have things that do not overlap with ethical values. Individuals exposed to the messages of unacceptable products or services are adversely affected. The most important ethical problem is the role of subliminal messages. The ethical dimension has become controversial as it is possible to use the hidden fears and impulses that people have unconsciously by using them for different purposes from commercial areas, advertising and marketing. Therefore, the channels from which the subliminal messages come from and what techniques are used must be determined and precautions should be taken. One of the areas where subliminal messages are mostly used is the sports sector. Stadiums and crowded sport halls bring together millions of people and provide the appropriate ground for the companies to send real-time subliminal messages at the same time. Thus, sports have become a popular means of

Ö. Ercan (✉)
Faculty of Sports Sciences, Sinop University, 57000 Sinop, Turkey
e-mail: ozgercan@windowslive.com

© The Author(s), under exclusive license to Springer Nature Singapore Pte Ltd. 2021
T. Dutta and M. K. Mandal (eds.), *Consumer Happiness: Multiple Perspectives*,
Studies in Rhythm Engineering,
https://doi.org/10.1007/978-981-33-6374-8_5

meeting companies' subliminal messages with consumers. In this chapter; it is aimed to provide a theoretical contribution to the studies in this field by dealing with the subliminal messages sent to the subconscious of consumers, the methods and techniques used for this aim, the reasons why they are preferred by the companies, the ethical dimension of the subliminal messages and the effects of subliminal messages in the sports arena.

**Keywords** Advertising ethics · Customer subconscious · Sports marketing · Subliminal message

## Introduction

The brain has a complex structure that amazes with its perfect structure. It is an extraordinary and complex system that directs people and ensures their survival. The brain is the control center that controls all the movements of the body, consciously or unconsciously (Ambler, Ioannides, & Rose, 2000; Keleş & Çepni, 2006; Perrachione & Perrachione, 2008).

Even in the light of modern science, some of its structures are still undiscovered. One of these structures is the subconscious. According to Sigmund Freud, who introduced the concept of the subconscious, consciousness consists of a structure that is both visible and invisible. The name of the invisible part is "subconscious". The behaviors affected by the messages received through the sense organs act involuntarily with the guidance of the subconscious (Özkalp, 1991).

On the other hand, brands try to motivate people act according to their own wishes, and they use different scientific methods and techniques in every field for this purpose (Demirtaş, 2004). These messages including psychological, sociological and neurological structures are sent to the subconscious by the help of these kinds of methods and techniques in line with the needs and expectations of people (Treisman, 1964).

Similar situations are applied to managing consumption habits and behavior. Consumers create their brand preferences, loyalties, brand awareness, purchasing intentions and behaviors, attitudes and behaviors toward the brand in the context of messages that they are not aware of but that are already subconsciously coded (Treisman, 1964).

The ability to link brain activities to human decisions and behavior has enabled cognitive neuroscience to be brought together with social sciences (Levallois, Clithero, Wouters, Smidts, & Huettel, 2012; Tüzel, 2010).

Cognitive neuroscience theories and practices have caused rapid and radical changes in many areas including traditionally expressed marketing, advertising, and sponsorship. Thanks to subliminal messages, the acceptance of consumers without resisting or questioning the messages they see and developing a behavior as determined by these messages have become the strategic goals of brands (Plassmann, Venkatraman, Huettel, & Yoon, 2015).

Within marketing activities in which there are fierce competition, leaving people with subliminal messages has brought discussion of the ethical dimensions of these messages in marketing practices. Subliminal messages have sides that do not match ethical values (Rotfeld, 2007).

Exposing people to these messages by legitimating with subliminal messages in the use of products and services that are not welcomed in the community or using these kinds of messages other than marketing purposes and marketing components are some situations in which there are ethical violations. Therefore, the source of subliminal messages is of great importance. Measures should be taken to prevent people from being exposed to these messages by determining which channels and techniques are applied (Shiv & Yoon, 2012).

Although subliminal messages penetrate the subconscious of individuals, these messages are not applied to a single person. These messages are sent to the subconscious of many people at the same time. It is carried out by selecting the appropriate place and convenient time where crowded groups of people come together (Hubert & Kenning, 2008).

Sports organizations are one of the most suitable grounds for subliminal messages to attract more people at the same time. Sports organizations have great roles in reaching the goals of the brands and have become unlimited in this context nowadays. Many brands serve their subliminal messages, whether or not their products or services are related to sports, during the sports organizations and create brand sympathy in the masses (Ratten, 2016).

It is obvious that sports are the most effective communication platforms between the target audience and the brand. The brand phenomenon cannot be placed so well in another event, in the memory (subconscious) of the athletes or the supporters (consumers). Of course, this reality attracts the attention of all kinds of brands, whether it is accepted or not. Therefore, the effects of subliminal messages on sports marketing are indisputable (Fett, 2011).

In this chapter, subliminal messages sent to the subconscious, subliminal messages of consumers, the methods and techniques used for this purpose, the reasons for being preferred by companies, the ethical dimension of subliminal messages and the effects of subliminal messages in the sports sector are introduced to make a theoretical contribution to the studies in this field.

## The Power of Subconscious

The subconscious refers to the area under the threshold of consciousness. There is no awareness reflex to consciousness in this area (Küçükbezirci, 2013). The threshold of consciousness differs from person to person. At the same time, the threshold of consciousness also varies for the same people, influenced by their interests, needs, mood, alertness and so on. In other words, for a person, the subconscious can be conscious for another time for the same person or for the same time for a different person (Fullerton, 2010; Strauch et al., 1976).

In psychology, the subconscious is a place where hidden emotions and actions are stored. The subconscious in the psychology dictionary is the place where there are materials and processes that exist below the level of consciousness and are not conscious but can be easily raised to the level of consciousness if desired (Karaca, 2010).

There is no order and general will in the subconscious. Subconscious acts completely with impulses. The energy that the subconscious possesses strictly adheres to the principle of pleasure and acts with the impulsive requirements found in the principle of pleasure. The energy dominated by these instinctive impulses is naturally completely unaware of judgments, good, bad and morality values. It does not listen to the laws of logic, and it experiences violent excitement, which does not contradict and prevent each other (Yücel & Çubuk, 2013).

The subconscious possessed by the individual saves messages unconsciously and without filtering anything and then reveals them as a form of behavior later on (Key, 1981; Bahrami, Lavie, & Rees, 2007). The memories and records that have been accumulated constantly cause changes in behavior and thought in people over time. These changes may be both positive and negative (Kılıç, 2011; Talbot, Duberstein, & Scott, 1991).

Sigmund Freud's experiments on his patients have been effective in the development of the subconscious concept and its scientific validation. Freud measured some of the patients' reactions regarding unconsciousness and stupor. He faced a certain resistance when he tried to bring some situations to the consciousness of the patient. However, the patient contended that he had no knowledge of these conditions. Thus, the "unconscious resistance" situation has been determined (Özkalp, 1991).

On the other hand, the Swiss scientist Jung stated that the subconscious is in an unchanging, stationary structure. For Jung, the subconscious is uninterrupted and very different from consciousness. It continues its subconscious functions even when we deal with different activities like reading, speaking or writing. With special methods, it can be revealed that the subconscious works uninterruptedly. Occurring under the consciousness, this event occurs as a dream at night or as a small and strange imbalance during the day (Karaca, 2010).

95% of the messages perceived by the sense organs are processed subconsciously. In addition, the same percentage of the decisions is made without any awareness. Even the most logical and rational decisions are made as a result of emotional accumulation (Baysal & Akalın, 2011). People do not even know the reasons of many choices they make. When they are asked why they like some things, they cannot give an exact answer because the subconscious is the basis of the decisions taken, especially the buying behavior decisions (Baysal & Akalın, 2011; Marangoz & İşli, 2018).

Subconscious advertising is a marketing communication tool born in this process. The modern communication method of the companies that try to make the target audiences prefer themselves to gain superiority to their competitors is subconscious advertising. In subconscious advertisements, stimulants below the lowest level of perception based on subliminal perception are sent to the consumers' emotions, thoughts and purchasing behaviors to affect their purchasing behaviors on purpose.

The message in subconscious advertisements is given indirectly, not directly (Carter, 1986; Öztürk, 2007).

Considerable attention is paid to the form of subconscious persuasion. Scientific data on subliminal advertisements are not sufficient since they are forbidden or wanted to be banned by some governments and industry organizations (Çakır, 2007; Darıcı, 2014; Öztürk, 2007; Taylor, 2012).

Increasing competition intensity and conscious behaviors of consumers force companies to use subliminal advertising techniques. The most frequently used subconscious advertisement techniques are those that create sexual connotation for the consumer and are intended to carry out the purchase action. These unethical advertisements made by companies cause negative attitudes by consumers toward the product and the brand. Subconscious advertisements mostly affect the consumer by placing hidden sexual messages in advertisements (Özdemir, 2009).

The subconscious is sensitive to the concepts of birth, death and sexuality and is affected mostly by these concepts. The use of these concepts in messages sent to the subconscious is also because of their high level of influence (Kılıç, 2011).

For this reason, subliminal advertising applications are used for many products or services.

## Conscious Versus Subconscious: Which Among the Two Generate More Experience in the Consumers

A consumer makes choices among thousands of product stacks and as a result of these choices makes the purchase decision for the product. While deciding, conscious or subconscious behavior affects the decision to buy. Consciousness is awareness of the individual both for himself and his environment. The subconscious, on the other hand, covers the activities that our consciousness carries out unconsciously (Touhami, 2011).

Therefore, consumers provide the information they need to make a decision during the purchase from two sources through internal and external search. Intrinsic search is the search of information in the memory that is relevant to the decision. This kind of search benefits from the experiences in the subconscious. For this reason, brands try to be preferred when making a choice in the memory of the consumer, to remain in the mind by creating sympathy and thus to influence the subconscious of people. When purchasing an advertised product, consumers unconsciously act with an emotion, impulse or desire (Öztürk, 2014).

When the information that emerges as a result of the experiences is not enough, consumers try to obtain data from around. This process is called external search. If consumers have enough experience, the expectation for external search does not show itself. In case of the data obtained by the consumers in the light of their experiences are not sufficient in quantity and content, external search is activated (Levallois et al., 2012).

Advertising plays an important role in the decision-making process of the consumer as an external source in the information periods of the individuals. In the evaluation of advertising activities, it is useful to determine the pre-and post-broadcast reviews and the results related to them, to determine what information the consumer needs and to determine the most accurate expression to be sent to the targeted consumer classes in the messages to be transmitted. Thus, the consumer will be able to respond to his needs and become more conscious in the process of choice through open messages (Aktuğlu, 2006).

Although the internal sources, which are the consumer's information source, depend on the memory, the information in this memory is created by obtaining from external sources (Odabaşı & Barış, 2002). For this reason, many companies work to place their products or services in the subconscious of consumers permanently with subliminal messages.

The main aim with subliminal messages is the acceptance of the situation. For example, when the packaging of some products is seen, it feels as if they have been looking for that product for many years. In this process, the subconscious comes into play and convinces that the desired product has been found. However, there is no question of persuading for one of the two products in this situation. The selection has been directed by the effect of the subconscious (Tığlı, 2002).

Subliminal messages may be hidden in mass media such as newspapers and television, sometimes in the packaging of products, in the advertisements of those products, or in posters hanging on billboards. But, they all serve the same purpose. Whether the product is beneficial to consumers or not, whether the product is quality or poor quality, expensive or cheap, the aim is to sell the product to consumers (Gratz, 1984; Ceylan & Ceylan, 2015).

Therefore, individual and non-individual factors affect consumer behavior. Among individual factors, needs, motives, perceptions, attitudes, experiences, self-concept, and value judgments can be listed. Marketing management members use psychological techniques to understand and evaluate consumers' emotional responses to goods and services, and what motives them to buy (Zurawicki, 2010).

Non-individual factors can be specified as culture, profession, family and reference groups (Tenekecioğlu, 2003). All of these factors play an active role in unconscious behavior and accumulation. Hence, in subliminal advertising applications, stimulants below the lowest level of perception based on subliminal perception are sent to affect the emotions, thoughts and buying behaviors of consumers. Subliminal advertisements are given indirectly, not directly (Öztürk, 2007). These messages are dictated to the sense organs by some methods and techniques.

The most commonly used methods in subconscious advertising are product placement, 25th square technique, embedding technique where shapes or texts are placed on objects and tactotop methods.

Product placement is described as an advertisement without indicating that there is a branded product or an advertisement related to the product in movies, series or related places, and that the server, the player or the people appearing on the screen carry a branded product. In a different reference, product placement is explained as providing a product or desired brand to be displayed on TVs or cinemas (Demir, 2008).

Advertisers want their products to appear in movies, but they do not want to make their products the focus of the movie. The effect of product placement arises in this way. Otherwise, the spectators have the feeling that they are watching a long advertisement, rather than thinking that they are watching a long movie. From the moment, the audience begins to think that what he or she is watching is advertising, the product placement technique will be no different from other classic advertisements (Demir, 2008).

Another method used to address the subconscious in television or cinema is the 25th frame technique. By the help of this way, the subconscious threshold of the audience is crossed. The audience has the opportunity to perceive the images as motion thanks to the 25th Frame (Güler, 2008). 25th square technique has been used in time, but its effect is less than other methods, and it is a technique that is no longer used by cinema technology today (Darıcı, 2012).

The effect of this technique remains very light among other methods. This technique was used at times when the role of cinema and television in our lives was important, and computer use, Internet, and digital technology were not developed. With the developing technology, the reason of being old fashioned of this method is that someone who has a computer at home can easily capture a photo frame placed with this method. There are difficulties in the application of this technique due to the unique features of the television channels (Darıcı, 2012).

In some cases, the subliminal advertisement can be not just the product, but only the marks or the name of the brand. As a different example, images related to the product can be transferred to the program or film digitally (Erdem, 2015).

Embedding technique, inserting text or shape on objects, is the most frequently used subconscious advertisement application. It is a technique that occurs in printed and display advertisements and messages as symbols, texts and shapes placed in messages or advertisements, which sometimes can be noticed with the naked human eye but sometimes not noticed. The purpose of this technique is to reach people's subconscious. This technique is effective in reaching the subconscious. Figures, symbols and texts can be seen by looking closely and carefully to this technique (Aydemir, 2014).

Embedding is a very common technique in advertisements. Both pictures, figures and words are successfully placed in the ads. The purpose of this technique is to adopt the message to be made by addressing the subconscious (Eldem, 2009). Embedding is done by giving the same texture, same darkness and contrast settings as the ground. This process is also called image feeding. Image feeding can be done on the whole screen or on a small frame on the screen. The message to be given to the subconscious is combined with the advertising floor in this kind of advertisement. Words such as sex, buy; images such as sexually evoking images; or elements of fear, such as dry heads, faces are intentionally placed on the advertisement's ground to not be noticed (Darıcı, 2012). Sex text or other subconscious words are put in pictures and photographs in a way that is not noticeable with the naked eye. These drawings are made on the background, as well as on hair strands and dress wrinkles (Eldem,

2009). In some advertising applications, the subconscious impulse is applied by tactoscope-related methods that transmit subliminal message at high speed (Kelly, 1979).

Tools used to create impulse in the experimental studies of subconscious visual perception are very important factors. Various selected tools affect the test results. The tactoscope is used very often in experiments. In the 1960s, the tactoscope device was further developed, and the effectiveness of some advertisements was aimed to be increased with images sent to the subconscious in a time such as one thousandth of a second (Brown, 1958; Kılıç, 2011).

The tactoscope device acts like a slide or a projection with a conventional camera system. With the increase of computer technology and its widespread use, the tactoscope device has started to lose its importance and usage today. However, it is still used in market researches, for advertisements of products to appeal us, making the product names and their packages remembered, and making pictures in more real sizes than computers (Kılıç, 2011).

Although the effect and ethical aspect of subconscious advertisements are discussed worldwide, it is prohibited in many countries. Subconscious advertising is an advertising technique that uses a lot of different technologies than conventional advertising (Aytekin, 2009).

## How Ethical is the Use of Subliminal Messages

Ethics is the whole set of concepts which determine the good or the bad, principles, moral duties and responsibilities of individuals or groups that must be followed in the society. Shortly, ethics is expressed as spiritual principles and values that regulate the behavior of the individual and society. In other words, it is a set of criteria that evaluate and basically treat behavior.

Everybody in the society should do their duty completely, fully and accurately for the benefit of humanity (Zengin & Şen, 2006). Therefore, the responsibilities of our daily activities are the basis of ethics (Çakır, 2007).

In this context, advertising ethics can be defined as the advertising sector actors and general society-based evaluations and inquiries for the production of the advertising products created in accordance with the general moral rules of the society, personal rights, the prevention of unfair competition and the protection of consumer rights (Elden, Ulukök, & Yeygel, 2006).

Ethical principles differ from society to society and from individuals to individuals. Norms considered correct in one society can be perceived as bad in the other since cultural norms and values are different in every society. Therefore, it is not easy to determine ethical limits. However, there should still be a basis for ethical values. Generally, the basis of ethical values is determined by the situation, time and evaluation of possible behavior results (Sungur, 2007).

However, it cannot be deduced from this that all legal acts are ethical. Although many product or service advertisements (tobacco, alcohol, etc.) are not ethically correct in the community, they may not have a legal restriction (Katzper, Ryback, & Hertzman, 1978).

Advertisers are moving away from ethical principles in order to convince consumers by making use of various advertising practices. Therefore, a lot of ethical problems have been encountered in many advertisements recently (Snyder, 2008).

Issues that can be taken into account in advertising ethics in the process of informing the consumer can be listed as misleading ads, comparative ads, witnessed ads, hidden or subconscious ads, and the use of children, female and male sexuality during the transmission of the advertising message (Aktuğlu, 2006; Fisher, 2012). Therefore, the ethics of subliminal messages require thinking more than the norms that require compliance with the standards of society.

Persuasion is at the core of subliminal messages. This technique is also considered a kind of mind control (Karabulut, 2002). Exposing people to these messages is not ethically correct. With the messages sent to their subconscious, the consumers are encouraged to buy without their will. However, consumers should purchase goods or services for their personal use.

When subjected to subliminal messages, the customer cannot understand the underlying factors in his buying behavior. Consumer behaviors are shaped by the effects of many socio-cultural factors such as the individual's needs, motives, personality, perceptions, education, attitudes and beliefs, the culture of the society in which he is located, the social class to which he belongs, and the family. When the subconscious messages of the brands are combined with these factors, the buying behavior becomes completely different. Therefore, brands must adhere to ethical values when determining their goals (Nebenzhal & Jaffe, 1998).

Today, the effects of subconscious advertisements may not be fully understood, but legislators should pay due attention to this issue and take necessary measures. Otherwise, it may be too late when the subconscious advertisement technique is proved (Yolcu, 2005).

## Subconscious Advertising Practices in Sports Events

The characteristics of sports consumers are different from other consumers. The fact that there are common conscience and common goals in sports, bringing people together with the phenomenon of success, provides an open and usable environment for all the emotions that ads want to give toward their goals and objectives (James, 2011).

Professional sports are one of the main shows of media culture. James (2011) states "sports combine race, nationalism, fame and star power, crime and scandalous demonstrations and raise their symbols to a divine position". Today, sports consumers make up a large part of the society. People learn about competition, success, values of a society and the way of action through sports. On the other hand, sports heroes are the highest paid and wealthiest people in the society in the eyes of consumers (Kapar, 2010). Therefore, they are the symbol of dreams about a good life. Sports consumers identify with these values while consuming sports competitions and go one step further toward specialization and criticism. They try to actively take part in any stage of the sport (James, 2011).

The phenomenon of mass consumption of sports is important for subconscious advertisements. Subconscious advertisements always have an interesting influence on the audience (Kardeş, 1991). Brands that have been using the effect of sports to society in recent years are trying to load subconscious ads in the memory of sports consumers in a fun-based and encouraging way, whether or not they directly offer sports-related products or services.

Especially the companies that produce cigarettes and alcohol, or gambling and betting companies, which are not accepted in most societies, seem to be sympathetic to consumers in subconscious advertising applications and sports competitions (McKelvey, 2004).

In addition, sports provides an opportunity to escape to people who are in the middle of the tight progress caused by capitalism and enables them to be slaves of advertisements without their knowledge (Talimciler, 2008).

With subconscious advertisements, sports, sports bulletins, match broadcasts, sports news programs, sports documentaries, sports magazine programs, and sports training programs encourage the masses, stimulate their feelings, make them angry, and at last prompt them to spend money to calm down (Erdoğan & Alemdar, 2011).

# References

Ambler, T., Ioannides, A., & Rose, S. (2000). Brands on the brain: Neuro-images of advertising. *Business Strategy Review, 11*(3), 17–30.

Aktuğlu, I. K. (2006). Tüketicinin bilgilendirilmesi sürecinde reklam etiği. *Küresel İletişim Dergisi, 2*, 1–20.

Aydemir, M. (2014). Medyada şiddetin dönüşümüne bilinçaltı etki yöntemleri ve "mutlu şiddet" ilişkisi. *Karamanoğlu Mehmetbey Üniversitesi Sosyal Ve Ekonomik Araştırmalar Dergisi, 2014*(3), 166–170.

Aytekin, P. (2009). *Reklamda Etik (Türkiye'de Televizyon Reklamlarının Etik Açıdan İncelenmesi)* (Doktora Tezi). Celal Bayar Üniversitesi, Manisa.

Bahrami, B., Lavie, N., & Rees, G. (2007). Attentional load modulates responses of human primary visual cortex to invisible stimuli. *Current Biology, 17*(6), 509–513.

Baysal, A., & Akalın, N. (2011). Pazarlamanın Bilinçaltı. *MediaCat Aylık PazarlamaDergisi, 202*, 60–68.

Brown, J. I. (1958). Teaching reading with the tachistoscope. *Journal of Developmental Reading, 1*(2), 8–18.

Çakır, V. (2007). Avrupa Birliğine Uyum Sürecinde Türkiye'de Televizyon Reklamlarına Yönelik Düzenlemeler. *Selçuk Üniversitesi İletişim Fakültesi Dergisi, 5*(1), 198–209.

Carter, R. (1986). Whispering soft nothings to the shop thief: How "reinforcement messaging" works. *Retail and Distribution Management*, January/February, 1986.

Ceylan, İ. G., & Ceylan, H. B. (2015). Ambalaj Tasarımında Bilinçaltı Mesaj Öğelerinin ve Nöropazarlama Yaklaşımının Kullanımlarının Karşılaştırılması. *Electronic Turkish Studies, 10*(2), 123–142.

Darıcı, S. (2014). *Bilinçaltı Reklamcılık Teknikleri ve Tüketicinin Yönlendirilmesi* (Yayımlanmamış Yüksek Lisans Tezi). Gelişim Üniversitesi, İstanbul.

Demir, M. (2008). *Televizyon ve Sinema Dünyasının Yeni Yıldızları: Ürünler ve Ürün Yerleştirme Tekniğinin İzleyiciler Üzerindeki Etkileri* (YayımlanmamışYüksek Lisans Tezi). Fırat Üniversitesi, Elazığ.
Demirtaş, H. A. (2004). Temel ikna teknikleri: Tutum oluşturma ve tutum değiştirme süreçlerindeki etkilerinin altında yatan nedenler üzerine bir derleme. *Gazi Üniversitesi İletişim Fakültesi Dergisi, 19,* 73–91.
Erdem, Ö. (2015). *Televizyon Ve Sinemada Gizli Reklam ve Subliminal Mesaj* (Yayımlanmamış Yüksek Lisans Tezi). Marmara Üniversitesi, İstanbul.
Eldem, Ü. İ. (2009). *Bilinçaltı Reklamcılık ve Tüketici Davranışları Üzerindeki Etkisi* (Yayımlanmamış Yüksek Lisans Tezi). Maltepe Üniversitesi, İstanbul.
Elden, M., Ulukök, Ö., & Yeygel, S. (2007). *Şimdi Reklamlar* (2nd ed.). İstanbul: İletişim Yayınları.
Erdoğan, İ, & Alemdar, K. (2011). *Kültür ve İletişim*. Ankara: Erk Yayınları.
Fett, M. (2011). *Neuromarketing in sports—How emotions strenghten the consumers' perception of a brand* (Bachelor's thesis). University of Lugano, Lugano.
Fisher, T. D., Moore, Z. T., & Pittenger, M. J. (2012). Sex on the brain?: An examination of frequency of sexual cognitions as a function of gender, erotophilia, and social desirability. *Journal of Sex Research, 49*(1), 69–77.
Fullerton, R. A. (2010). "A virtual social H-bomb": The late 1950s controversy over subliminal advertising. *Journal of Historical Research in Marketing, 2*(2), 166–173.
Gratz, J. E. (1984). The ethics of subliminal communication. *Journal of Business Ethics, 3*(3), 181–184.
Güler, H. (2008). *Algıların Ötesi: Bilinçaltı Reklamcılık Bilinçaltı Reklamcılığın Tüketici Davranışları Üzerindeki Etkileri* (YayımlanmamışYüksek Lisans Tezi). Kocaeli Üniversitesi, Kocaeli.
Hubert, M., & Kenning, P. (2008). A current overview of consumer neuroscience. *Journal of Consumer Behaviour: An International Research Review, 7*(4–5), 272–292.
James, J. D. (2011). Attitude toward advertising through sport: A theoretical framework. *Sport Management Review, 14*(1), 33–41.
Kapar, S. (2010). Resimde Sembolik İmgelemi Oluşturan Psikolojik Etkenler. *Sanat Dergisi, 15,* 43–46.
Karabulut, B. (2002). *Algı Yönetimi*. İstanbul: Alfa Basım Yayın.
Karaca, H. E. (2010). *Resimde Bilinçdışı Anlatımın Rastlantısal ve Deneysel Süreci* (YayımlanmamışYüksek Lisans Tezi). Gazi Üniversitesi, Ankara.
Kardeş, S. (1991). Gençliğin Spor Ayakkabı Satın Almasında Tercihlerin Belirlenmesi. *Pazarlama Dünyası, 26,* 20–27.
Katzper, M., Ryback, R., & Hertzman, M. (1978). Alcohol beverage advertisement and consumption. *Journal of Drug Issues, 8*(4), 339–353.
Keleş, E., & Çepni, S. (2006). Beyin ve öğrenme. *Journal of Turkish Science Education, 3*(2), 66–82.
Kelly, J. S. (1979). Subliminal embeds in print advertising: A challenge to advertising ethics. *Journal of Advertising, 8*(3), 20–24.
Key, W. B. (1981). *The Clam plate orgy, paperback*. New York: New American Library.
Küçükbezirci, Y. (2013). Bilinçaltı Mesaj Gönderme Teknikleri Ve Bilinçaltı Mesajların Topluma Etkileri. *Electronic Turkish Studies, 8*(9).
Kılıç, Y. (2011). *Subliminal Hipnoz*. İstanbul: İkinci Adam Yayınları.
Levallois, C., Clithero, J. A., Wouters, P., Smidts, A., & Huettel, S. A. (2012). Translating upwards: Linking the neural and social sciences via neuroeconomics. *Nature Reviews Neuroscience, 13*(11), 789–797.
Marangoz, M., & İşli, A. G. (2018). Bilinçaltı Reklamcılık ve Tüketicilerin Satın Alma Niyetine Etkisi. *Pamukkale Journal of Eurasian Socioeconomic Studies, 5*(1), 15–33.
McKelvey, S. M. (2004). The growth in marketing alliances between US professional sport and legalised gambling entities: Are we putting sport consumers at risk? *Sport Management Review, 7*(2), 193–210.

Nebenzhal, I. D., & Jaffe, E. D. (1998). Ethical dimensioins of advertising executions. *Journal of Business Ethics, 17*(7), 805–815.
Odabaşı, Y., & Barış, G. (2010). *Tüketici Davranışı* (9th ed.). İstanbul: MediaCat Akademi.
Özdemir A. H. (2009). *Televizyon Reklamlarında Etik ve Marka Sadakati İlişkisi Üzerine Bir Araştırma: Ankara İli Örneği* (Doktora Tezi). Gazi Üniversitesi, Ankara.
Özkalp, E. (1991). *Psikolojiye Giriş*. Eskişehir: Eylül Yayınları.
Öztürk, C. (2014). *Bilinçaltı Reklamcılık ve Göstergebilimsel Reklam Analizleri* (Yayımlanmamış Yüksek Lisans Tezi), Beykent Üniversitesi, İstanbul.
Öztürk, Ö. (2007). *Aldatıcı Televizyon Reklamlarına Karşı Tüketicinin Korunması* (Yüksek Lisans Tezi). Gazi Üniversitesi, Ankara.
Plassmann, H., Venkatraman, V., Huettel, S., & Yoon, C. (2015). Consumer neuroscience: applications, challenges, and possible solutions. *Journal of Marketing Research, 52*(4), 427–435.
Perrachione, T. K., & Perrachione, J. R. (2008). Brains and brands: Developing mutually informative research in neuroscience and marketing. *Journal of Consumer Behaviour: An International Research Review, 7*(4–5), 303–318.
Ratten, V. (2016). The dynamics of sport marketing: Suggestions for marketing intelligence and planning. *Marketing Intelligence & Planning, 34*(2), 162–168.
Rotfeld, H. J. (2007). Mistaking a marketing perspective for ethical analysis: When consumers can't know that they should want. *Journal of Consumer Marketing, 24*(7), 383–384.
Shiv, B., & Yoon, C. (2012). Integrating neurophysiological and psychological approaches: Towards an advancement of brand insights. *Journal of Consumer Psychology, 22*(1), 3–6.
Strauch, I., Schneider-Düker, M., Zayer, H., Heine, H. W., Heine, I., Lang, R., & Müller, N. (1976). The impact of meaningful auditory signals on sleeping behavior. *Archiv für Psychologie.*
Sungur, S. (2007). Bilinçaltı Reklamcılık Ve Toplumsal Etkileri. *İstanbul Üniversitesi İletişim Fakültesi Dergisi, 29.*
Snyder, W. S. (2008). The ethical consequences of your advertisement matter. *Journal of Advertising Research, 48*(1), 8–9.
Talbot, N. L., Duberstein, P. R., & Scott, P. (1991). Subliminal psychodynamic activation, food consumption, and self-confidence. *Journal of Clinical Psychology, 47*(6), 813–823.
Talimciler, A. (2008). Futbol Değil İş: Endüstriyel Futbol. *İletişim Kuram ve Araştırma Dergisi, 26*(2), 89–114.
Taylor, E. (2012). *Self-hypnosis and subliminal technology.* Hay House, Inc.
Tenekecioğlu, B. (2003). *Genel İşletme.* Eskişehir: T.C. Anadolu Üniversitesi Yayını 1268. Açık Öğretim Fakültesi Yayını 704.
Tığlı, M. (2002). Bilinçaltı Reklâmcılık, *İstanbul Üniversitesi İletişim Fakültesi Dergisi, 25.*
Touhami, Z. O., Benlafkih, L., Jiddane, M., Cherrah, Y., Malki, H. O. E., & Benomar, A. (2011). Neuromarketing: "Where marketing and neuroscience meet". *African Journal of Business Management, 5*(5), 1528–1532.
Treisman, A. M. (1964). The effect of irrelevant material on the efficiency of selective listening. *The American Journal of Psychology, 77*(4), 533–546.
Tüzel, N. (2010). Tüketicilerin Zihnini Okumak: Nöropazarlama ve Reklam. *Marmara İletişim Dergisi, 16.*
Yolcu, E. (2005). Bilinçdışı (Bilinçaltı) Reklam Tartışmaları ve Çalışmaları. *İstanbul Üniversitesi İletişim Fakültesi Dergisi, Istanbul University Faculty of Communication Journal, 22.*
Yücel, A., & Çubuk, F. (2013). Nöropazarlama ve bilinçaltı reklamcılık yaklaşımlarının karşılaştırılması. *Niğde Üniversitesi İktisadi Ve İdari Bilimler Fakültesi Dergisi, 6*(2), 172–183.
Zengin, B., & Şen, L. M. (2006). Seyahat acentelerinin pazarlama faaliyetlerinde karşılaşılan etik sorunlar ve çözüm önerileri. *Pazarlama Dünyası, 5,* 54–63.
Zurawicki, L. (2010). *Neuromarketing: Exploring the brain of the consumer.* Berlin: Springer Science & Business Media.

# Subliminal Messaging: Moving Beyond Consciousness

**Ratul Sur**

**Abstract** Despite the growing interest among the scientific community regarding the power of subconscious, the current research did not find any evidence of its superiority over consciousness in generating positive consumer experience. This study shows that researches in subliminal messages have been set to work towards a set of pre-defined results and can only be used to generate some insignificant changes in social behaviour. Such behaviour is elicited only in laboratory conditions with specific situational variables. Interpretation of the existing corpus of the literature shows that subliminal messages can create negative experience which leads to hostile behaviour like derogatory comments on an African by an American citizen when the latter was primed with negative subliminal messages. Positive priming on the other hand showed weak presence in behaviour. However, research in the field of subliminal messages is required to inspect whether it is capable of improving mental health as indicated by few researches. Further exploration is required to prevent subliminal abuse. As indicated in the current study, subliminal messages when used in commercials are not capable of making a significant increase in sales figures when compared to supraliminal messages. Such messages and their wide-spread broadcast are not ethical because of the advertiser's inclinations to use lascivious, disparaging or satanic stimuli which can lead to fatal outcomes like alleged suicide of a 10-year old boy. Positive experience or happiness is a subjective feeling and is generated by supraliminal messages which has been shown in the study to rely heavily on consciousness.

**Keywords** Supraliminal · Subliminal priming · Social behaviour · Consumer experience · Mere exposure · Threshold level

R. Sur (✉)
University of Kalyani, Kalyani, India
e-mail: ratulsur@gmail.com

© The Author(s), under exclusive license to Springer Nature Singapore Pte Ltd. 2021
T. Dutta and M. K. Mandal (eds.), *Consumer Happiness: Multiple Perspectives*, Studies in Rhythm Engineering,
https://doi.org/10.1007/978-981-33-6374-8_6

# Introduction

## *Mind and Its Antecedents*

Of all the faculties that constitute the human body, the mind has shown to be one of the most complex and widely diverse systems. In 1949, Gilbert Ryle coined a term 'ghost in the machine' to denote the complex functioning of the human mind and its observed influence on human body and behaviour. This raises an important question: 'does the human body functions differently from the mind?' So we will start by answering this basic question.

The origin of the word 'mind' has both Old Germanic and Anglo-Saxon ancestry. In Old Germanic, the word has its roots in 'minna' which means 'memory'. On the other hand, the Anglo-Saxon root refers to the word 'gemynd' which means 'memory'. So, we can sum it up that mind is a faculty which acts as a repository of functions and hence is a separate faculty. At a later point of time, Rene Descartes forwarded his theory of bifurcation of body and mind. However, his theory was long suppressed, and it took almost 12 years to publish his work. Having established mind as a separate faculty, we will do a brief review of the functions of the mind.

However, a complete review of the functions of mind will be a mammoth task and is beyond the scope of this study. To complete our task, we will stick to the theories of Karl Deutsch (1998), who had confined the operations of the mind in seven discrete areas. These are:

1. Abstracting,
2. Communicating,
3. Storing,
4. Subdividing,
5. Recalling,
6. Remembering,
7. Replaying.

As we go about discussing the functions of mind, pertaining to the topic in hand, we will discuss each of the aforesaid functions with due importance while we propel our way to a serene nook of our mind which is least dawdled by our conscious dispositions. The astounding fact about the human mind lies in the fact that a major part of it is inaccessible and unknown to us. Yet, its implications leave far-fetching results. Let us take a feisty example. Noted psychologist Dijksterhuis and Nordgren (2006) once stated this example in one of his works. He said that one of his colleagues had received two job offers at the same time. Both the job offers being equally good, she found it difficult to choose the correct alternative. Dijksterhuis gave her a simple task to perform. He told her to write all the attributes of both the jobs and then to put 'tick' marks against the favourable one and 'cross' marks against the rest. The job that receives the most number of tick gets the nod and the other will be rejected. After some time, this lady came to Dijksterhuis and said that she feels that more number of tick marks are coming on the wrong side. Dijksterhuis asked her how she came to know which was the right side without completing the task which left her without an answer. To this, Dijksterhuis answered that her unconscious had already decided which job to accept. As such, the unconscious exerts an immense effect on human behaviour. Having said that we will move towards the epicentre of the study, i.e. subconscious mind and its influence.

## *Manifestations of the Subconscious in the Society*

Though modern psychological research bears some evidence that subliminal messages and their application are a modern event, it has its roots in the mediaeval ages and has spread throughout the world.

Subliminal messages or subconsciously perceived stimuli have been a function of the society and are always conveyed by someone who has been regarded by society to be a superior. That superiority is conferred upon the individual by the positional system, economical prosperity, administrative status or educational intellect. Such messages from a superior individual can now social suggestibility (Sidis, 1898) in the context of subconscious psychology. A brief review of social suggestibility, as such, will help us to understand how the society gave birth to a laughing new-born son.

The law of suggestibility lays down two streams of consciousness. As evident, these two streams of consciousness are the waking consciousness and the sub-waking consciousness (Sidis, 1898). Subjects under systematized anaesthesia reveal the presence of a subconscious waking, and research findings of such subconscious waking will be dealt in the next section. For now, we will review various social events to see how the underlying metaphors of such events drive individuals to unpursued goals.

Suggestibility is an inherent virtue of the society that gradually permeates into human beings. What is required here is a complete disaggregation of the consciousness for social suggestibility. Such suggestibility under a strong, consistent belief gets metamorphosed into a mental epidemic and henceforth grows exponentially. Since the mediaeval ages, the spectre of witchcraft loomed at large in various parts of Europe. Demonophobia cast its ominous shadow not only over the Holy Church but had also eclipsed the waking consciousness. Boris Sidis in his famous work 'The Psychology of Suggestion: A Research Into the Subconscious Nature of Man and Society' has given an animated description of the mental epidemic. Such mental epidemic in recent time has been called the 'Subliminal Messages' and has been regarded as a spectre which is haunting America. One of the widely used methods used to eliminate witchcraft was to tie the suspect to a table or chair in an uncomfortable manner and leave them to an observer for 24 h without food and water. It was believed that the witch will be visited by one of her imps during this period to suck her blood. The imp which was most likely to take the form of a moth or a fly was intentionally admitted to the room by making a hole in door or window. If a moth or fly enters the room which is a normal event, the suspect would be condemned a witch and killed.

Beyond the set of apparently unrelated events, there is a hidden metaphor. The message that such social events bore was clear, deep and pertinent. While the Pope was keen in the micromanagement of the society, the witch hunters kept themselves busy in the simony. The ever-present social message appealed directly to the subwaking consciousness and sent a message to the individuals—'witches exist today and they can emerge as parallels of churches'. This message crossed the threshold of consciousness and started to make behavioural changes in people. It reached such an extent that even among such horrifying consequences, a young girl from Germany confessed that she can cause sterility in cattle by merely uttering some magic words.

As such, under enormous socio-static pressure of the political system, repression of the economic system and the strict dictums of the religious organizations, individuals tend to lose control over their relations. The cleavage between the conscious and the subconscious widens (Sidis, 1898). The personal self sinks and the suggestible subconscious surfaces get exposed and get tuned to the apparently unheard tones of the epidemic. Subsequently, it becomes an intangible part of a historical tragedy and becomes the content of a social spectre.

## *Personality and Its Importance in the Creation of the Subconscious*

Thought is never a single. Here, in this section, we will see how a thought is born in the human mind and how impressions are carved out of both the conscious and subconscious counterparts.

Each thought is formed out of a multitude of other thoughts. In this case, the particular thought which enters the human mind is the one which identifies its own ego with that of a person. Any message communicates with the psyche of an individual in terms of cognitive pulses. In this regard, it must be noted that every pulse carries a message, but it only becomes aware of it when it is about to die and needs to pass on the message to the next pulse. In this regards, Sidis (1898) remarks that 'each thought is thus born an owner, and dies owned, transmitting whatever it realizes as itself to its later proprietor'. It can be explained by the following diagram.

a in Fig. 1 is a single pulse which bears the identity 'A', and it has inherited it from a previous pulse. In b and c, we see that 'B' has inherited its identity from 'A' and 'C' has inherited its identity from 'B'. This is the underlying process of memory creation which also donates full-heartedly to the formation of personality. As such, this synthesis is not aware of its mechanism, and hence, the formation of personality is an unconscious process. The reason we are talking about personality in connection to subliminal messages is the fact that the subconscious lies below the crust called personality. The tougher the crust is, the harder it is for the message to communicate with the subconscious. The synthesis of 'A', 'B' and 'C' in the diagram

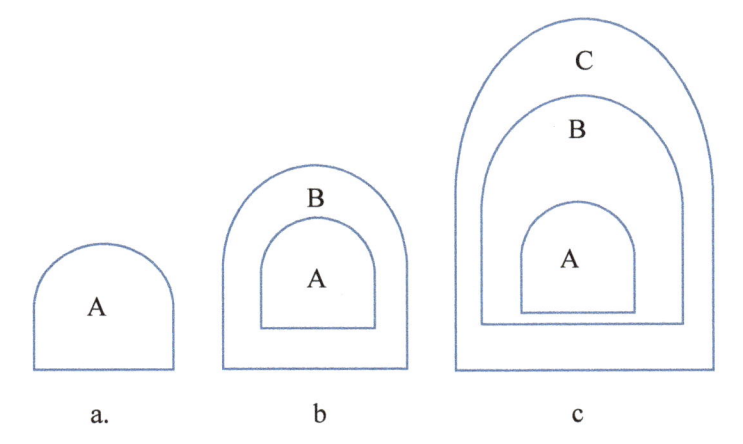

**Fig. 1** Creation of thought and its subsequent transmission

above has created the personality. But what happens when the message cracks the crust and communicates with the subconscious self. In the next few sections, we will see how a subconscious self is created and how it creates behavioural changes in the individuals.

## *Mechanism and Stages of Consciousness*

The subconscious self is a constellation of particles of conscious. In this regard, Sidis (1898) comments that synthesis and catalysis are at the heart of the subconscious. We have already discussed in the preceding section what a synthesis of thought is and how that contributes to the development of personality. Here, in this section, we will review the process of catalysis and see how both synthesis and catalysis together create the subconscious.

We have already said that the subconscious is a constellation of particles of consciousness. Now, each constellation can now be broken into very minute moments of consciousness. Such moments of consciousness are no more a part of the constellation and are not recognized by the rest of the consciousness. These moments of consciousness coalesce under a strong influence of suggestibility. Such conditions favourably catalyse the dissociation of the primary from the secondary consciousness, and the cleavage of the consciousness as mentioned in the last section further widens and may be enhanced by the presence of a powerful external stimulus, e.g. a shocking political image. Under such circumstances, the voluntary muscles of the human body paralyse momentarily, and they are compelled to fixate all their attention to the stimuli. The stimuli from this point become more powerful and gradually permeate through various levels of consciousness and reach the subconscious. The further a message permeate, the difficult it becomes for consciousness to recognize

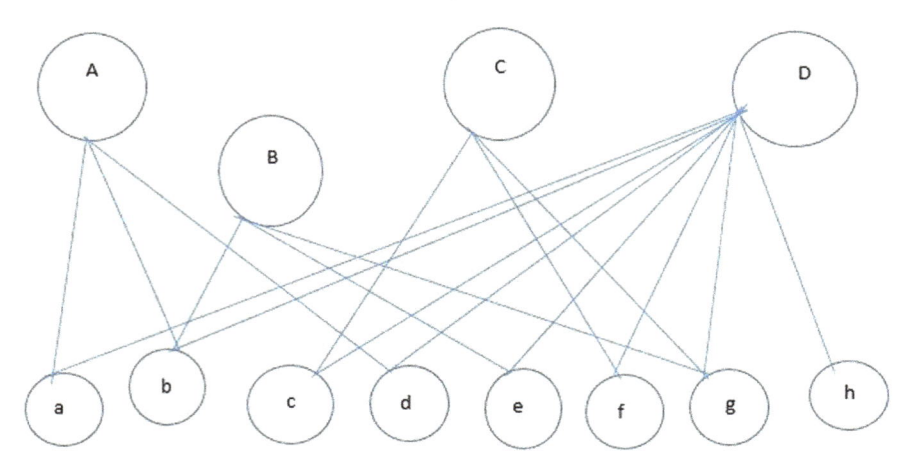

**Fig. 2** Synthesis of the elements of consciousness

its presence. Thus, a message falls from plane of reorganization to the plane of synthesis and from the plane of synthesis to desultory plane. The recognitive plane is the surface which identifies a message, and the plane of synthesis is the surface where thoughts coalesce with each other before dying (as seen in Fig. 1), and desultory is the lowest form of consciousness.

Since the subconscious is a constellation of moments of consciousness, each unit of this constellation lends its identity to dissociated particles of consciousness. A diagrammatic explanation will make it clear.

In Fig. 2, A, B, C and D are constellation or collections of consciousness whereas, a, b, c, d, e, f, h are dissociated particles of consciousness. Here, A has passed on some content of identity to a, b and d, and thus, A can be said to have synthesized A, B and D. Such synthesis is most basic and is called primary synthesis or synthesis of apprehension. However, the total of creation of subconscious self is actually more intense which is explained by the following diagram:

In the secondary stage of synthesis, we see two important elements E1 and E. E1 represents the beats of consciousness which arise from a stimulus. E1 then gives rise to successive beats of consciousness which are represented by concentric circles in Fig. 3. After successive synthesis, it gives rise to E. E can now be called synthetic moment–consciousness and has two important qualities:

1. It can change its synthesized moment–consciousness, i.e. A, B, C, D.
2. It can change its moment contents, i.e. a-j. So, if it can synthesize A, A will also change a. Also, it can directly synthesize b without changing B.

As such, E1 and E are critical elements and are usually present as a stimulus in a message. Figure 3 is the process by which subconscious self emerges. Conclusion of this discussion leads us to a basic question: How does this affect our lives? Does it bear any connection to our study? If so, how? Of all the prominent ways by which such subconscious manifests itself in our lives in by dreams. So we will start the next

Subliminal Messaging: Moving Beyond Consciousness                                        107

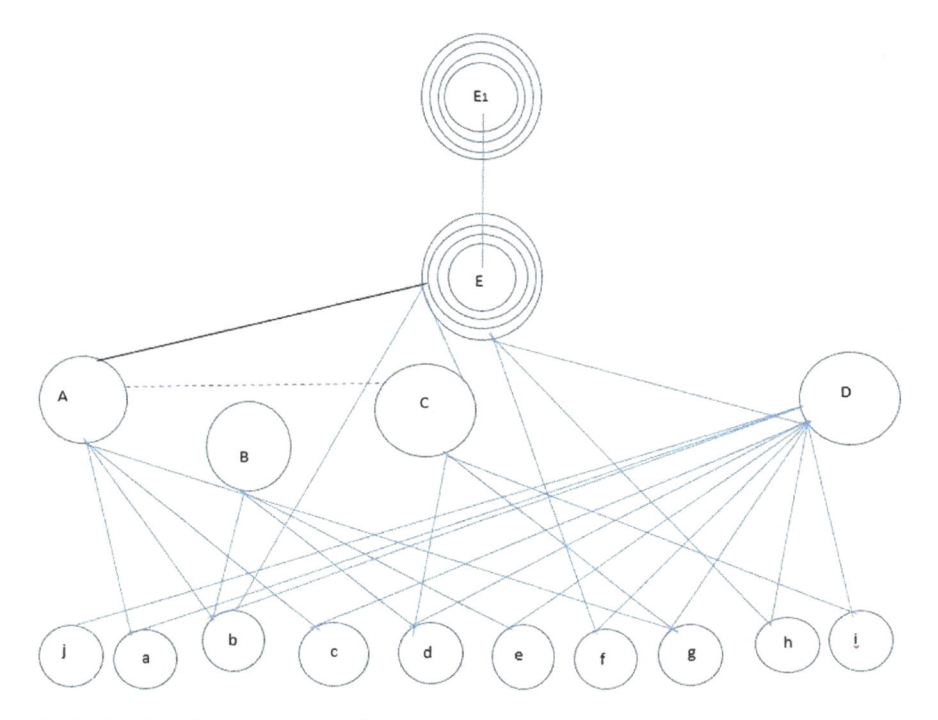

**Fig. 3** Creation of subconscious self

section with a brief introduction of dreams and in turn that will establish a connection with the complex psychological operations and our behavioural differences.

## *Moving Beyond Consciousness*

In all of the previous sections, how elements of consciousness get dissociated and get endowed with the identity of the moments–consciousness. Now, we will review the case of 'Dream of Irma's Injection' and that will give a realistic feel of the moments–consciousness while setting a prelude for the psychoanalysis of the subliminal self in our daily lives.

During July 1895, Freud had a dream about one of his patients Irma whom he had treated for hysteria. In this dream, Freud saw that he is receiving many guests in a large hall; among them, Irma was also a guest. Irma complained Freud of her pain in the neck, stomach and abdomen. Freud examined Irma who was reluctant to an examination looked pale and bloated and found that she has a spot inside her mouth. He quickly called Dr. M, who also looked pale and had a clean chin, limped to him and diagnosed the infection and told that it will be cured in a few days. Also, he found an infiltrated portion on Irma's left shoulder.

Freud from this point picked up each element of this dream and analysed it. The dream opens with a scene where Freud is receiving many guests which is a dissociation of the event where his wife expressed her wish to celebrate her birthday, which was to come in a few days. Irma was indeed his patient, but somehow, she did not follow Freud's prescribed medication which led to some medical complications. Such medical complication again got dissociated and gave birth to different psychic contents. The white spot that Freud found in Irma's mouth represents another event whose dissociation gave rise to this psychic content. This is an event where Freud met one of the Irma's friends who was diagnosed with diphtheria by Dr. M and Irma's reluctance was synthesized by the shy behaviour of one of the Freud's patients who was a governess and had false teeth. This made her reluctant to oral examination also. The sloth and uneasy movement of Dr. M was deduced from one of the Freud's brothers who had a clean chin and sometimes limps because of arthritis. The infiltration on Irma's left shoulder derived its psychic contents from his own rheumatism. Hence, we can say that every single beat of consciousness has its own subconsciousness counterpart, and the contents of this consciousness are shown in the example. Diagrammatically, it would look like this in analogy to Figs. 2, 3 and 4.

Here, conscious reality contained the events of the last few days and even weeks. But the mind detained only some important elements and the other elements withered away. In this case, the information contained in the memory exists either both as semantic memory or as episodic memory. For example, the incident of the governess of Irma's friend's diphtheria was stored as episodic memory, while wife's birthday and knowledge of brother were stored as semantic memory. But, as these events passed away, such elements fell from the plane of recognition to the plane of synthesis.

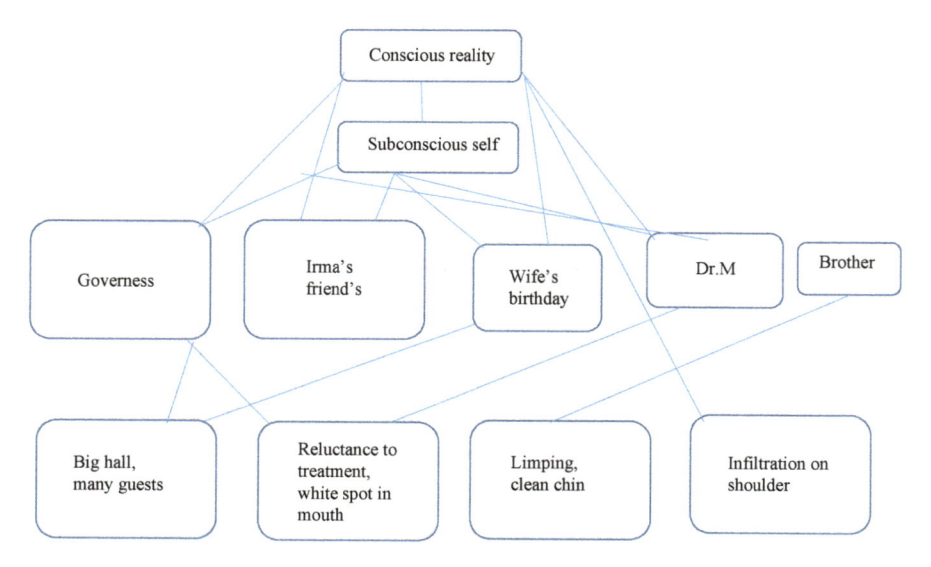

**Fig. 4** Analysis of 'Dream of Irma's Injection'

Here, such elements gave rise to the psychic contents of the elements: 'big hall', 'many guests', 'reluctance', 'white spot in the mouth', 'limping', 'infiltration on the shoulder'. It should be noted that the element 'governess' synthesized 'reluctance', 'white spot in the mouth' because this governess had false teeth and hence was reluctant to an oral examination. Also, 'Irma's friend's diphtheria' synthesized 'white spot' because Dr M who was examining Irma's friend detected diphtheria by a white spot. At the same time, Dr. M's limping movement and his clean chin were synthesized by Freud's brother who had a clean chin and suffered from arthritis.

It must be noted here that the subconscious manifests itself in the form of a dream. Does it prove that it has no influence in a waking state? Hence, it must be asserted here that the stimuli contained in all the episodes, i.e. wife's birthday and Irma's friend, occurred in the natural course of events and had no social or commercial underpinning. But when such carefully crafted stimuli are communicated, as we will see in later sections, undesirable changes are noted in the presence of suitable cues.

## *The Nature of the Subliminal Plane*

The nature of the subliminal planes will be discussed in two parts:

1. Metaphysical description of the subliminal plane, its nature, characteristics and primacy over human consciousness,
2. A more psychological and applied approach towards the subliminal plane which will talk about its creation and utility as a marketing tool.

This section will deal with the metaphysical description of it. In the opening section, we have discussed Rene Descartes' theory of bifurcation of body and mind. This, however, is in the alignment of the Buddhist concept of body and mind. We, as individuals, tend to perceive ourself as a fixed entity. This entity, as we perceive it to be is a communion of body and mind. This is a divorce from the fluid nature of life. Life is in constant motion, and hence, the human self is not fixed. The moment we try to deny this universal truth, our suffering begins. The world is dynamic, and we must align ourselves to the direction of time. Such a bifurcation of body and mind can have immense potential. To illustrate this in the light of reality, the experiment of Benson (1982) carries a special significance. In 1981, HH Dalai Lama verbally instructed three Buddhist monks to perform gTum-Mo meditation which increased the temperature of fingers and toes by 8.3 °C. The bifurcation of body and mind becomes more pronounced. It is during this time that the 'prana' of the soul is withdrawn from the normal states of consciousness and entered the 'central channels', thereby igniting the heat of the body (Benson, 1982). Though it might sound metaphysical, but when we go to the next section, we will see that this bifurcation is not something spiritual and happens in normal waking state with normal individuals. For now, we will explore further the nature of the subliminal plane which will pave the way for the next paragraphs.

To put it in a better metaphysical sense, I would like to talk about the cognitive layer. In meditative states, the practitioner turns rigid to the lucid external world, but as normal individuals, who do not practice the yogic process, the nature of the subliminal self cuts a different scenario. With the reality changing every moment, the consciousness gives birth to several cognitive layers starting from most gross to the most subtle. Here, what we see that with the minute change in the external reality, our consciousness creates a few subtle layers, more subliminal layers of consciousness. These layers of consciousness essentially exchange energy between themselves. It is more like an exchange of information between a client and a business.

It appears that we sometimes have unexpected moments which the cognitive mind reacted, but the entire event did not communicate only to the gross layers of consciousness, but some subtle stimuli also communicated to more subtle subliminal layers. It feels less defined than thought, and when we try to describe it, it slips away, yet we can feel that something significant has changed. The rigid nature of human psychology with reference to the external world is the cause of sorrow, whereas the subtle changes of the layers of consciousness pave the way for healing and pleasure. Psychologically, this healing is carried out by our ego defence mechanism. An ego defence mechanism is a process of 'excluding the unpleasant impulses from our consciousness'. According to Buddhist psychology, hidden at the core lies the subtle centre of 'knowledge'. It must be noted that this knowledge is different from cognitive learning. In this case, the learning happens through feelings, emotions and ideas without becoming identified with them. This, in turn, creates a separate personality, rather a point of consciousness having its own emotionality (Johansson, 1985). Such a personality is characterized by its clarity which lies enshrouded in the mists of our individual belief. The more we tend to believe in the rigidity of life and its reality, the more obstacles we get in understanding the subliminal personality. More the intensity of belief, more the lack of clarity.

## *Subliminal Messages: Does It Make Us Happy? a Conclusive View of Different Researches*

We have already mentioned that the changes in the subtle layers of consciousness create a feeling of healing. It must be noted that this healing is by large a process which is beyond our consciousness. The concurs with the Freudian theory that human beings are born to draw pleasure (Freud, 1920), and they do so by a proper balance of the id ego, superego and ego (Freud, 1936). However, in this section, we will address the focal point of the study, i.e. the potential of the conscious versus the subconsciousness in creating happiness. In all the previous sections, we have discussed the subconscious from a metaphysical point of view and from a psychological point of view. From this section, we will treat the subject from the perspective of business management, more specific from the perspective of subliminal messages and their application in advertisement and other social communication. So, before we start our meta-analytical discussions, we will see what motivated advertisers to take up this type of communication techniques.

Let us take the American case. By the end of World War II, the life of the Americans started getting better. As per the Govt. reports, the unemployment rate dropped from 14.6% in 1940 to 1.2% in 1944 (Sur, 2015). At the same time, GNP grew by 132.42%. It was found that the GNP grew from $91.9 billion in 1939 to $213.6 billion in 1945 (Washinton DC, Govt. Printing Office, 1961). It was definitely a dream-like situation where real income grew by 68% with the fear looming in the market that public saving might go down. Over time, this fear came to be true. This can be substantiated by an example. It was found that the purchase of refrigerators, then considered to be a luxury item, grew from 44% in 1942 to 80% in 1950 (Sur, 2015). Such consumerism prompted the advertisers to communicate to the consumers on a more personal level. Advertisers started to exploit the subconscious and the whole new horizon of subliminal advertising as unfurled. The situation became more dramatic after James Vicary's seminal experiment on subliminal priming in 1957. James Vicary flashed the words 'Hungry? Eat popcorn' and 'Thirsty? Drink Coca Cola' for a very brief time on the cinema screen in a movie theatre. Following this experiment Vicary claimed that the sale of popcorn and coca-cola grew in those theatres where the message was flashed. He demanded that he has successfully primed the audience and made them behave in the way he wanted and claimed it to be a successful priming experiment. Later on, it was revealed that the results of the experiment were doctored and that Vicary had inflated the outcomes. However, the importance of the experiment to date lies in the fact that it made the consumers aware of this 'devil' (Vokey & Read, 1985). It was the beginning of a spectre. A spectre is ever present, ever absent, but the fear looms large. People started to explore subliminally primed, hidden message or backmasked messages in ad jingles. Judas Priest had been alleged to have used backmasked messages. In their 'better by you, better by me' was alleged to contain backmasked words like 'try suicide', 'do it', 'let's be dead', which prompted suicide tendencies. Led Zeppelin was found to have used subliminal messages 'master satan', 'serve me', 'there's no escaping it' in there, and it created havoc in 1981. In a similar case, in 1985, two young men Raymond Belknap (18) and James Vance (20) committed suicide after listening to Judas Priest. Even, in 2019, the ny post reported having found hidden subliminal messages in YouTube kid's videos which are feared to have created suicidal tendencies. In one of such potentially risky video, a YouTube prankster was shown in a kid's video in which this prankster known as Filthy Frank demonstrated how kid's can harm themselves by the signing off words 'End it'.

As such, in the coming few paragraphs, we will see how the subconscious self is exposed in the presence of subliminal stimuli, how the particles of consciousness dissociate and give birth to the subconscious and how behavioural changes are triggered by such subliminal stimuli.

To start with the works of Zajonc (1968, 1980, 2001) needs to be mentioned. In such works, Zajonc elaborated the nature and functioning of 'mere exposure'. This refers to a minute exposure of a stimulus, e.g. 1/3000th of a second (as we saw in

Vicary's experiment). It was found that the respondents pointed to the right stimuli when asked to choose from a set of stimuli with 61% accuracy. Though they did not see anything more than a flicker, their answers were well above chance occurrences.

To start a delving discussion on subliminal messaging, it is imperative to understand what type of stimuli it is and how it works. The word 'subliminal' has two parts—a prefix 'sub' meaning below or beneath added to 'liminal' or 'limen' meaning 'threshold' (Sur, 2015). Takes together, it means that subliminal message is a message which contains a stimulus which passes beneath the threshold of our consciousness. Such a message enters the psyche without the consumer being aware of it. As per traditional belief, such a message can make certain behavioural changes, as desired by the propagator and makes a long-lasting effect on the consumer. Here, onwards, we will do a strict analysis of some eminent researches on subliminal messaging and draw meaningful insights of them while re-interpreting the results of them.

As mentioned in the earlier paragraph, the nature of the stimuli has to be subtle. This means that in most of the cases, such messages are hidden, e.g. if we look at the brand logo of Baskin Robbins, we will see that the way 'B' and 'R' are written, and it resembles the numerical digit '31' which represents the 31 flavours that Baskin Robbins serves to its customers. A close look at Amazon's brand logo reveals smiley with an arrow from 'a' to 'z' denoting that Amazon sells from 'a' to 'z'. FedEx has very skilfully used a subliminal message in the logo. It has been found that there is an arrow sign placed between 'E' and X' which denotes the speed and ability of the company to deliver the consignment.

At this point of time, we can put forward two arguments:

1. If the stimuli are so subtle and hidden, then how can we expect that it will be noticed by all the consumers?
2. Even if we are propagating something subliminally, how can we make sure that it will have its desired effect on the consumers?

Moore (1982) put forward two important arguments in this connection. He said that subliminal stimuli are so weak in nature that they are not perceived equally by all consumer. Concurrently, Engel and Blackwell (1982) commented that a fractional stimulus could be easily swamped by other strong stimuli. Regarding the outcome of such messages, it has been found that a message 'telephone now' flashed in a TV programme 3–5 times elicited behavioural changes like urges to eat, to take off shoes, driving safely and even to buy an electric frying pan (Sur, 2015). However, the stimulus discussed so far is visual stimuli, and we will also consider the case of masked auditory stimuli with diverse implications.

In one of the eminent experiments, Henley and Dixon (1974) examined the laterality differences as one of the important variables in auditory subliminal messages. This study reported a marked emergence of subliminal stimuli. This study encouraged future studies when Mykel and Daves (1979) replicated the experiment of Henley and Dixon. In their experiment, Henley and Dixon presented the subliminal stimuli to their respondents in the presence of musical stimuli. Here, I raise a few issues. Music has a proven effect on human behaviour. Gueguen and Jacob (2010) found in a study that when romantic music was played in a flower shop; both male and females respondents spent more time and spent more money compared when pop and rock music was played. This proves that usage of a supraliminal stimulus adjacent to a subliminal stimulus cannot be a right experimental condition, and as Moore (1982) pointed out, it might go unnoticed completely. In a similar experiment, Gorn, Goldberg, and Basu (1993) made their respondents test a pair of speakers after listening to the music out of the speakers. One of the speakers played a piece of unpleasant music, and the other speaker played a piece of sad music. It was found that the speaker that played pleasant music was evaluated higher than the speaker that played mellow music. This indicates that music can induce mood in consumers which are then translated in certain behavioural changes.

The reason for such outcome roots from the affective response—a subjective feeling of well-being or a feeling of rue or remorse. Such an affective response arises when affect induces social judgment and decision (Forgas, 1992) and acts as a source of information to form a judgment. This process is beyond the scope of consciousness (Freud, 1922) and hence sets in quickly. This also explains the mysterious results of Zajonc (1968), where the respondents successfully recognized the stimuli on the screen 1/3000th of a second. Here, the response developed an affective feeling and hence was able to recognize the stimuli. The experiment and Henley and Dixon were again replicated by Benes, Gutkin and Decker (1990) by using subliminal stimuli along with the music. The respondents were provided with the stimuli along with the music which was either mellow or frenetic. The stimuli consisted of subliminally propagated word related to 'water' or 'family'. After the stimuli were presented to the respondents, they were asked to write all the random words coming to their minds. This was followed by a questionnaire from where the respondents were to choose from a set of stimuli, the words that they think about. It was found that when mellow music was played, most of the respondents reported some family words in the presence of mellow music. This indicated that subliminal messages work best under some mellow mediation. It can also be deuced from here that subliminal messages are not capable of creating happiness. To further the discussion, let us review another important case.

Silverman (1976) found that subliminal presentation of the stimulus 'DESTROY MOTHER' resulted in an increased feeling of depression, whereas a control stimulus 'PEOPLE THINKING' was unable to evoke such a feeling. This can be taken as a piece of evidence that even subliminal messages can address our subconscious wish, and it is capable of creating some depressed feelings. It can also evoke feelings and wishes which can address to the suppressed wishes. Based on Silverman's findings, Saegert (1987) commented that such a finding might have enough potential for subliminal advertising. However, on the basis of earlier findings, I raise a few questions on Silverman's arguments. Silverman found pronounced effects of depression in schizophrenics and depressive patients, but he did not find any effect of subliminal messaging on normal individuals. Secondly, the 'wish' of the respondents, which both Silverman and Saegert talks about, is a vague proposition. Every individual has a set of wishes of which they do not have complete knowledge themselves. At such a juncture, it is almost impossible both for the advertiser and the researcher to explore those wishes and to trigger them. Thirdly, Silverman proposed that only subliminal messages heighten a wish. This is again a very difficult task because different people have different threshold limits and the nature of subliminal stimuli is very much dependent on that. Currently, psychological research claims that stimuli being propagated below threshold limit can heighten some feelings (Marcel, 1983). Here, also there is a problem. Any message to qualify for a subliminal message must be in the range of 20–110 ms (Marcel, 1983) or for the minimum of 10–25 ms (Fowler, Wolford, Slade, & Tassinary, 1981). Such a threshold value is too small for everyone and might go unnoticed totally, leading to low impact (Beatty & Hawkind, 1989). To stress on the usefulness of subliminal messages, Chakalis and Lowe (1992) conducted an experiment in which they presented subliminal memory improvement messages in audiotapes. This subliminal message was embedded in a pleasant audiotape called 'Garden of Dreams'. The visual stimuli consisted of faces of common people with their name and profession. The visual stimuli were presented for 10 s. It was found that the respondents in the experimental conditions remembered the face-name-profession in a better way than the control group. Based on the findings, the researchers claimed that memory performance can be better in a subliminal condition. There are two possible explanations for this experiment. Firstly, the visual stimuli were presented for 10 s which does not qualify as a subliminal stimulus. Secondly, the subliminal stimuli were masked, and surprisingly, no investigation for subliminal stimuli was done. Thirdly, the experimental results were very much similar to the findings of Kunst-Wilson and Zajonc (1980) in which the respondents recognized the stimuli correctly even when they were presented for 1 ms, and the respondents saw nothing more than a flicker. The reason forwarded by Kunst-Wilson was because of development of an affective response. It suggests that there is a 'capacity for making affective discrimination without extensive participation of respondents'. Hence, we can deduce that subliminal medium is not an appropriate medium for instilling happiness in consumer because whatever happiness it creates, it is beyond the scope of consciousness.

**Alternative Research**

As an alternative to motivational research, research on affective response or feeling-based should be advocated. We have seen in the previous paragraphs that whatever the subliminal event is suspected, it was an affective feeling that had set-in, and hence, an elaborate is required. Kotler (1974) in one of his eminent works stressed heavily on the importance of feeling-based variables as determining factor in advertisement research. Such variables included factors like smell and lighting which can be regarded as spatial aesthetics and regarded them as a silent language of business. Unfortunately, such affective states have been regarded as cold states (Forgas, 1995). A few works that have been done in this regards are pertaining to the relationship between incidental affect and task-related affect (Sur, 2015), the influence of affect on eating capacity (Garg, 2007), temporal existence of affect (Pocheptsova & Novemsky, 2010) or compensatory consumption.

# References

Beatty, S. E., & Hawkins, D. I. (1989). Subliminal stimulation: Some new data and interpretation. *Journal of Advertising*, 4–8.
Benes, K., Gutkin, T., & Decker, T. (1990). The effects of mellow and frenetic music on reported cognitions resulting from auditory subliminal messages. *Journal of General Psychology*, 83–89.
Benson, H. L. (1982). Body temperature changes during the practice of g Tum-mo yoga. *Nature, 295*, 234=236.
Chakalis, E., & Lowe, G. (1992). Positive effects of subliminal stimulation on memory. *Perceptual and Motor Skills*, 956–958.
Deutsch, K. (1998). *Perception and cognition at century's end: history, philosophy, theory.* USA: Academic Press.
Dijksterhuis, A., & Nordgren, L. F. (2006). A Theory of unconscious thought. *Perspectives on Psychological Science*, 95–109.
Engel, J., & Blackwell, R. D. (1982). *Consumer behavior.* London: The Dryden Press.
Forgas (1995). Mood and Judgment: The Affect Infusion Model. *Psychological Bulletin*, 117(1):39–66.
Forgas, J. P. (1992). Affect in social judgments and decisions: A multiprocess model. *Advances in experimental social psychology*, 227–275.
Fowler, C. A., Wolford, G., Slade, R., & Tassinary, L. (1981). Lexical access with and without awareness. *Journal of Experimental Psychology: General*, 341–362.
Freud, S. (1920). *A general introduction to psychoanalysis.* New York.
Freud, S. (1922). Beyond the pleasure principle. *International psychoanalytical library, 4.*
Freud, A. (1936). *The Ego and the Mechanisms of Defence.* Karnac Books.
Garg, N. (2007). The influence of incidental affect on consumers' food intake. *Journal of Marketing*, 194–206.
Gorn, G. J., Goldberg, M. E., & Basu, K. (1993). Mood, awareness, and product evaluation. *Journal of Consumer Psychology*, 237–256.
Guéguen, N., & Jacob, C. (2010). Music Congruency and consumer behaviour: an experimental field study. *International Bulletin of Business Administration*, 56–63.
Henley, S., & Dixon, N. (1974). Laterality differences in the effects of incidental stimuli upon evoked imagery. *British Journal of Psychology*, 529–536.
Johansson, R. E. (1985). *Analysis of Citta*, 34.
Kotler, P. (1974). Atmospherics as a marketing tool. *Journal of Retailing*, 48–64.
Kunst-Wilson, W., & Zajonc, R. (1980). Affective discrimination of stimuli that cannot be recognized. *Science*, 557–558.
Marcel, A. J. (1983). Conscious and unconscious perception: Experiments on visual masking and word recognition. *Cognitive Psychology*, 197–237.
Moore, T. E. (1982). Subliminal advertising: What you see is what you get. *Journal of Marketing*, 38–47.
Mykel, N., & Daves, W. F. (1979). Emergence of unreported stimuli into imagery as a function of laterality of presentation: A replication and extension of research by Henley & Dixon (1974). *British Journal of Psychology*, 253–258.
Pocheptsova, A., & Novemsky, N. (2010). When do incidental mood effects last? Laybeliefs versus actual effects. *Journal of Consumer Research.*
Ryle, G. (1949). *The concept of mind.* Chicago: University of Chicago Press.
Sidis, B. (1898). *The psychology of suggestion: A research into the subconscious nature of man and society.* Andesite Press.
Silverman, L. H. (1976). Psychoanalytic theory: the reports of my death are greatly exaggerated. *American Psychologist*, 621–637.
Sur, R. (2015). Impact of subliminal messages on consumer behavior: A fresh look into the future. *International Journal of Interdisciplinary and Multidisciplinary Studies*, 56–59.

Vokey, J. R., & Read, J. D. (1985). Subliminal messages: Between the devil and the media. *American Psychologist*, 1231–1239.

Zajonc, R. (1980). Feeling and thinking: Preferences need no inferences. *American Psychologist*, 151–175.

Zajonc, R. B. (1968). Attitudinal effects of mere exposure. *Journal of Personality and Social Psychology*, 1–27.

Zajonc, R. B. (2001). Mere exposure: A gateway to the subliminal. *Current Directions in Psychological Science*, 224–228.

# Customer Happiness: The Role of Cognitive Dissonance and Customer Experience

Anil V. Pillai

**Abstract** Creating a superior customer experience is a definitive means to customer happiness. The customer experience though must have meaning and context. However, when the delivered experience is at odds with the customer's closely held beliefs and values, this creates cognitive dissonance, and instead of the happiness, it is supposed to generate, unhappiness and disengagement results. This chapter explores a unique perspective on cognitive dissonance, customer experience, their linkage and possible outcomes in order to drive customer happiness.

"Happiness is the overall experience of both pleasure and meaning" Tal Ben Shahar.

Based on this excellent definition of happiness, we gather, in order to deliver customer happiness, three elements needed to be in alignment.

The customer has to derive functional fulfilment; the functional objective of that interaction has to be met, and the customer must derive a "pleasurable" "experience", from the entire interaction journey with a brand and such lastly the experience has to hold meaning, must have "context" which is aligned to the customer. This meaning and context have to be in alignment with the customer's beliefs and thinking and value system.

First some clarity on terminology; "Happiness" is often defined as both a state of mind and an emotion.

"Satisfaction", on the other hand, is far more transactional, a state where a customer's needs are met. It is an outcome of the knowledge that a customer's demands and desires are made possible.

The converse of happiness, the state of being unhappy, therefore comes about when there is dissonance between what one desires and what is experienced, between what one's moral, ethical and social beliefs and contexts are and what the brand or the organization espouses.

---

A. V. Pillai (✉)
Pune, India
e-mail: ap@terragni.in

© The Author(s), under exclusive license to Springer Nature Singapore Pte Ltd. 2021
T. Dutta and M. K. Mandal (eds.), *Consumer Happiness: Multiple Perspectives*,
Studies in Rhythm Engineering,
https://doi.org/10.1007/978-981-33-6374-8_7

To take a deeper look at what causes this state of unhappiness, we go back in time and look at an elegant theory that Leon Festinger proposed. A theory that has since been rather unfortunately consigned back to the pages of history or at best summarized into a rather pithy, unfair label; "buyer's remorse".

This has occurred simply due to a lack of a deeper understanding of Festinger's seminal work in cognitive dissonance.

Cognitive dissonance is one of the most acute drivers of unhappiness. Think of it, each time you have felt unhappy, what really happened? What occurred was a tussle between your world view, your closely held beliefs, your value system and that of an occurrence in reality.

According to Festinger's (1957) theory of cognitive dissonance, people prefer consistency over inconsistency; Festinger (1957) also suggests that dissonance is an experience. Cooper (2007) also posited that cognitive dissonance is experienced as tension or unpleasant mental state as a result of dissonance aroused by two inconsistent cognitions.

The attribute of magnitude significantly differentiates cognitive dissonance from other theories of inconsistency. Higher the discrepant information, higher is the discomfort; greater the magnitude of dissonance, higher the motivation to reduce it (Cooper, 2007).

Dissonance magnitude is proportional to discrepant cognition and inversely proportional to consonant cognition, each weighted by its importance (Cooper, 2007). Change in attitude is one of the common approaches to reducing cognitive dissonance (Cooper, 2007), as it is easier to change attitudes for people because it is a private construction, not a public committed behaviour.

Several limiting conditions have been identified that arouse dissonance. For the arousal of cognitive dissonance, decision freedom is a necessity (Linder, Cooper, & Jones, 1967). The magnitude of commitment is another concern that appears to limit the occurred dissonance, e.g. when a person feels committed to his or her attitude-inconsistent statement, such as one is publicly identified with the statements (Carlsmith, Collins, & Helmreich, 1966) or when one cannot retract them at later date (Davis & Jones, 1960). Nel, Helmreich, and Aronson (1969), who were the first to spot the aversive or unwanted consequence for the arousal of dissonance, later discovered that the mere implications that an unwanted result might occur also permit counter-attitudinal behaviour to lead to dissonance (Cooper, 2007). According to Cooper (1971), foreseeable aversive consequences lead to dissonance; unforeseeable consequences do not.

New-look model of dissonance proposed two aspects of dissonance theory: one that generates the dissonance arousal and the other that motivates people to change (Cooper & Fazio, 1984). "Dissonance process begins with behaviour. People act. And as a result of those actions, consequences ensue. Those consequences can be positive, negative or aversive" (Cooper, 2007, p. 74).

Harmon-Jones (1999) suggested that people normally would take a stance towards the environment that they normally operate in, wherein they could act in a manner devoid of ambivalence and conflict. "New look version of dissonance theory predicts that aversive consequences lead to dissonance; the inconsistency with a *but only* modifier holds that attitude-inconsistent behaviour leads to dissonance, but only when an aversive result is produced" (Cooper, 2007, p. 83).

The theory of self-affirmation, proposed by Steele (1998), Steele and Liu (1983), indicates that most people would like to see themselves as good and honest. Both self-affirmation and the new-look model agree that dissonance reduction is not just about restoring consistency. Self-consistency theory (Aronson, 1968; Thibodeau & Aronson, 1992) also holds the concept in the consistence of Steele's self-affirmation theory. When an individual's behaviour is inconsistent with their self-concept, it gives rise to dissonance. This is the key to understanding why customers find themselves unhappy even when a product or service they have procured has met their functional needs. Brands and organizations that understand this aspect will also realize that customer happiness goes far beyond merely meeting a customer's functional needs. This will also explain why customers defect even when brands have met a customer's ostensible needs. As per Stone and Cooper (2001), there were two major classifications of standards that a person can use to assess the meaning of the consequences of his or her behaviour; personal and normative. People use a personal standard to judge the outcome of their behaviour. When such outcomes are perceived as undesirable, then people experience ideographic dissonance (self-esteem moderation). If people use the normative standard to judge their behaviour outcomes, nomothetic dissonance (no self-esteem moderation) occurs because consequence behaviour falls into a class that people realize is consensually negative. Self-standard model (SSM) of dissonance (Stone et al., 2001) allows us to continue the metamorphosis of dissonance theory. It expands on the new-look model (Cooper et al., 1984) which suggested a radically different motivational basis for dissonance from the one Festinger had introduced in 1957.

Dissonance can also occur when other people make choices or when otherwise act in a way that brings about the aversive event; it will open up a way to experience a vicarious emotion and will result as vicarious cognitive dissonance (Cooper & Hogg, 2007; Norton, Monin, Cooper, & Hogg, 2003).

Rosenfeld, Kennedy, and Giacalone (2001) illustrated that post-decision enhancement occurs instead of pre-decision moderation. People who have already played are more confident about winning the "gumball guess" lottery game than those asked just before they started playing it. The effect of post-decision cognitive dissonance on attitudes towards the chosen alternative is also supported by their experiment. Sharma (2014) explored the factors—"family status, religious value, customs, belief", etc.—as key factors contributing to cognitive dissonance in the customer buying process.

Thus, cognitive dissonance deeply influences how humans perceive experiences.

## Customer Experience

Late 1999, Bernd Schmitt, re-looked at the essence of traditional approaches to marketing and involving sensory, affective, cognitive, experience, actions and relations in the field of marketing and came up with a rather elegant definition of customer experience.

Focus on customer experience: experience is an outcome of the interactions and encounters of the customers and the brand; experience also engages the brand and the organization to the customer's lifestyle and places individual customer actions and the context of the customers' purchase in the broad social perspective (Schmitt, 1999).

As per Schmitt (2003), customer experience management (CEM) connects with the customer at every encounter and demands for integration of different elements of the customers' experience.

To my mind, experience is a set of recollected memories. Here is an experiment I often try out in my workshops and research interactions. If I were to tell you to close your eyes and recall the experience of the best childhood meal that you ever had, what would you do? You would lean back, relax and cast your mind's eye to a set of memories, the fragrant aroma of the meal, the sights and sounds around the dining table, the crinkle in your mother's eyes as she served you, the texture of the morsel as you first tasted it. The slipping of the glass from your hand as it crashed onto the floor, the glare of your father!

Experiences are multisensory, and they are recollected, and because they are recollected, they need not necessarily be an accurate representation of reality.

In an organizational setting, there occur twin challenges, one of designing an experience which is contextual, relevant and has rich meaning and the other of profitably executing such an experience in a consistent manner across the customer journey, across all touchpoints and on each day of the year.

As functional differentiation between products narrow, after all, one mobile phone is almost as good or bad as the other, experience takes centre stage. Organizations that recognize this are now moving on to create experiential positioning. "The experiential positioning depicts what the brand stands for. It is equivalent to the positioning statement of traditional marketing, but it replaces that vague positioning statement with an insightful and useful multisensory strategy component that is full of imagery and relevant to the buyers and users of the brand" (Schmitt, 2003). The experiential positioning that organizations thus choose is a promise, a promise of an experience that the customer will get when she interacts with the brand. Thus expectations are built. If such expectations, therefore, fall short in actual deliverables, the customer will experience disappointment leading to disengagement.

The implementation theme chosen by the company showcases the summary of the core messages of style and content in the brand experience, the customer interface and future innovations (Schmitt, 2003).

Meyer and Schwager (2007) projected the understanding of customer experience; it is a subjective response to direct or indirect contact with a firm. Direct contacts are initiated by the customers usually with the course of purchase, etc.; indirect contacts are the unplanned interactions with experience of a company's product, brands or services. As mentioned earlier, it is important to understand the difference between customer experience management and customer relationship management. Customer experience management (CEM) captures what customers think, feel, perceive about the organization or the brand; it also captures insights across the customer journey and may be monitored by surveys, depth interviews or observational studies. These observations about customers are used by the business or functional leaders to attain the expectations of their customers by providing a better product, service and contextual, pleasurable memories. While customer relationship management (CRM) captures what companies know about their customers or customer action after the fact, monitored by market research or automated tracking of sales, this research could be used by functions such as marketing, sales and field services to enhance efficiency and effectiveness. Companies check the past, potential and present patterns of interaction with customers to enhance the understanding of customer experience that they deliver. Each pattern has a different method of generating and analysing data (Meyer et al., 2007).

The implementation of superior customer experience would require superlative teams of employees who are trained and empowered to deliver such an experience profitably and consistently.

Thus, employee experience necessarily becomes critical, because customer experience will not improve until employee experience becomes a serious priority for organizations (Meyer & Schwager, 2007).

Pre-experience is the first stage of customer experience management (CEM) where the customer starts consumption with imaging, searching, planning and budgeting. Brand attributes develop customer relationships. They also shape attitudes which influence behaviour. Multichannel interaction consists of design, deployment, coordination and evaluation of multiple channels that the company and its customers use to interact with each other (Fatma, 2014). A sense of comfort and quality of service encounters positively impacts the perception of satisfaction among customers. Customer satisfaction depends on the service outcome provided in the physical environment; the social environment, word of mouth, online application and media, it provides a connection between customers and the community.

When customers have a positive experience, it creates enjoyment, helps them in developing new skills and fantasizes about how the experience was. "The CEM has three types of consequences: first, customer satisfaction which is an important antecedent of loyalty, second customer loyalty" (Fatma, 2014). Positive customer experiences and relationship management have a positive impact on "customer loyalty among customer is derived due to the experiences which they undergo" (Garret, 2006) and third customer equity. "Customer equity is made up of three drivers; brand equity, retention equity, and value equity" (Bejou & Iyer, 2006; Rust, Zeithaml, & Lemon, 2000).

Rawson, Duncan, and Jones (2013) stated that while touchpoints are important, it is the entire customer journey that creates an impact. The interaction between customer and organization during and after purchase is a critical moment for developing a positive experience, but a narrow focus on maximizing satisfaction distorts customer experience. A major topic of argument was that customer journeys aligned with customer expectations see a positive impact on customer retention and increased competitiveness.

Findings suggest that "companies need to combine top-down (used for fixing few glaring problems in a specific journey); judgment driven evaluation and bottom-up (add some research into customer experience of their journey). Companies can use such ideal customer journeys to redesign the entire customer experience" (Rawson et al., 2013).

Spiess, T'Joens, Dragnea, Spencer, and Philippart (2014) explored ways of utilizing big data insights and integrating this with process automation that has to do with key customer touchpoints which in turn improves or influences the customer experience.

Using insights of customers' "behaviour", "wants" and "needs", the brand provides superior experience through enhanced perception of their company supported by the ease of doing business that satisfies and delights customers (Spiess et al., 2014).

So where does this leave customer happiness? The path to customer happiness as is now clear leads from providing a *superlative* customer experience. It can be seen from all the above analysis that a well-executed customer happiness strategy has the immense potential to deliver customer happiness. The challenge comes when, despite the best effort, many a well-intentioned experience fails to deliver happiness. And here is where we go back to the definition of happiness. The two key concepts emerge, " Pleasure and Meaning".

Both pleasure and meaning require a third key component, that of context. When customer happiness is designed and executed minus of context, we end up spending enormous amounts of resources and create dissonance, not happiness.

An experience thus designed devoid of the context of the consumer will run into the "dissonance wall". The dissonance wall is defined as that cognitive friction that inhibits a customer from truly engaging with a brand on account of the inconsistency of beliefs between the customer and the brand. Each element of the brand experience thus has to be designed with great care to ensure that such dissonance is eliminated.

We have seen earlier that the presence of two or more inconsistent cognitions leads to the arousal of dissonance. This dissonance manifests as uncomfortable tension. Left unattended such tension over time takes on the form of pain and friction. Such friction has drive-like properties and can result in serious unhappiness.

Let us paint a scenario, if all around me there is strife, there is uncertainty, and there is looming recession and I indulge in a luxury purchase because I have a limited period discount of 50%. The product (functional fulfilment) is superlative and so has been the pleasure across all touchpoints, and I also got a great deal to boot. Yet, I am just not happy, why? There are two cognitions that are inconsistent with each other. One is aligned to frugality in solidarity with my suffering fellow human beings; the other is aligned to my hedonistic need and not wanting to miss out on a great opportunity.

The interesting thing with dissonance is that dissonance has a magnitude. This marks it distinctly different from other theories of inconsistency. The more the gaps, the more the discrepancy between the two differing cognitions, the higher the order of magnitude of the dissonance. The higher the discrepancy, the greater the gnawing within me, the higher the discomfort and the more my pressing need to eliminate this difference.

Brands have an opportunity here. Those brands that can recognize such dissonance and can address it will be in a unique position to generate happiness. This is what a good contextual customer experience must deliver.

Such an effort requires superior insight gathering skills that go beyond standard, cookie-cutter structured questionnaires or that bane of all insight gathering methods, the focus group! The ability to ask "why" and keep going down that path is key and so are techniques rooted in consumer neuroscience to help elicit emotional contexts and deep-rooted biases and attitudes.

Going back to our example of the luxury purchase, what could the brand have done?

If the brand had the insight into such possible dissonance, perhaps the brand could have lowered the dissonance wall, by promising to contribute 10% of the product price to a charity of the customer's choice. And stop at that? Not in the least. Promising to contribute to charity helps attenuate the discordant inconsistency, but to create real, ongoing happiness to the customer, the brand could go further. How about, every 3 months send pictures to the customer on how her 10% is bringing a smile to little children and how they are progressing.

So you see, contextual customer experience that is designed to deliver true customer happiness never ends when the customer walks out of the store, nor with simply resolving to customer complaints. It goes far beyond that. It contributes to a continual engagement between customer and brand.

When faced with unhappiness arising from discordant cognitions, humans have one of two responses; one would be to affix blame at a source outside of themselves; in this case, the brand and the other response would be to change one's own attitude, a reconciliation of sorts. "So what if there is a recession, I was right in buying this luxury product, after all, I got a 50% discount and moreover I have pumped money into the economy".

A brand thus has a choice, either create cognitions that minimize the gap between opposing cognitions (10% of your bill goes to charity) or create communication which assists the customer to shift attitudes (Thank you for shopping with us, your purchaser helps keeps our shop assistants in employment and contributes to the economy). Brands may choose either approach or a combination as the situation dictates. Innovation led organizations, with deep insight into fundamental attitudes, could go one step further. They could use the right mix of approaches on differing contextual customer segments.

Dissonance reduction thus includes focusing on encouraging beliefs that offset the dissonant belief or behavior, decreasing the importance of contradictory belief. A possible approach would involve changing the conflicting belief, such that it is consistent with other beliefs or behaviors, through credible sources which customers trust. At the same time, word-of-mouth communications need to increase emphasis on selecting a likeable source and finding out the sources which match customer's earlier belief. Unexpected sources sometimes may have more impact on customer's attitudes to overcome cognitive dissonance (Sharma, 2014).

George et al. (2009) revealed that the difficulty in overcoming dissonance is significantly associated with the level of involvement in the process and decision making. As per George et al. (2009), the cognitive dissonance with a low involvement purchase is easier to overcome than with a higher involvement purchase. Empirical results of this study also concluded that the more involved people are, less likely are they, to reject the previous cognition as compared to their less involved counterparts.

It is unfortunate however that brands that focus on designing experiences to deliver happiness skimp on very important learning, humans abhor inconsistency. It is not that this is merely a matter of preference that they would merely prefer consistency over inconsistency. Humans will consistently and actively seek out ways and means of eliminating inconsistency. It is that fundamental a drive. Brands, human beings, life partners and organizations that promise such consistency win our trust and win our hearts.

Therein lies the key to happiness.

## References

Aronson, E. (1968). Dissonance theory: Progress and problems. In R. P. Abelson, E., Aronson, T. M. Newcomb, M. J. Rosenberg, & P. H., Tannenbaum (Eds.), *Theories of cognitive consistency: A sourcebook* (pp. 5–27). Chicago, IL: Rand McNally.

Carlsmith, J. M., Collins, B. E. & Helmreich, R. L. (1966). Studies in forces compliance: I. The effect of pressure for compliance on attitude change produced by face to face role playing and anonymous essay writing. *Journal of Personality and Social Psychology, 4*(1), 1–13.

Cooper, J. (1971). Personal responsibility and dissonance: The role of the foreseen consequences. *Journal of Personality and Social Psychology, 18,* 354–363.

Cooper, J. (2007). *Cognitive dissonance: 50 years of a classic theory.* Sage Publications.

Cooper, J., & Fazio, R. (1984). A new look at Dissonance theory. In L. Berkowitz (Ed.), *Advances in experimental social psychology* (Vol. 17, pp. 229–264). Orlando, FL: Academic Press.

Cooper, J., & Hogg, M. A. (2007). Feeling the anguish of others: A theory of vicarious dissonance. In M. P. Zanna (Ed.), *Advances in experimental social psychology* (Vol. 39). San Diego, CA.

Festinger, L. (1957). *A theory of cognitive dissonance.* Stanford, CA: Stanford University Press.

Harman-Jones, E. (1999). Toward an understanding of the motivation underlying dissonance effects: Is the production of aversive consequences necessary? In E. Harman-Jones & J. Mills (Ed.), *Cognitive dissonance: Progress on a pivotal theory in social psychology* (pp. 71–103). Washington, D.C.: American Psychological Association.

Linder, D. E., Cooper, J., & Jones, E. E. (1967). Decision freedom as a determinant of the role of incentive magnitude in attitude change. *Journal of Personality and Social Psychology, 6,* 245–254.

Norton, M. I., Monin, B., Cooper, J., & Hogg, M. A. (2003). Vicarious dissonance: Attitude change from the inconsistency of others. *Journal Personality and Social Psychology, 85,* 47–62.

Rosenfeld, P., Kennedy, J. G., & Giacalone, R. A. (2001). Decision making: A demonstration of the post-decision dissonance effect. *Journal of Social Psychology, 126*(5), 663–665.

Rust, R. T., Zeithaml, V. A., & Lemon, K. N. (2000). *Driving customer equity: How customer lifetime value is reshaping corporate strategy.* New York: The Free Press.

Schmitt, B. (1999). Experiential marketing. *Journal of Marketing Management.*
Schmitt, B. (2003). *Customer experience management.* New York: Wiley.
Sharma, M. K. (2014). The impact on consumer buying behaviour: Cognitive dissonance. *Global Journal of Finance and Management, 6*(9), 833–840.
Spiess, J., T'Joens, Y., Dragnea, R., Spencer, P., & Philippart, L. (2014). Using big data to improve customer experience and business performance. *Bell Labs Technical Journal, 18*(4), 3–17. https://doi.org/10.1002/bltj.21642.
Steele, C. M., & Liu, T. J. (1983). Dissonance processes as self-affirmation. *Journal of Personality and Social Psychology, 45,* 5–19.
Stone, J., & Cooper, J. (2001). A self-standards model of cognitive dissonance. *Journal of Experimental Social Psychology, 37,* 228–243.
Thibodeau, R., & Aronson, E. (1992). Taking a closer look: Reasserting the role of the self-concept in the dissonance theory. *Personality and Social Psychology Bulletin, 18,* 591–602.

# Advertising: A New Visual World (Re-Conceptualization of Advertising Through Creative Design)

**Aleksandra Krajnović**

**Abstract** We live in a time that demands more creativity in all spheres of life and business, so creativity is imperative and necessity of the twenty-first century. In this chapter, the author argues the thesis according to which the advertisements and branding directed "not more only on product/service, but also in the wider context of the identity of an institution, company, brand or product, or even further—to its very existence and transformation in time." (Golub, in Exhibition by Boris Ljubičić: The singularity of plural—Symbol, sign, logo, brand, 2018) The author in this chapter explores the questions: What is the point of transforming creative forms of advertising versus traditional ones? What are some new ways of celebrity endorsement in advertising? What are the specifics of creative advertising in the areas of: products/services—corporate identity—creating the logo of the brand? Do advertisements nudge consumers to a positive experience? Are creative experiments better than traditional ones? (creativity vs. traditional) Do they impress upon the brain differently? Through a qualitative analysis of some mostly award-winning, internationally relevant design projects in creative advertising, the author presents (new) conceptualizations of these projects, but also explores the emotions that these innovative/experimental forms of advertising seek to produce.

## Introduction

We are living in the world of constant changing. New era of economics and economic behavior has come. In this chapter, we address the issue of a new approach, but also of a new paradigm in marketing and advertising. However, the focus is not only solely on the use of new tools in advertising, but also on the complete transformation of the concept of advertising and overall promotion, which today aims, like never before, to capture the consumer's emotion and move "straight to the heart."

---

A. Krajnović (✉)
Department of Economics and Management, University of Zadar, Zadar, Croatia
e-mail: akrajnov@unizd.hr

Rifkin (2005) argues that "wealth is no longer considered physical capital but human imagination and creativity." Furthermore, today, we live in an era of *experience economy* and *attention economy*. Content in the digital age has grown increasingly abundant and immediately available, so the attention becomes the limiting factor in the consumption of information.

Kapferer (1992) has previously argued that "brand is not a product but an essence of the product and thus becomes a metaphysical category." Therefore, the very essence of advertising and branding changes, through the innovative creation of advertising messages that are increasingly moving from the rational (cognitive) sphere to the emotional one. This proves the thesis that we live in the era of reconceptualization of advertising in which marketing/branding and art have never been closer, thereby moving the border of both advertising and art through creative design.

With the first organized forms of marketing, which relied on the theory of utility, the goal was to sell the product/service at any price, and thus, the aim of every successful marketing department was to optimize the marketing mix—4P marketing: product, price, place, promotion. However, in the increasingly competitive environment, following the impact of globalization and technological advances, but also given the strong concentration in many sectors, it has become clear that 4P optimization alone is not enough since it can be easily achieved, at least through imitation, by other competing companies. Therefore, we have introduced new, subtler methods that propose reaching a potential customer from the opposite side to the approach taken so far, from the emotional aspect. These theses are all scientifically supported by a new field called behavioral economics. It is a direction based on the assumption that *homo oeconomicus* is not as rational being as it was previously thought given that economic decisions are also being made based on a person's internal drivers of which one is often not even aware of. In the marketing field, this area has recently been further explored under the term neuromarketing, providing astounding results and demonstrating that as a rule, one is not aware of why one likes and buys certain products/brands, and not others. This finding, of course, calls into question the whole paradigm of marketing, advertising and market research per se. In such a context, the issue of promotion, and in particular of advertising, is gaining in importance. The focus is on finding the ways how to market a product, how to "target" a potential customer without actually seeming, at least not directly and aggressively, while keeping in mind Drucker's famous statement that the goal of modern marketing is to "make selling superfluous." The aim of marketing is how to know and understand the customer so well that the product or service fits him and sell itself (Drucker 1973, Swaim 2013). How do we tackle this problem? How do we find out what is the best way to approach the customer? Which emotion, or more specifically, what marketing appeal should we "use" in an advertising message to "move" a customer? Can only modern technology be helpful, or can we use classic marketing tools in this new marketing paradigm? These are all questions that will be further addressed in this chapter. The aim is first of all to answer the following questions:

How to understand the need to change advertisements? Do advertisements nudge consumers to a positive experience? Emotion versus cognition: Does celebrity endorsement work better than presentation of plain facts? Creativity versus traditional: Are creative experiments better than traditional ones? Do they impress upon the brain differently?

The chapter consists of several sections. The first section provides a brief overview of the new concept and the new marketing paradigm. As it was described earlier, it is a new model based on the a school of thought of behavioral economics, which places emotions rather than rationality at the center. In other words, it is the so-called inbound marketing, which emerged as a contrast to the somewhat outdated, outbound marketing model. In the second part, we explore the scientific and professional theses that confirm the need to make a turn in theorizing the concept of advertising, which is the very basis of the concept of inbound marketing. The third section presents the results of the research, followed by a discussion and a conclusion confirming the key hypothesis of the article about a necessary turn in the conceptualization of marketing advertising.

The methodology of this research can be divided into *three parts*. In the first part, we review the conceptual frameworks of the research topic. The second part presents a qualitative analysis of the marketing appeals of selected high quality, internationally award-winning advertising projects and campaigns. This analysis deals with the aspect of "artistic vision" of marketing campaigns, but also with the expected public perception, as it was conducted on the example of two public exhibitions of marketing projects of the most famous advertising agencies in the Republic of Croatia, which create projects of international importance. The third part summarizes several case studies dealing with celebrity endorsement to determine whether or not this form of promotion is declining and, if not, whether it has retained its existing approach or whether there has been a change in concept. The final section is a review of the (new) role and meaning of visual communications in marketing and beyond, followed by a conclusion. It should be noted that the only possible approach to this complex issue is a multi- and interdisciplinary approach, which goes far beyond the boundaries of economics and moves significantly into domains of psychology, sociology, anthropology, design science, medicine, ethics and other scientific disciplines.

## Inbound Marketing Versus Outbound Marketing, Behavioral Economics and the Emergence of Neuromarketing

The contemporary age is an age marked by revolutionary changes in fields of marketing and branding. Marketing is becoming increasingly focused on individuals and their needs, relying on specific, new approaches (Fig. 1). Marketing is an applied science that tries to explain and influence how firms and consumers behave in markets. Marketing models are usually applications of standard economic theories, which rely on strong assumptions of rationality of consumers and firms. Behavioral

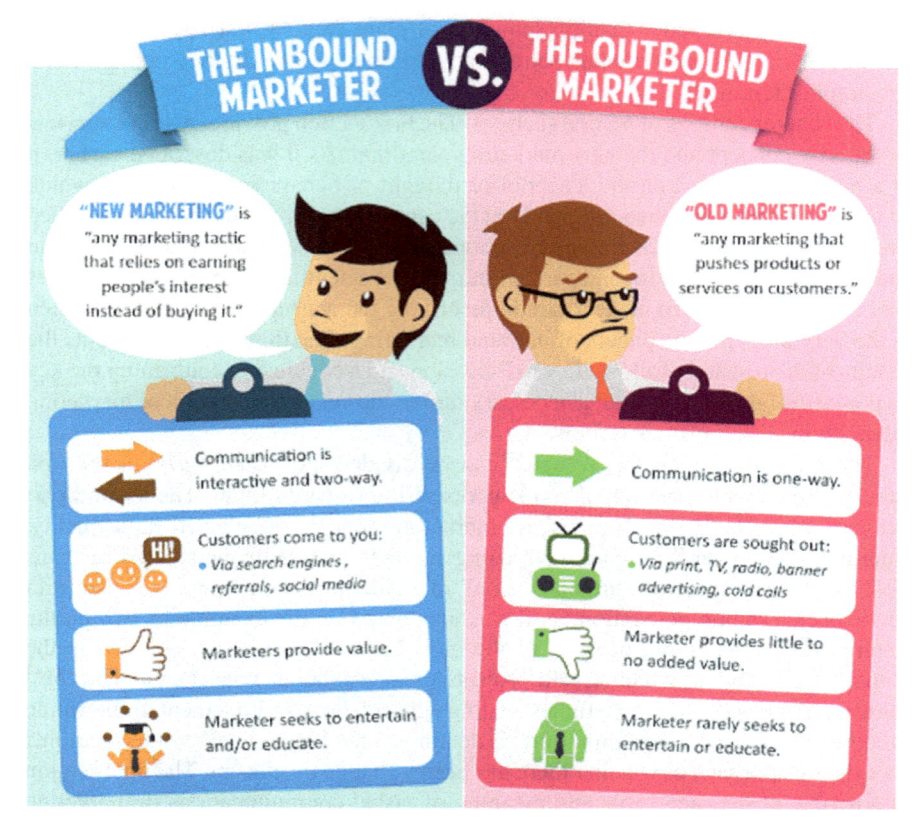

**Fig. 1** Differences between inbound and outbound marketing. *Source* Drell (2011)

economics explores the implications of the limits of rationality, with the goal of making economic theories more plausible by explaining and predicting behavior more accurately while maintaining formal power (Teck, Lim, & Camerer, 2006). Behavioral economics is a relatively new school of thought and is increasingly attracting the attention of many authors. It seeks to understand the model of *homo economicus* decision-making process that provides insight about contradiction in the previous interpretation that decisions are made rationally, not emotionally. Primary source of behavioral economics is in psychology, but is mostly applied for better understanding of the consumers particularly in the area of decision-making, while its basic feature connects these two sciences by explaining how different terms and knowledge of scientific disciplines influence the behavior of consumers and their decision on the choice (Krajnović et al., 2008).

Behavioral economics connects theoretical frameworks of rationality of classical and neoclassical economics and the way on which consumers make decisions, including personal feelings. At the same time, its contribution to the modern understanding presents departure from rationality and recognition of the existence of "mistakes" of human behavior in the decision-making process. In fact, these "errors" are not random, but are generated by applying heuristics or intuitive psychological mechanisms that produce bias. The neoclassical theory in its simple terms describes the behavior of consumers, given that the model has been perceived in such a way that man behaves like a robot, which will be in decision-making process followed by self-interest and rely only on the costs and benefits (Henrich et al., 2001). Introduction of behavioral economics in the study of human behavior emphasizes the importance of emotions and their interaction with society (Bakucs, Ferto, & Szabo, 2010).

Nowadays, some companies continue "marketing tradition" of "pushing" a certain aggressiveness to the customer, which is reflected through their own names of strategies and tactics that have dominated in the marketing by the end of the twentieth-century century: „push " marketing, „offensive marketing tactics " etc. But because of this aggressiveness, effectiveness of this promotional model systematically began to decline. This is supported also by contribution of the modern technology which provided new tools which allow users to block displays of advertising messages on cell phones and computers. Therefore, marketing is needed to find out new solutions. New marketing stopped to push product/service to the customer and created new tactics which aim is to deserve customer attention. That new marketing model has been often called *inbound marketing* and traditional marketing model as *outbound marketing*.

However, it would be a mistake to label classic marketing as a "boring, aggressive and undesirable" form of marketing communication in comparison with new advertising models. In that regard, it is interesting to look at research that dealt with the first forms of advertising—the first printed posters and other promotional tools. Discussing the role of design in advertising, research has shown that in these first forms of advertising, there was no significant difference between art and commercial advertising, i.e., the two areas intertwined. It is believed that this approach in marketing and design began in the Art Deco era (Art Deco 2011). Following Kotler's early claims that marketing is both "art and science," (Kotler & Keller 2008) the inbound and outbound marketing theses should be taken as an assessment of the general concept; that is, inbound marketing should be viewed as the "return to the old days" when the role of creative designs in advertising campaigns was much more prominent.

These changes also prompted the emergence of a new field of marketing, with a distinctly interdisciplinary approach called neuromarketing. The idea behind this new approach was to introduce the neuroscientific methods into marketing activities, with the aim of providing marketers with "insight into the human brain" and discovering unknown and hidden facts and data that are important to customers. With the advent of neuromarketing, marketers have begun to look more closely into the relationship between certain marketing elements and customer behavior. The aim was to apply this knowledge in marketing activities and to create a product that, even at a subconscious level, would stimulate certain emotions in customers, facilitate its association with the brand/product and thus encourage them to buy that product. Neuromarketing relies on the fact that many decisions, as many as 70% of them, are made at a subconscious level and that many people cannot logically explain why they made a particular decision. It is a specific area of behavioral economics, a new area that proves that economic decisions, not just those involved in the buying decision process, often lack a rational basis, as was the case in classical economic theories, as mentioned earlier (Krajnović, Sikirić, & Hordov, 2019).

It is understandable that the marketing experts have been focusing on "deeper" and better familiarization with the consumers, their emotions, desires and needs—in short, focusing on discovering their inner world. In an attempt to get the consumers "know" better, marketing experts have started using new methods, tools and scientific disciplines. It can be openly said that they have shifted the area of operation from "visible to invisible." Instead of traditional research of all conscious, rational and explicable elements involved in the purchase process, they have turned to exploring unconscious, irrational, inexplicable and emotional elements (Krajnović, Sikirić, & Jašić, 2012).

Neuromarketing offers the ability to disclose information about purchase decisions and previously unknown customer preferences. In order to transform this data into valuable information, marketers are looking for ways to relate the collected data to customer preferences, their selection process and their behavior, all to achieve the set marketing goals. For marketers, neuromarketing has been a remarkable novelty and a revolution in market research precisely because it offers, what was previously unimaginable in marketing, the ability to discover hidden information in the mind of consumers (Krajnović et al., 2019).

## *Neuromarketing and Ethical Dilemmas*

Wilson, Gaines, and Hil (2008) raise concerns about ethical issues involving neuromarketing and state:

"Marketers seek to influence the intricate processes of evaluation and selection by consumers, sometimes reverting to tactics and technologies that redirect decision makers without their explicit permission. ... Relevant issues for our discussion are whether and to what extent marketers are willing to engage in activities that lack transparency." Schull (2008) states that there are numerous scientists who disagree with the whole concept of neuromarketing. Actually, they oppose the increasing efficiency of marketing campaigns on these grounds. They believe that the factors not directly affecting the quality of a product must not influence a potential buyer in any way. They insist that the consumers should be encouraged to buy a product exclusively by emphasizing its quality. However, Sinnott-Armstrong, as cited in Wieckowski, argues that "Neuroscience will never be a threat to our free will." and adds: "A lot of the concern about neuromarketing is based on a perceived exaggeration of its power versus that of other kinds of marketing. All marketing is about influencing people—of course you want to induce them to buy your product instead of another product. But when neuroscience enters the picture, people worry that they're not just influenced, they're *forced* to do certain things. And though the influence of these techniques will grow as they improve, forcing people to do things is just not on the horizon" (Wieckowski 2019).

## Literature Overview and Conceptual Framework

Dibb, Simkin, Pride, and Ferrell (1995) affirm that "if you ask a number of people what marketing is, they will answer the question in various ways." Marketing involves more activities than most people might imagine. Although a single definition of marketing does not exist, there are a few definitions that have been widely approved. The American Marketing Association (AMA) defines marketing as "a process for creating, communicating, delivering, and exchanging ideas, goods and services, establishing their prices, promotion and distribution, in order to satisfy the goals of customers, clients, partners, and society at large." This definition emphasizes the marketing focus on planning and performing marketing activities in order to satisfy the requirements of customers. … Sheth and Parvatiyar (1995) state that marketing as a discipline has initially stemmed from economy, "due to economists who lacked interest in exploring the market behaviour, in particular with regard to the function of intermediary persons."

The classic marketing approach is based on Maslow's well-known hierarchy of needs, which Kotler modifies adding the categories of desire and demand. According to this model, the marketing objective is to satisfy the following: need → desire → demand. Today, as the concept of marketing is changing, one can speak of a new model: attention → gratification → happiness. While the classic model is based on the Attention, Interest, Desire, Action (AIDA) approach in promotional messages, the other mentioned categories are still being explored. Table 1 provides an overview of the selected authors who explored the effects of marketing messages in terms of gratification and happiness categories, although it seems that in this area of research, there is a lack of literature and a lack of a solid conceptual framework.

Gratification theory is a field of mass media theory that began to develop significantly in the 1980s (Palmgreen, & Rayburn II, 1985) and is nowadays stepping into the field of digital marketing, including research in social media marketing and online shopping. In the case of the happiness category, research is mainly focused on the broader context of sentiment analysis, while some neuromarketing studies place emphasis on exploring connectivity and design interpolation in marketing communication.

**Table 1** Categories of gratification and happiness in marketing messages—selected authors and research areas

| Category | Theorists | Research area |
|---|---|---|
| Gratification | Palmgreen and Rayburn II (1985) | Alternative gratification/expectancy-value models to predict satisfaction with television news |
| | O'Donohoe (1994) | Marketing and non-marketing uses of advertising by young adults |
| | O'Donoghue and Rabin (2000) | Economic analysis of immediate gratification |
| | Xueming (2002) | Influences of marketing messages on various online consumer behaviors |
| | Huang (2008) | The impact of use and gratification on e-consumers' acceptance of B2C Web sites |
| | Chung and Austria (2010) | The effect of social media gratification and attitude toward social media marketing messages |
| | Lee et al. (2013) | Emotional shopping motivators |
| Happiness | Johnston Laing (2004) | Advertising art and marketing design |
| | Sikirić (2011) | The conscious and subconscious reaction of the recipient to marketing appeals |
| | Stroe and Iliescu (2014) | Decisional process of the buyer |
| | Chabrol and Furrer (2014–15) | Sentiment analysis of guerilla marketing communication |
| | Zaušková et al. (2016) | The interaction between religious tourism and marketing communication |
| | Navas-Loro et al. (2017) | Sentiment analysis—emotions toward a brand |
| | Mrduljaš (2019) | Marketing design, marketing communication project and its wider social resonance |

*Source* Author's Research

## Research Results

Below are the results of the first part of the research, presented in Tables 2 and 3 and in accompanying images. The results refer to two multimedia exhibitions held in 2019 in the Croatian cities of Pula and Rovinj, which are examples of the most successful marketing communication projects in that area. These are marketing campaigns of two advertising agencies of international importance, based on the Republic of Croatia—Studio Sonda and Bruketa&Žinić&Grey Agency. This approach was chosen in order to emphasize the artistic significance of campaigns, expressed mainly through creative marketing design. In that regard, an attempt was made to find out what constitutes the success (innovation) of the marketing appeal of the displayed

Table 2 Selected advertising campaigns on display at the Museum of contemporary art of Istria—Studio Sonda

| Campaign | Product—service/client | Creative idea/marketing appeal* |
|---|---|---|
| Fashionable conversations | Exhibition catalog /HDLUI (Croatian Association of Artists of Istria) | Problematizing body issues, representational messages of fashion, clothing, and art, and their interrelationships within the gallery and social contexts. That is why the symbol of the exhibition was made by combining two motifs: a hanger as a recognizable object related to fashion and clothing items, and lips that evoke the idea of conversation or communication. The catalog is printed on tailor-made paper and also functions as a B1 poster. Thanks to the perforation on the margins, the edge of the poster can be detached as a tailor's bullet, while the graphic imitates the cutting sheet. |
| Closed on Sundays* | Poetry collection, Drago Orlić: closed on Sundays /self-initiated project, 2015 | The conceptual solution to the poetry collection "Closed on Sundays" questions whether the contemporary poet must present his poetry as consumer goods to be noticed at a time when poetry is becoming completely marginalized. The poets' words are being packed and weighed in kilos, playing with the allusion to everyday consumerism, which today becomes almost the only form of communication understood by modern man. In the Istrian dialect, the term "doprta butiga" or in Croatian "otvoren dućan" is mainly used when one has an open fly on their trousers. The collection is called "Closed on Sundays," but the fly on the poet's trousers (the portrait represents a real photography of the author) is open, like an open shop. In other words, poets have no free Sundays, no matter how much they wish for it. |

(continued)

Table 2 (continued)

| Campaign | Product—service/client | Creative idea/marketing appeal* |
|---|---|---|
| Agrocal—the secret of the most beautiful garden | Fertilizer—product packaging /Holcim Lafarge, 2016 | Agrocal powder is a completely natural and environment-friendly source of calcium and magnesium for rapid and efficient calcification and increase in soil fertility, leading to a long-term increase in agricultural yields. After the initial large packages intended for farmers and large crops, Holcim decided to offer the product to urban gardeners as well, in a small package of 4 kg. The solution was to pack the product in an eco-friendly, paper-readable, one-color package. The product is then placed in a wooden box that gardeners can remodel and use, for example, for seedlings, tool storage, etc. Also, various motivational messages are cut into the box to encourage gardeners to buy the entire collection. In designing the concept, the guiding principle was functionality in the sales space: the packaging was designed so that the boxes could be stacked on one another, looking like an independent entity. This ensures that the product does not have to contend for shelf space in agrarian space. Another very important functional element is the possibility of easy transport. |
| On the line of fire—monograph* | Poreč regional museum, public fire station—Poreč fire protection center, 2016 | On the line of fire: history of the Poreč fire department is a monograph that provides, through 254 pages, a chronological overview of the history of voluntary and professional firefighting in the Poreč area from the nineteenth century to the present, as well as a list of all firemen who are part of that history. The cover is made of the fingerprints of all employees who were employed by the Public Fire Station—Poreč Fire Protection Center at the time the book was created. |

(continued)

**Table 2** (continued)

| Campaign | Product—service/client | Creative idea/marketing appeal* |
|---|---|---|
| Radenska—can | Radenska mineral water—product packaging, Radenska, 2019 | Radenska is natural, carbonated mineral water that originates from the source Kraljevi vrelec in Slovenia and has been on the market since 1869. In 2019, the well-known brand wanted to bring innovations in the Slovenian market and find a way to reach a younger audience by creating a concept that would adequately brand to present in foreign markets. And so, the idea of a can was born, as new packaging for Radenska. The goal was to get closer to the ubiquitous trend of leading a healthy life, but in a way that is acceptable to the modern consumer: The concept aims to show how drinking water, in addition to being healthy, can be a lot of fun! This is demonstrated by the playfulness of the new packaging, but also by the new way of consuming water: Drinking from a can is in itself a special and fun ritual (opening sound, stronger bubble concentration than in traditional PET packaging, etc.) The can is sold in the following markets: Austria, Italy, Slovenia, USA. |
| *On paper*—exhibition catalog* | HDLUI (Croatian Society of Artists of Istria), 2009 | The "On Paper" exhibition seeks to acknowledge the many works on paper that exist in the shadow of what is often understood as primary art production and which are often understood as sketches of something else or as works of lower artistic value. The exhibition catalog is a departure from the usual presentation of the catalog and emphasizes the material of which the catalog is made: paper. While in the classic catalog view the user pays more attention to the information, in this view the focus is placed on the information carrier, on paper as a medium. The work is dislocated to such an extent that it undoubtedly draws attention to its purpose, form and to its incredible potential in creative use. As the exhibition seeks to emphasize that paper sketches have no lower artistic value than completed works, crumpled paper that at first glance looks like a waste, in this case, has risen from being discarded to something of value—the exhibition catalog. |
| *Water Guide*— A book that saves water | Water guide—water supply in the history of Poreč, monograph, Elena Uljančić, Vekić, | The mentioned book sheds light on the history of water supply in the Poreč area and talks about the importance of this valuable resource. However, the plan was to create a book whose technical design would also embody the idea of saving water. Thanks to the choice of paper types and techniques that allow for less water consumption in the printing process; the production saved 7355 L of water compared to standard non-recycled paper and conventional printing techniques. By comparison, one person consumes an average of 50 L of water a day for meal preparation and personal hygiene, so we can say that this project has achieved sufficient water savings for approximately five months of an individual's normal life. The book further uses no additional graphic elements, conveying the message at its core—concrete water savings. Thus, the work that speaks about the importance of conserving water resources, by its very design, states that saving water is possible and important at every step. |

(continued)

**Table 2** (continued)

| Campaign | Product—service/client | Creative idea/marketing appeal* |
|---|---|---|
| BOOKtiga Award—Award to the most read author, BOOKtiga Used Books Fair | Poreč City Library, 2014-present | The BOOKtiga Used Books Festival annually awards the most widely read Croatian book in Istria. The task was to create an award for the author. Instead of a classic prize or a statue, we wanted to give the authors something special, give them an award that will testify that their work has affected the lives of those who give the award, readers (Sonda, website 2020). In order to achieve this, the actual copy of the award-winning book "travels" for one-month through to the libraries of Istria, inviting readers to write their impressions in. Filled with reader comments, the book is then wrapped in a specially designed cover and returned to the author as such. Following the concept of the festival (second-hand books), the prize itself is the same as the prize-winning book—it is one copy of that book—second-hand, colorful, worn out, and most importantly readable, with reader comments on the margins of the book, which is its biggest value for the author. |

*Note* Descriptions of displayed projects from exhibitions are cited with minimal intervention by the author (Shown in Fig. 2)
*Source* Materials from the exhibition "On the Power of Creativity," Croatian Designers Association, Studio Sonda and Museum of Contemporary Art of Istria, Pula, 2019

**Table 3** Selected advertising appeals on display at the Adris Gallery in Rovinj—Bruketa & Žinić & Grey Agency—B2B, medium: annual report

| Medium—annual report | Marketing appeal—creative concept |
|---|---|
| 2009: Beyond Expectations | The team that worked on Adris's annual report messed up a bit because the covers turned out to be smaller than the book pages. The idea was actually to illustrate results that exceeded all expectations, and so the book outgrew its cover |
| 2011: In Good Hands | The Adris Group is a company whose employees and its owners are shareholders. The cover of the annual report entitled " In Good Hands" was printed with a thermally sensitive color, so when you hold it in your hands, the warmth of your palms reveals green floral details, which metaphorically shows how things change and evolve in good hands. The creative concept is complemented by ten short autobiographical stories in which employees-shareholders offer their personal perspectives about the Adris Group's impact on their lives, as well as their contribution to Adris' results |
| 2013: A book that grows on its own | Adris Group is one of the leading Croatian and regional companies. The company grows in the good hands of its employees, and many of them are also owners—shareholders. Thus, Adris' annual report for 2013 literally grows in the hands of its readers |
| 2014: 4 books in 1 | In 2014, Adris Group acquired a majority stake in Croatia osiguranje. A new story began that would ensure the company's long-term growth, development and sustainability. The annual report is therefore comprised of four books in one, interconnected by the well-known Croatia osiguranje plaque, which passes through a hole cut into these various volumes, creatively linking all of the company's assets |
| 2015: The Success Factory | The annual report focuses on the motive of the staircase that symbolizes growth. After becoming part of the Adris Group, each new company grows and becomes more successful than before. Thus, Adris is a success factory |
| 2017: Tested for Challenges | Only companies that constantly adapt to circumstances can develop, grow and be resilient in the face of the challenges. The publication is literally resistant to challenges. It cannot be torn, crumpled, wetted or set on fire. Regardless of which test is subjected to, the book remains intact, illustrating how the Adris Group's results are stable and solid, and the company can tackle every situation |

campaigns. We will refer on the exhibition entitled "On the Power of Creativity" and on the promotion of the same-named book (texts written by Marko Golub), jointly published by the Croatian Designers Association, Studio Sonda and MSU (MCA) of Istria, held at the Museum of Contemporary Art of Istria (Croatia) in Pula from September 20, 2019, to October 31. The second exhibition, held at the same time at the Adris Gallery in the Croatian town of Rovinj, is entitled One Croatian Story. This is an exhibition of communication projects by the agency Bruketa & Žinić & Grey, which was held during the Weekend Media Festival and which represents the "concise Croatian story of creativity" as a necessary tool for success in the global market. The exhibition consists of 24 award-winning creative advertising projects. In addition to the projects exhibited at the aforementioned exhibitions, two more case studies of the interesting and award-winning projects of the mentioned agencies are presented (Fig. 2).

Below are three more interesting cases. The first refers to Studio Sonda and the other two to Bruketa & Žinić & Grey.

**Case study 1. The Happiest Eggs—Visual and Verbal Identity for The happiest Perfa Eggs—Studio Sonda**

The first example demonstrates how the concept of happiness in marketing has a much wider context. In this example, eggs are described as "happy eggs" because they were laid by "happy hens", hens which are free, healthy and coming from free-range systems. It is a great example that shows how it is possible to link ecology to the notion of happiness through marketing communication. The following is a description of the project as downloaded from the Studio Sonda Web site:

"The producer *Perfa eggs* has put a new product on the market: eggs of free hens. Up to now, the already existing *Perfa Eggs* are promoted on the market under the name and slogan *Happy Eggs*. But as they become, through the new production system, even happier, since the hens enjoy free walks in the fresh Zagorje air and are fed on natural or bio-food, the new products are now carrying the names: *The happiest eggs* and *The happiest eggs BIO*. A new packaging has also been designed. By upgrading the input of the present visual expression, while simultaneously nurturing the tradition of the region the eggs come from, illustrations for the new packaging are inspired by the creative opus of naïve art. The landscapes in which the hens live are made, according to the product itself, with egg-shaped shapes, emotionally transferring the beauty and calmness in which *The happiest eggs* are created. The illustrations, besides on the packaging, are applied also to other *Perfa Eggs* business and promotional materials." (https://sonda.hr/?s=Happy+eggs&post_type=post).

**Case study 2. Untouched by light: the worlds first sparkling wine made, sold and tasted in complete darkness—Bruketa&Žinić&Grey**

The following example is about promoting different, new experiences, surprises, fun and education in marketing advertising. This is a campaign called "Untouched by ligth: the worlds first sparkling wine made, sold and tasted in complete darkness," which is described on the Web site of the Bruketa&Žinić&Grey agency that created it: "A new sparkling wine brand called *Untouched by Light*, released by

**Fig. 2** Marketing campaigns: "On Paper," "Closed on Sundays," "On the Line of Fire" and "BOOKtiga"—Studio Sonda

Radgonske Gorice recently, claims that best things happen in the dark. According to Professor Emerita Ann C. Noble[1][1], the exposure of wine to daylight or artificial lighting results in what is commonly known as light-struck aromas. It decreases the intensity of citrus aromas, but increases the intensity of cooked cabbage or wet dog aromas. For that reason Radgonske Gorice made a bold decision, after a suggestion by Bruketa&Žinić&Grey, to turn to the dark side and produce their new sparkling wine. Untouched by Light is made from Chardonnay grapes picked at night and aged

---

[1]Noble, Ann C., Sensory study of the effect of flourescent light on a sparkling wine and its base wine, American Journal of Enology and Viticulture, Vol. 40, No. 4, 1989.

in pitch darkness of a cave to be sold in (sun)lightproof bottles and preferably served in the dark." (https://bruketa-zinic.com/2019/08/28/untouched-by-light-the-worlds-first-sparkling-wine-made-sold-and-tasted-in-complete-darkness/)

**Case study 3. The book "Hope" among the world's 50 best book covers, part of *Yale* and *Columbia* libraries**

The American Institute of Graphic Arts (AIGA) has published the book *Hope* one of the world's 50 best book covers. The list has been created as part of the 50 Books I 50 Covers annual international competition in collaboration with Design Observer. By entering this list, *Hope* becomes part of the *AIGA collection* at the *Rare Book and Manuscript Library* within *Columbia University's Butler Library* and the *Robert Haas Arts Library* at *Yale University*.

*Hope* was made in collaboration with *The Adris Foundation* and *Bruketa&Žinić&Grey* agency, and it presents science, education and arts projects funded by *The Adris Foundation* in the past ten years, and its cover reacts to touch. "Gathered in this book those projects give hope that Croatia has a good future, this is why the more the book is used, the more hope there is – the book is sensitive to touch," says Creative Director Davor Bruketa. At first blank, the covers start showing the letters NADA, which means hope in Croatian, after touched by the reader. Besides, the project has a cultural value because it gathered 24 illustrators, some of the best ones in Croatia. (https://bruketa-zinic.com/2018/11/12/the-book-hope-among-the-worlds-50-best-book-covers-part-of-yale-and-columbia-libraries/).

The three case studies described show the third part of the research. It is particularly interesting since it presents results of a B2B campaign as a creative solution that transforms the company's annual report, in this case, Adris Group, into a creative medium with a marketing message that varies from year to year.

## Celebrities in Modern Advertisements

Bergkvist and Qiang Zhou (2016) affirm that there are six areas of research on celebrity endorsements: celebrity prevalence, campaign management, financial effects, celebrity persuasion, non-evaluative meaning transfer and brand-to-celebrity transfer. Part of the research focuses on celebrity endorsement with a focus on psychological processes based on celebrity endorsement effects. In the article "Impact of Celebrity Endorsement on Brands Marketing Campaign" it is stated that: " The brand value added by celebrities is almost immediate and when a brand links a contract with the celebrity, a sudden stock rise can be visible as soon as the information is made public. However, there is a need to have a relevant celebrity to endorse the brand as having a celebrity that is not well matched to the brand will not influence consumers at all. If consumers are not convinced with the celebrity, it will lead to a negative impact on the brand." An example is presented in the same article: a campaign by GOQ ii, a healthcare platform which associated with Bollywood actor

Akshay Kumar to launch their *India Health Quiz*. Known for being naturally fit even at the age of 51, Akshay Kumar is one of the top fitness icons in the country. Hence, he spreading awareness about health and fitness has received a great response.

A study by University of Arkansas suggests that consumers between the ages 18 and 25 develop their identities and personalities based upon their favorite celebrity. Celebrity endorsement resonates more with Gen Z and millenials as per a report by Neilson in 2015. The right combination of imagery, heartfelt storytelling, creative presentation and playful wit is what it takes to entice an audience to support the campaign and get motivated to purchase. While celebrity endorsements are used to attract consumers, the influence on purchasing decision is still ambiguous. As per a survey in the book Contemporary Ideas and Research in Marketing, 85% of people surveyed said that celebrity endorsements enhanced their preference for the product, but only 15% said that celebrities had an impact on their buying decisions. This sheds light on the current market scenario where consumers are unsure about the influence of celebrity endorsements with 51% of consumers claiming that they have little to no impact on their purchasing decisions.

The presence of celebrity is impactful, but it has to be considered as just a portion of a multi-channel marketing approach to create a wider impact on the audience. Celebrity endorsement alone does not guarantee success; it is a mix of several other factors such as price of the product, durability and quality which will be considered before making the decision to buy the product and help the brand in reaching their ultimate goal, sales. In her research from 2019, Pavlović came to the same conclusion and confirmed that celebrities have indeed a big influence, both positive and negative, on product sales. According to Pavlović:

"People have different views and opinions. What is good for some people is not good for others. Advertising, therefore, requires careful selection of the right person to advertise in order for the product/brand to receive positive reviews and sales to grow. (Pavlović 2019)"

In his professional paper entitled "Top 5: Celebrities and Celebrities in Commercials," Hrastovčak presents, in a popular manner, what he considers to be the five most prominent case studies of using celebrities in advertising:

1. **Matthew McConaughey—Lincoln Motor.** You all know Matthew McConaughey and his films very well, but what you may not have known is that he recently appeared in an advertising campaign for Lincoln Motor. The campaign featured several subtle commercials that showcase McConaughey's charisma, and the most attention was drawn to one in which a famous actor comes across a bull on the road!
2. **Robert Downey Jr.—HTC**. When Iron Man and the mobile phone manufacturer HTC team up, the result is everything but the ordinary commercial. In a two-minute video by the Taiwanese company, Robert Downey Jr. attempts to discover the meaning of the HTC acronym, which leads to humorous and somewhat bizarre situations. The ad is part of HTC's big "Change" campaign, launched two years ago to help strengthen its presence in the US market. In other words, the engagement of the popular Hollywood actor was no coincidence.

3. **Nicole Kidman—Chanel**. What does a $40 million ad look like? Like a short movie starring Nicole Kidman! Promotional video for Chanel No. 5 attracted a great deal of attention ten years ago thanks to a huge budget and famous actress, but also thanks to the famous director Baz Luhrmann, who previously collaborated with Nicole Kidman on the movie Moulin Rouge.
4. **Scarlett Johansson—SodaStream**. The SodaStream commercial raised a lot of controversies after a short line spoken by Scarlett Johansson was censored at SuperBowl early last year. In the commercial, the popular actress enjoys the aforementioned drink and then turns to the camera and says: "Sorry, Coke and Pepsi!" Given that Coca-Cola and Pepsi are the major advertisers of the Super Bowl, such jokes on their account obviously could not pass. Scarlett's controversial line was eventually replaced by another, and the original ad became a viral hit on the Internet.
5. **Britney Spears, Beyoncé, and Pink—Pepsi**. Not one, not two—but three pop divas! Pepsi made a big hit in 2004 hiring Britney Spears, Beyoncé and Pink in their expensive marketing campaign to sing their version of Queen's "We Will Rock You" in the Roman arena. However, the three singers are not the only celebrities in the commercial—there is also Enrique Iglesias as Roman Emperor! This is nothing unusual for Pepsi, who has a long tradition of hiring celebrities for endorsements, and many of us still remember campaigns from the 1980s starring Michael Jackson (Hrastovčak 2015).

However, it is becoming increasingly important to incorporate both the celebrity endorsement strategy and influencer marketing, and the similarities and differences between the two marketing tactics are also being more and more explored. In the article "How Influence Marketing Differs from Celebrity Endorsement" Geepert states:

"Celebrity endorsements have always been an effective method of piquing public interest in a product or service. But since the advent of social media marketing and advertising, influencer marketing has also emerged as a viable way to build a brand's reputation."

The following is an example of the innovative use of celebrity endorsers, a case study that illustrates a new approach in the area, which we could almost categorize as the "next generation of celebrity endorsement." As part of the "Men All Over the World Are Crying" advertising campaign, the agency hired well-known men from the world of skiing to recount their most intimate moments. However, it is interesting to note that this is a campaign that was launched to promote the world's lightest skis for women. The official Facebook page for Elan Skis reads: "Now we know why Filip Flisar, Davo Karničar, Bine Žalohar, and Glen Plake were crying…" (www.2016.weekendmediafestival.com/en/program/balcannes/).

On the official Web site of Studio Sonda, it is stated: "Elan's Delight women's skis marketing campaign "Men all over the world are crying" by Studio Sonda was awarded the prestigious prize "Project of the Year" at BalCannes 2016. Christian Mayer—World cup alpine ski racer, Olympic and World Cup medalist, Filip Flisar—World Champion and overall World Cup winner in Ski cross, Davo Karničar—The first person to ski down from Mt. Everest, Matthias Myer—Professional free-skier and Big mountain explorer, Bine Žalohar—Professional free-skier, competitor & instructor and Glen Plake—Pioneer of extreme skiing in USA, share with us their most intimate moments. Real heroes, confessing having cried for reasons which many would consider banal. But who dares to decide what is important to this guys, well known for their abilities and courage? They proudly say: "Do not judge other people's reasons. Support true feelings!" (https://sonda.hr/men-all-over-the-world-are-crying/).

## The Role of Creativity and Design in Contemporary Marketing Advertising

Previously mentioned exhibition is accompanied by an invaluable publication feat, a trilingual catalog "Jedna hrvatska priča/Una storia Croata/The One Croatian Story." In the catalog Maroje Mrduljaš inter alia recounted: "This "Croatian Story" has released the latent creative energy of the local context and has been recognized internationally through numerous awards, publications, exhibitions and collections. The global success of these works is aided by their authenticity and the fact that they could hardly have originated elsewhere. The collective effort put into these works by both Bruketa & Žinić (&Grey) and the Adris Group, with numerous authors involved, feeds back into the corporate culture of the company. (…) When it comes to the contribution of Adris Group and Bruketa & Žinić (&Grey) to the public domain, their "extended freedom" of the design scene and contributions to visual culture should certainly be emphasized, but for the contemporary Croatian context it is even more important to use corporate channels to communicate basic social values in times when they are increasingly being brought into question." (https://bruketa-zinic.com/2019/09/24/the-one-croatian-creative-story-told-at-adris-gallery/) In the introduction to that book, P. Grubić states: "The Adris Group is consciously exposed to the risk of creativity. We believe that the time of traditional solutions in the economy is passing and that, even though too slowly, or even dramatically too slowly, the Croatian economy is moving towards a creative economy in which design will be crucial, if not decisive, for increasing competitiveness and adding value to companies in the global market. In order to be as efficient as possible in this business, we must, despite the risk, be open to new knowledge, innovative and creative solutions and accept the cooperation between entrepreneurs and designers. The presented Rovinj fragments will undoubtedly contribute to the growth of the design culture in our country and to its socio-economic and cultural affirmation. It will certainly, in counteraction with the unbearable *aesthetic of ugliness*, stimulate that constant energy of visibility…".

Previously described advertising agency, Studio Sonda, an independent, creative studio situated in a small village Vižinada, which "advocates the idea that design and communications must be logical, honest and intuitive. In an inspiring environment overlooking the green hills, they create globally recognized projects in the area of design and communications." Many of them have received numerous national and international awards such as Cannes Lions, Red Dot Best of Bests, ADC gold, IF, World Star Packaging, LIA, Pentawards, Creativity Annual, ID, Rebrand, Plus X Award, Grand Prix of the Croatian Design Society, AIGA, Golden Drum, Cannes Lions Shortlist, ICOGRADA, etc. They are members of professional associations HURA (Croatian Association of Advertising Agencies) and HDD (Croatian Design Society). (https://www.glasistre.hr/kultura/u-muzeju-suvremene-umjetnosti-istre-o-moci-kreativnosti-studija-sonda-od-petka-u-puli-597609).

The curator of previously mentioned exhibition, Marko Golub, discussing the projects of the Studio in the introduction of the book "On the Power of Creativity" notes: "There are no small or big projects for them, no less important ones or more important ones. Furthermore, some of the smaller projects that are local in character are also the ones that went on to have the widest impact...." He further comments on how marketing agencies, in this case, Sonda, are contributing to the development of the local and wider social community: "... but their most important contribution to the development of a design environment in Istria (Author's comment: Croatian region where the agency operates) is that, through their work for local producers, tourism brands, cultural and festival events, they helped to increase awareness of clients and users about the importance and impact of design and visual communications, and thus, they greatly paved the way for design in the Istrian region. Dislocation from the center, or rather a firm insistence that the center can be wherever you choose to place it, is also one of the principles by which this studio is run. It is also a starting position in asking questions: how we are connected and what do we have to say to each other. Devoting its attention to designing messages that will be relevant to the specific local situation, context, and community, Studio also came up with concepts that proved to be important in today's general understanding of visual culture and communications, design and advertising." It is interesting to point out that one of the basic assumptions of Studio Sonda's authors can be summarized in a claim that "the best trigger for creativity ... is a meaningful and fulfilling life."

## Conclusion

It may be concluded from the above discussion that modern marketers are in search of new ways to reach the customer. One way they are trying to achieve this is by providing gratification and attracting the attention of consumers, although more recently, they seem to be trying to elicit an emotional response from consumers to the point of even offering them happiness. The very concept of happiness, which is studied in economics mainly through the index of perception of happiness and as an "addition" to GDP, is the essence of a new marketing approach. However, this approach raises certain ethical dilemmas as to whether such marketing, in fact, intensifies consumerism, since it encourages people's emotional attachment to material objects.

Without going further into the psychological definitions of happiness, this chapter tried to answer the question of how creative ideas and artistic designs in marketing really can evoke the emotions of a potential customer. However, the reaction does not necessarily imply "happiness in the narrow sense of the word," but rather it is about triggering various emotional reactions commonly associated with fun, humor, tradition, ecology, pride, humanitarian action, engaging the user in interactive (equal) communication with the sender of the message, etc. Thus, the answer to the question of whether modern marketing is actually "manufacturing happiness" may be considered affirmative if we bear in mind a broader understanding of the concept of happiness. For instance, the previously mentioned case in which an author receives his or her book as a reward along with reader comments is an example of creating happiness in two ways: Both the author of the book and the readers are being rewarded. In the example of a monograph on firefighting, a white book cover painted only with the black fingerprints of local firefighters arrived immediately after the firefighting intervention, that is, a picture of the dangerous work of brave firefighters cannot leave anyone indifferent. Therefore, it can also be concluded that the best and most innovative marketing appeals are indeed "manufactured" to provoke the surprise and other strong emotional reactions of the recipient. But these strong and often deep and/or long-lasting emotional reactions, which bind the recipient of the message even more to the promoted brand, relate to the full range of emotions and emotional reactions that the marketing appeal has generated. This segment certainly needs further exploration, especially by psychologists and sociologists.

When it comes to celebrities in advertising, they are still used in marketing, but in slightly different ways. Today's innovative marketing models use celebrities in their appeals in a whole new and unusual way. For example, in previously mentioned ski advertising campaign, the world's most famous skiers are crying; in another case, a famous poet is photographed on the cover with an unzipped fly, which is a provocative message about the hard work of poets who work even "on Sundays." Another example of Studio Sonda's creative campaigns is the poster of a famous Croatian pop singer who is half-covered with a poster for a book fair, ironically showing through marketing communication how much easier it is to persuade people to visit a pop singer concert than a book fair. It should also be noted that celebrities have some sort of competition in marketing, and these are influencers that are emerging as new marketing as well as a sociological phenomenon. Therefore, we can conclude that celebrities are still used in marketing advertising, but no longer so that they are "placed" above others, ordinary people, but now they become vulnerable, fragile, they make mistakes, send messages, they are imperfect, etc. In other words, marketing wants to bring them closer to ordinary people, to make them almost one of us.

And finally, it can be said that in modern marketing, the question is: Who are the real celebrities? In the creative marketing appeals of the Sonda Studio, for example, these are professional and amateur firefighters, and in the creation of the Bruketa&Žinić&Grey agency, they are employees of the Adris Group, which designed one of its annual reports so that the cover of each copy is a photograph of one company employee. Thus, it can be said that new marketing expands and erases boundaries, breaks prejudices, develops new mental models, creates new communication paradigms, but also new aesthetics versus ubiquitous aesthetics of ugliness, as characterized by Predrag Grubić in the introduction of the book One Croatian Story. In the end, this discourse can be concluded with a photograph of a successful project of contemporary marketing visual (as well as multimedia) communication that summarizes all the postulates of the new contemporary paradigms of advertising. It is warm, nice, fun, cheerful, but also educating, supports what is good, ecological and humane (Fig. 3).

Marketing appeal: How can one describe that Cromaris fish are better than the competition because of the location and method of farming? Cromaris fish are not like other fish. Because of the attention that goes into cultivating them, they are more than commons. They are well-raised fish.

**Fig. 3** Advertising products in a creative way—fish from organic farming (Cromaris, Croatia, Bruketa & Žinić & Grey agency, 2019) *Source* Jedna hrvatska priča—Una storia croata—The One Croatian Story, Rovinj

**Acknowledgements** The author gratefully acknowledges for the kind permission to: "Adris Grupa d.d." Rovinj—Croatia, "Bruketa&Žinić&Grey d.o.o." Zagreb—Croatia, Muzej suvremene umjetnosti Istre/Museum of Contemporary Art of Istria, Pula—Croatia, "Studio Sonda" Vižinada—Croatia

# References

Art Deco. (2011). Documentary film, director and screenwriter Sirotić, G., production by HRT (Croatian Radio Television).

Bakucs, Z. L., Ferto, I., & Szabo, G. (2010). *Institutions in transition—Challenges for new modes of governance* (pp. 3–5). Germany: IAMO Forum.

Bergkvist, L., & Qiang Zhou, K. (2016). Celebrity endorsements: A literature review and research agenda. *International Journal of Advertising, 35*(4), 642–663. https://doi.org/10.1080/02650487.2015.1137537.

Bruketa&Žinić&Grey d.o.o., offical website: https://bruketa-zinic.com. Accessed February 3, 2020.

Chabrol, J. B., & Furrer, O. (chair). (2014–2015). *Marketing de guérilla: Analyse et recherche sur un sujet souvent utilisé mais peu délimité, Travail de Bachelor*. Université de Fribourg, Chaire de Marketing.

Chung, C., & Austria, K. (2010). Social media gratification and attitude toward social media marketing messages: A study of the effect of social media marketing messages on online shopping value. In *Proceedings of the Northeast Business & Economics Association* (pp. 581–586).

Convinceandconvert.com. https://www.convinceandconvert.com/digital-marketing/influence-marketing-differs-from-celebrity-endorsement/. Accessed February 3, 2020.

Dibb, S., Simkin, L., Pride, W. M., & Ferrell, O. C. (1995). *Marketing—europsko izdanje*. Zagreb: MATE d.o.o.

Drell, L. (2011). *Inbound marketing vs. outbound marketing* (INFOGRAPHIC) October 30, 2011, Mashable (online). https://mashable.com/2011/10/30/inbound-outbound-marketing/?europe=true. Accessed February 1, 2020.

Exhibition. On the power of creativity (images). In M. Golub, & K. Milićević Mijošek, (Eds.) organized by HDD (CDA) & MSU Istre (MCA of Istria), in collaboration with Studio Sonda.

Glas Istre: Web stranica. https://www.glasistre.hr/kultura/u-muzeju-suvremene-umjetnosti-istre-o-moci-kreativnosti-studija-sonda-od-petka-u-puli-597609. Accessed February 3, 2020.

Golub, M. (2018). In *Exhibition by Boris Ljubičić: The singularity of plural—Symbol, sign, logo, brand*. https://www.media-marketing.com/en/news/exhibition-by-boris-ljubicic-the-singularity-of-plural-symbol-sign-logo-brand/. Accessed February 3, 2020.

Henrich, J., Boyd, R., Bowles, S., Camerer, C., Fehr, E., Gintis, H. & McElreath, R. (2001). In search of homo economicus: Behavioral experiments in 15 small-scale societies. *The American Economic Review, 91*(2), 73–78. Available in: https://www.umass.edu/preferen/gintis/Anthro%20AER%202001.pdf. Accessed April 17, 2020.

Huang, E. (2008). Use and gratification in e-consumers. *Internet Research, 18*(4), 405–426. https://doi.org/10.1108/10662240810897817.

Hrastovčak, T. (2015) Top 5: Poznati i slavni u reklamama. https://planb.hr/top-5-poznati-i-slavni-u-reklamama/. Accessed April 17 2020.

In P. Grubić (Ed.), *Jedna hrvatska priča—Una storia croata—The one Croatian story*. (2019). Rovinj: Adris grupa d. d. & Bruketa&Žinić&Grey d.o.o.

Johnston Laing, E. (2004). *Selling happiness: Calendar posters and visual culture in early-Twentieth-Century Shanghai*. Honolulu and USA: University of Hawaii Press.

Kapferer, J. N. (1992). *Strategic brand management: New approaches to creating and evaluating brand equity*. Kogan Page.

Krajnović, A., Sikirić, D., & Jašić, D. (2012). Neuromarketing and customers' Free Will. In D. Gomezelj Omerznel, B. Nastav, & S. Sedmak, (Eds.), *MIC 2012: Managing Transformation with Creativity; Proceedings of the 13th International Conference, Budapest, 22–24 November 2012* (pp. 1143–1163). Koper: University of Primorska, Faculty of Management, Koper; Budapest: Corvinus University of Budapest.

Krajnović, A., Sikirić, D., & Bosna, J. (2018). Digital marketing and behaviorial economics. *CroDiM—International Journal of Marketing Science—The International Scientific and Professional Journal, 1*(1), 33–46.

Krajnović, A., Sikirić, D., & Hordov, M. (2019). *Digitalni marketing—nova era tržišne komunikacije*. Zadar: University of Zadar.

Kotler, P., Wong, V., Saunders, J., & Armstrong, G. (2006). *Fundamentals of marketing—4th* (European). Zagreb: MATE d.o.o.

Kotler, P. & Keller, K. L. (2008), Upravljanje marketingom, dvanaesto izdanje, Zagreb: MATE d.o.o.

Lee, M., Kim, Y., & Lee, H. (2013). Adventure versus gratification: Emotional shopping in online auctions. *European Journal of Marketing, 47*(1/2), 49–70. https://doi.org/10.1108/03090561311285457.

Medianews4u.com. https://www.medianews4u.com/impact-of-celebrity-endorsement-on-brands-marketing-campaigns/. Accessed February 3, 2020.

Mrduljaš, M. (2019). Jedna hrvatska priča. In P. Grubić, (Ed.), *Jedna hrvatska priča—Una storia croata—The One Croatian story*. Rovinj: Adris grupa d.d. & Bruketa&Žinić&Grey d.o.o.

Navas-Loro, M., Rodríguez-Doncel, V., Santana-Perez, I., & Sánchez, A. (2017). Spanish Corpus for sentiment analysis towards brands. In A. Karpov, R. Potapova, & I. Mporas (Eds.), *Speech and computer. SPECOM 2017. Lecture Notes in Computer Science*, vol. 10458. Cham: Springer

O'Donohoe, S. (1994). Advertising uses and gratifications. *European Journal of Marketing, 28*(8–9), 52–75(24). https://doi.org/10.1108/03090569410145706.

O'Donoghue, T., & Rabin, M. (2000). The economics of immediate gratification. *Journal Behaviorial Decision Making, 13*, 233–250. https://doi.org/10.1002/(SICI)1099-0771(200004/06)13:2%3c233::AID-BDM325%3e3.0.CO;2-U.

Palmgreen, P., & Rayburn, J. D., II. (1985). A comparison of gratification models of media satisfaction. *Communication Monographs, 52*(4), 334–346. https://doi.org/10.1080/03637758509376116.

Pavlović, S. (2019). Korištenje slavnih osoba u oglašavanju i njihova uloga u stvaranju lojalnosti marki, Master's thesis / Diplomski rad, University North, Varaždin, Hrvatska / Croatia

Rifkin, J. (2005). *Doba pristupa—Nova kultura hiperkapitalizma u kojoj je cijeli život iskustvo za koje se plaća*. Zagreb: Bulaja.

Studio Sonda (2020) https://sonda.hr Accessed April 17, 2020.

Schull. (2008). Decision and preference in Neuromarketing. *Neuroscience and Society*. http://ocw.mit.edu/courses/science-technology-and-society/sts-010-neuroscience-andsociety-spring-2010/assignments/MITSTS_010S10_paper4_js2.pdf. Accessed February 2, 2020.

Sheth, J., & Parvatiyar, A. (1995). *The evolution of relationship marketing*. International Business Review.

Sikirić, D. (2011). *Neuromarketing i slobodna volja potrošača*. Final paper, Zadar: University of Zadar.

Stroe, M. A., & Iliescu, E. M. (2014). Interdisciplinary perspectives on individual consumption decision, challenges of the knowledge society. In *Economics* (pp. 579–585). Bucharest: Titulescu University Editorial House. Available online: https://search.proquest.com/openview/4ccf534feba015f0e887fe1151399844/1?pq-origsite=gscholar&cbl=2036059. Accessed February 01, 2020.

Swaim, W.R., (2013). Peter Drucker on Sales and Marketing https://www.processexcellencenetwork.com/innovation/columns/peter-drucker-on-sales-and-marketing#:~:text=Drucker%20on%20Selling%20and%20Marketing&text=In%20Management%20(1973)%2C%20he,fits%20him%20and%20sells%20itself.%22. Accessed April 17 2020.

Teck, H., Lim, N., & Camerer, F. (2006). Modeling the psychology of consumer and firm behavior with behavioral economics. *Journal of Marketing Research, 43*(3), 307331.

Xueming, L. (2002). Uses and gratifications theory and E-consumer behaviors. *Journal of Interactive Advertising, 2*(2), 34–41. https://doi.org/10.1080/15252019.2002.10722060.

Wieckowski, A. G. (2019). When neuromarketing crosses the line, January 23, 2019. Harvard Business Review (online). https://hbr.org/2019/01/when-neuromarketing-crosses-the-line. Accessed February 1, 2020.

Wilson, R., Gaines, J., & Hill, R. (2008). Neuromarketing and consumer Free Will. *The Journal of Consumer Affairs, 42*(3), 389–410.

Zaušková, A., Grib, L., Hliboký, M., & Kyselica, P. (2016). Innovative use of marketing communication tools in promoting pilgrimage sites. *European Journal of Science and Theology, 12*(1), 223–230.

# Consumer Happiness and Decision Making: The Way Forward

**Tanusree Dutta and Manas Kumar Mandal**

**Abstract** In a competitive market, companies must think beyond consumer satisfaction. The key lies in consumer happiness. Disciplines have crossed their traditional boundaries to understand their consumer better. This is reflected in the growing popularity of the use of neuroscientific techniques to understand the consumer brain. On the other hand, the field of behavioral economics has also made progress in understanding human decision making and cognitive bias. In this chapter, we discuss about the cognitive biases that influence decision making (cognitive) and consumer happiness (affective).

Demand and supply are fundamental to any transaction. Consumers' demand for a product leads to an increase in the production of goods and shares a linear relation with the consumers who need fulfillment, satisfaction, and happiness, to name a few. Since resources available with consumers are limited to fulfill their needs and desires, happiness lies in choosing the best among the available options. However, research suggests several factors that make a consumer behave irrationally and make suboptimal decisions (Bernstein, 1998).

Disciplines of marketing, consumer behavior, psychology, economics, finance, and other allied subjects are actively involved in decoding the underlying factors of consumer happiness. Over a period, these disciplines leaped their traditional boundaries to understand the consumer better, resulting in the emergence of fields such as neuromarketing, consumer neuroscience, neurofinance, behavioral economics, etc. While consumer neuroscience, neuromarketing, neurofinance and other related areas relied on the use of neuroscientific tools to understand behavior, behavioral economics reflected a departure from traditional economics.

T. Dutta (✉)
Indian Institute of Management, Ranchi, India
e-mail: tanusree@iimranchi.ac.in

M. K. Mandal
Dept of Humanities and Social Sciences, Indian Institute of Technology, Kharagpur, India

© The Author(s), under exclusive license to Springer Nature Singapore Pte Ltd. 2021
T. Dutta and M. K. Mandal (eds.), *Consumer Happiness: Multiple Perspectives*, Studies in Rhythm Engineering,
https://doi.org/10.1007/978-981-33-6374-8_9

Traditional economics was anchored on '*homo economicus.*' A concept developed by John Stuart Mill (see Persky, 1995). Homo economicus is a theoretical construct that suggests calculated self-interest as an underlying motive governing human transactions (Urbina & Villaverde, 2019). Traditional economic perceived humans as rational beings, while behavioral economists like Richard Thaler, Daniel Kahneman, Amos Tversky, and others believed that psychological phenomena like biases, heuristics, etc., affect human decision making (see Chang, 2019). Our thoughts are often based on insufficient knowledge or information, limited processing capacity, and influenced by the context in which we make decisions and not the result of careful deliberation. Being the social animal that humans are, there is also a tendency to exhibit social preferences and follow social norms.

In this chapter, we discuss how cognitive biases influence the relationship between decision making (cognitive) and consumer happiness (affective). We argue that these biases are a reflection of the subtle and /or expressive feeling of consumer happiness.

## Happiness and Freedom to Choose: Does the Availability of Choice Reduce Happiness?

We are all familiar with the prevalent phrase in English, The more, the merrier. Logically too, the availability of choices should make us happy as we experience the freedom to choose. Economists and social scientists value freedom of choice equally. In an experiment conducted by Iyenger and Lepper (2000), they found some alarming results. Iyenger and Lepper (2000) had conducted three experiments in the laboratory and field setting. Results suggested that individuals are more likely to make a purchase when options are limited than extensive, which was very unlike common understanding. The first experiment of the study was conducted in an upscale grocery store. Consumers attending the store were presented with two options, a limited six option of jam vis-à-vis an extensive array of twenty-four flavors. The findings of the study suggested that increased alternatives appeared attractive to the consumer but harmed purchase decisions.

On the other hand, consumers exposed to the six alternative jam options for purchase were more committed to making a purchase decision. In the second experiment of the study, students were asked to write an essay from among six topics presented to them in comparison to another group, which was given 30 essay topics. The intrinsic motivation of participants of both the group was measured. Results suggested that intrinsic motivation was higher when choices were limited. In the third study, one group of participants could sample chocolate from an array presented to them while in the other group, the participants tested chocolate chosen for them. The response of the participants in both the situation was compared in terms of satisfaction with the choice process, choice related expectation, satisfaction with the chocolates that they had sampled, and subsequent purchase behavior. The results of this study also confirmed that though the availability of choices is alluring, it harms satisfaction and intention to purchase.

## *Neuromarketing Implication*

Literature suggests that the availability of options makes a consumer feel happy, but at the same time, it impairs decision making. Freedom of choice has an expressive value (Schwatrz, 2005). And, the availability of moderate alternatives works best in consumer decision making. But, if the decision-making ability of the consumer is to be safeguarded in a situation of multiple choices, it is advisable to have the preferred alternative among the other options. This assists the brain to handle additional information better.

## *The Science Behind the Availability of Option and Consumer Happiness*

The anterior cingulate cortex and the striatum are the two areas of the brain that are activated when there is choice overload. In the study conducted by Reutskaja et al. (2018), participants were provided 6, 12, and 24 options. Results suggested that in conditions of fewer and more alternatives (six and twenty-four, respectively), activity in the striatum and ACC was lower. However, twelve options represented an optimal number of choices. Another interesting finding of this study was that with the addition of a preferred alternative, the capacity of the brain to handle more options increased. Reutskaja, et al., (2018), in their research at the authors argued that though the brain perceives the existence of choice as attractive, in reality, the availability of options makes it difficult for the brain to evaluate and compare these choices. This is referred to as 'choice overload' and helps explain why we often leave a shopping mall providing a wide range of options without making a purchase decision. Multiple options delay the decision-making process.

## Loss Aversion

The concept of loss aversion is important to a cognitive psychologist, decision theorists, and behavioral economist. It refers to the tendency to avoid loss vis-à-vis in acquiring equivalent gain. Consumers are motivated to avoid the pain of loss and seek pleasure in reward. This term was developed by Tversky and Kahneman (1979) and has been examined in both risky and riskless decision-making situations (Novemsky & Kahneman, 2005; Barnberis, 2013). In riskless decision making, loss aversion has been used to represent endowment effect (Thaler, 1980)—a tendency to retain the object that one owns in comparison to acquiring a new item, and status quo bias (Samuelson & Zeckhauser, 1988)—sticking to a decision made earlier. In the risky decision making a loss, aversion is reflected in dispositional effects (Weber and Camerer, 1998)—tendency to retain assets which have lost value and do away with assets that have gained value and framing effects (Tversky and Kahneman, 1981)—evaluating options based on their positive and negative connotations, and many others (Camerer, 2005; Barberis, 2013, c.f Cheng & He, 2017). Homonoff (2017) assessed whether a tax of $0.05 levied as a penalty for the use of the plastic bag had a more significant impact in comparison to a reward of the same amount. He found that imposing the tax had a more significant effect than a reward of the same amount. This change in behavior helps validate the concept of classic loss aversion in action.

## Neuromarketing Implication

Marketers have used the classic loss aversion in terms of displaying 'limited time offer,' 'Stocks to end soon,' 'last left in stock' Consumers feel happy when they get to buy something that is exclusive, rare or unique. The principle of fear-of-missing-out (FOMO) acts stronger in such situations. To increase appeal to consumers and generate a feel-good factor, it makes sense to display the discounted price with a referent showing the baseline price. It makes the consumer feel good in making the purchase now rather than deferring it to a time when the discount may not be applicable. The impact of 'Black Friday' on consumer behavior is one classic example. 'Black Friday' essentially refers to the day after thanksgiving, which marks the unofficial shopping season for Christmas. Almost all stores come up with great offers and discounts to attract customers to their stores.

## Neuroscientific Basis Behind Scarcity and Consumer Happiness

Research suggests that sensitivity to loss is governed by negative emotions (Camerer, 2005). Thus, when exposed to situations of potential loss, increased activity in the amygdala or anterior insula is recorded (Kahn et al., 2002; Kuhnen & Knutson, 2005). Both the amygdala and ventral striatum get activated during loss more strongly than corresponding gains. This suggests a possible role of amygdala in the integration of value and emotional cues (Charpentier, et al., 2016). Scarcity of the situation makes a consumer realize that they may not get the product, and to avoid the feeling of regret, they rush to buy the product. Attaining the product makes them feel happy.

## Sunk Cost Phenomenon

According to this phenomenon, humans tend to continue the effort once an investment is made. This investment could be in terms of time, effort, and / or money. This may be since denial implies wasted effort, which is difficult to accept. Empirical evidence suggests that customers who spent more for a season subscription for a theater series watch more plays because of the sunk cost in purchasing season tickets (c.f. Arkes & Blumer, 1985). In other words, there is a tendency to spend good money after bad to justify one's efforts. Another example related to our reading habit is continuing to read a book that you find boring simply because you have finished reading half the book. Thaler (1980) explains this phenomenon with prospect theory (Kahneman & Tversky, 1979). One aspect of the prospect theory is the objective evaluation of gains and losses compared to the subjective evaluation of perceived gains and losses. The other feature which plays an important role is the certainty effect. Certainty effect plays a dual role, certain gains are overvalued, and certain losses are undervalued. Since an individual is aversive to certain losses, he tends to abide by certain gains.

### *Neuromarketing Implication*

Companies may use this phenomenon to introduce membership programs that require investment from their customers to get some membership benefits. The chances are that the customer would continue with the membership with the organization to gain the advantage of the money invested so as not to 'lose' the buy-in fee. This also would give time to the organization to device alternatives ways to keep the customer further engaged and help them think away and positively from the sunk cost phenomenon. However, such a technique may not work well with customers who are price sensitive.

### *Neuroscientific basis behind sunk cost phenomenon and consumer Happiness*

Haller and Schwabe (2014) used fMRI to examine the functioning of the brain during a novel decision-making task. Results suggested that prior investments reduced the influence of the ventromedial prefrontal cortex (vmPFC). This reduced influence of the vmPFC correlated with the sunk cost effect. Further to this, the activity in the dorsolateral prefrontal cortex (dlPFC) was related to prevent wastage of resources and correlated negatively with vmPFC activity. This suggested that investments made in the past bias the decision-making process in the brain and the interaction between vmPFC and dlPFC induce a tendency to give away good money after bad, explaining sunk cost phenomenon.

## Goal Gradient Hypothesis

The goal gradient hypothesis was proposed by American Psychologist, Clark Hull (1934). In his experiment on rats, Hull found that when put in a straight rally the rats exhibited a tendency to run faster when they were closer to the target (food). Extending the findings of goal gradient hypothesis to understand consumer behavior. Kivetz, Urminsky and Zheng (2006) found that when customers were given a coffee card with 12-stamps of which two were pre-completed, the purchase intention of the customers increased in comparison to those customers who received card with 10 stamps but with no pre-completion. This was because of illusory goal progress. Customers perceived a reduction in the psychological distance in the achievement of the reward.

### *Neuromarketing Implication*

The goal gradient hypothesis has been very successful in encouraging donation behavior, performing a community act, and the like. When consumers are given an illusionary feeling of progress toward reaching the set objective, they feel encouraged and motivated to perform the desired behavior. People who donate toward the end experience a greater sense of accomplishment, satisfaction, and the personal impact of being instrumental toward solving a social problem. The perception of impact increases as the distance toward the goal decreases generating more happiness (Cryder, Loewenstein & Seltman, 2013).

The Mesolimbic reward system gets activated in the brain when we are receiving a reward and engaging in charity (Moll et al., 2006). This suggests why the goal gradient hypothesis works best for charity and equally successful in other situations where the consumer perceives that the goal is approaching.

## Nudging to Happy Decision Making

Perception is humans are rational beings capable of choosing the best option of their interest. Reality is that cognitive limitations and biases influence human judgment and decision making. Resultant is that individual at times makes choices that are against their best interest. Thaler and Sunstein (2003) proposed the use of nudge as an influential way to improve the quality of human decision making without infringing on their freedom of choice. Therefore, it becomes important to plan alternatives so that individuals feel that their freedom of choice is not infringed yet they are 'nudged' to a choice for their best interest.

Many people are reluctant to opt for being a donor for organ after one's death. Goldstein et al., (2008), in one of their publications, report that in Germany no citizen is an organ donor by default and the percentage of citizens who have actively opted for it is as low as 12%, whereas in Austria, a neighboring country, citizens are by default set as organ donors until and unless they take active measure to opt-out. The percentage of people who have not opted out from the default option of organ donorship is as high as 99.98%. Even for simple things, the use of nudge has proved to be very useful. For example in an ATM, they first ask you to collect the card that you have inserted and then dispense the money, so that the customer does not forget to collect the card (c.f. Brüggemann, et al., 2017). There are several such examples of the successful use of nudge.

## *The Science of Nudge and Happiness: Neuromarketing Implication*

To help consumers make effective decisions, nudges are being used as a behavioral trigger. Use of terms such as 'bestsellers,' 'popular,' 'healthy,' 'new,' etc., are nudges that help a consumer to choose a particular product. They help the consumer to get over the problem of choice overload and assist subtly in streamlining the thought process.

The phenomenon of nudge is best explained by the dual process theory. Our brain has an automatic system and a reflective system. The automatic system takes care of most of our behavior. It is fast, unconscious, and capable of both parallel and associative thinking. On the contrary, the reflective system is slow, conscious, and capable of serial and analytical thinking and consumes a lot of energy. Interestingly, both the systems work together on most occasions. Thaler and Sunstein (2008, pp. 19–22) refer to them as automatic (System 1) and reflective (System 2). Nudges designed to affect System I thinking (automatic processing) usually make use of graphical warning or default rules however nudges designed to work effectively with System II thinking use statistical information and factual disclosures (Sunstein, 2016).

## Conclusion

Some of the biases that we have discussed in the chapter may have an evolutionary perspective for support. Well-being homeostasis is largely governed by biases related to the self. Additionally, humans tend to justify and confirm behavior. Therefore, designing for consumer happiness makes sense in utilizing these biases rather than working against it. Moreover, many of the biases have been explored in the western culture and very few across cultures. So, research conducted on the acultural and cultural differences on these biases may help marketers to improve on consumer emotional experience, increasing consumer happiness.

# References

Arkes, H. R., & Blumer, C. (1985). The psychology of sunk cost. *Organizational Behavior and Human Decision Processes, 35*(1), 124–140.

Barberis, N. C. (2013). Thirty years of prospect theory in economics: A review and assessment. *Journal of Economic Perspective, 27,* 173–195.

Bernstein, P. L. (1998). *Against the Gods: The Remarkable Story of Risk.* New York, NY: John Wiley & Sons.

Brüggemann, P., Günter, A., Lorenz, J-T., & Münstermann, B. (2017). A 'nudge' for the better in assistance claims journeys. McKinsey & Company

Camerer, C. (2005). Three cheers—psychological, theoretical, empirical—for loss aversion. *Journal of Marketing Research, XLII*(129), 129–133 (May 2005).

Chang, K.-P. (2019). Behavioral cconomics versus traditional economics: Are they very different? Available at SSRN: https://ssrn.com/abstract=3350088 or https://doi.org/10.2139/ssrn.3350088

Charpentier, C. J., De Martino, B., Sim, A. L., Sharot, T., & Roiser, J. P. (2016). Emotion-induced loss aversion and striatal-amygdala coupling in low-anxious individuals. *Social Cognitive and Affective Neuroscience, 11*(4), 569–579.

Cheng, Q., & He, G. (2017). Deciding for future Selves Reduces Loss Aversion. *Frontiers in Psychology, 8,* 1644.

Cryder, C. E., Loewenstein, G., & Seltman, H. (2013). Goal gradient in helping behavior. *Journal of Experimental Social Psychology, 49*(6), 1078–1083.

Goldstein, D.G., Johnson, E.J., Hermann, A. & Heitmann, M. (2008). *Nudge your customers toward better choices.* Decision making, Harvard Business Review.

Haller, A., & Schwabe, L. (2014). Sunk costs in the human brain. *NeuroImage, 97,* 127–133.

Hull, C. L. (1934). Rat's speed-of-locomotion gradient in the approach to food. *Journal of Comparative Psychology, 17,* 393–422.

Iyyenger, S. S., & Lepper, M. R. (2000). When choice is demotivating: can one desire too much of a good thing? *Journal of Personality and Social Psychology, 79*(6), 995–1006.

Kahn, I., Yeshurun, Y.I.., Rotshtein, P., Fried, I., Ben-Bashat, D., Hendler, T. (2002). The role of the amygdala in signaling prospective outcome of choice. *Neuron, 33,* 983–994.

Kahneman, D., & Tversky, A. (1979). Prospect theory: An analysis of decision under risk. *Econometrica, 47*(4), 263–291.

Kunhen, C. M., & Knutson, B. (2005). The neural basis of financial risk taking. *Neuron, 47*(5), 763–770.

Moll, J., Krueger, F., Zahn, R., Pardini, M., de Oliveira-Souza, R., Grafman, J. (2006). Human fronto–mesolimbic networks guide decisions about charitable donation. *Proceedings of the National Academy of Sciences 103*(42), 15623–15628.

Persky, J. (1995). Retrospectives: The ethology of homo economicus. *The Journal of Economic Perspectives, 9*(2), 221–231.

Reutskaja, E., Lindner, A., Nagel, R., Andersen, R., A. & Camerer, C. F. (2018). Choice overload reduces neural signatures of choice set value in dorsal striatum and anterior cingulate cortex. *Nature Human Behaviour, 2* (12) 925-935.

Samuelson, W., & Zeckhauser, R. (1988). Status quo bias in decision making. *Journal of Risk and Uncertainty, 1,* 7–59.

Schwartz, B. (2005). *The paradox of choice: Why more is less.* Harper Perennial: US.

Sunstein, C. R. (2016). People Prefer System 2 Nudges (Kind Of). *Duke Law Journal, 66,* SSRN: https://ssrn.com/abstract=2731868 or http://dx.doi.org/https://doi.org/10.2139/ssrn.2731868

Thaler, R. (1980). Towards a positive theory of consumer choice. *Journal of Economic Behavior and Organization, l,* 39–60.

Thaler, R., H., & Sunstein. C. R. (2003). Libertarian paternalism. *American Economic Review, 93* (2), 175-179.

Thaler, R. H., & Sunstein, C. R. (2008). *Nudge: Improving decisions about health, wealth, and happiness*. Yale University Press.
Tversky, A., & Kahneman, D. (1981). The framing of decisions and the psychology of choice. *Science, 30thJan*, 453–458.
Urbina, D. A. & Ruiz-Villaverde, A. (2019), A Critical Review of Homo Economicus from Five Approaches. *American Journal of Economics and Sociology, 78*, 63-93.
Weber, M., & Camerer, C. F. (1998). The disposition effect in securities trading: An experimental analysis. *Journal of Economic Behaviour and Organization, 33*, 167–184.

## *Webinks*

Reference for Daniel Putler Homonoff (2017)
https://thedecisionlab.com/biases/loss-aversion/
Ran Kivetz, Oleg Urminsky, Yuhuang Zheng. "The Goal-Gradient Hypothesis Resurrected: Purchase Acceleration, Illusionary Goal Progress, and Customer Retention."
https://www8.gsb.columbia.edu/articles/ideas-work/goal-gradient-hypothesis-resurrected-purchase-acceleration-illusionary-goal-progress-and

# Paying a Price to Get a Value: Choose Wisely

**Vijay Victor and Elizabeth Dominic**

**Abstract** Of the traditional marketing mixes that are used to influence consumer purchase decisions, the development of a pricing strategy that retains a loyal customer base without harming profitability remains the most elusive. Advancements in behavioural economics have demonstrated that the economic choices of human beings may not always be necessarily rational. In fact, previous studies have shown that the assumption of rationality is overstated in economic theories. Even under normal circumstances, the consumers tend to confuse price and value, often ending up purchasing high-priced products on the pretext that higher price is associated with a higher value. The inherent elements of irrationality that persist in consumer behaviour offer opportunities for the sellers to develop and apply new heuristics and priming techniques for setting prices that may coax the consumers to make a purchase. Everyone gets influenced by one or another cognitive distortion at some point. On this account, understanding the cognitive biases in consumer decision-making process is key to developing an effective pricing strategy. This chapter gives a detailed account of how well the consumers respond to subtle nuances in price offers and other price-related priming strategies through the lens of behavioural economics.

**Keywords** Pain of paying · Pricing strategies · Consumer behaviour · Behavioural economics

## Introduction

Rational human behaviour is an assumption long held in economic theories. Rationality implies a thought process which is mainly based on the assessment of objective

---

V. Victor (✉)
Department of Economics, CHRIST (Deemed to be University), Hosur Road, Bangalore, Karnataka 560029, India
e-mail: vijay.victor@christuniversity.in

E. Dominic
Saintgits Institute of Management, Kerala 686532, India
e-mail: elizabeth.dominic@saintgits.org

© The Author(s), under exclusive license to Springer Nature Singapore Pte Ltd. 2021
T. Dutta and M. K. Mandal (eds.), *Consumer Happiness: Multiple Perspectives*,
Studies in Rhythm Engineering,
https://doi.org/10.1007/978-981-33-6374-8_10

facts. In simple words, a rational decision is a sensible decision that is not influenced by emotions (Hammond, 1997). It was assumed that the choices of a rational consumer are that which maximise his/her level of utility or satisfaction (Bernoulli, 1738). However, this belief was defied by two psychologists in the 1970s. Daniel Kahneman and Amos Tversky proved that at times humans make preposterously irrational decisions that are completely against logic. They proved it by exposing the heuristics involved in human decision-making which results in many cognitive biases that may finally yield suboptimal decisions (Tversky & Kahneman, 1974).

An emerging field of study analysing the inextricably intertwined relationship between emotions and reason in the process of decision-making is neuroeconomics. It follows an interdisciplinary approach in studying how people make decisions considering the neural, social and psychological aspects of decision-making (Zeki, Goodenough & Zak, 2004). The use of medical imaging techniques such as functional magnetic resonance imaging (fMRI) in neuroeconomics researches has enabled the researchers to objectively assess the development of preferences rather than the traditional methods of close observation of the decisions made by consumers which is largely subjective in nature (Loewenstein, Rick & Cohen, 2008). The recent developments in these fields of studies are advancing researches in many niche areas in behavioural sciences which require more objective and precise results.

This chapter gives a detailed account of the various researches conducted in the field of neuroeconomics related to changes in consumer behaviour when subtle nuances in price offer and other price-related strategies are made. The chapter is divided into six broad sections, and the first section opens with a description of the concept of 'pain of paying' which is the theoretical foundation used by many studies to expound the discrepancy observed in the spending behaviour of consumers. Section 2 discusses the psychology of price tags and priceless products. It mainly focuses on the behavioural changes observed under price and product priming techniques. Consumer preference towards bonuses and discounts is explained in the following section where the effectiveness of both techniques for different products and situations is given. Section 4 details the difference in the pain of paying while using credit cards and cash. How credit cards alleviate the pain of paying is elucidated using relevant theoretical and literature support. The positive experience generated by freebies is given in the section which follows. The last section explains the popularity of buy now pay later practice and how it influences the pain of paying.

## Pain of Paying

Economists have long maintained the view that the choices of consumers are mainly influenced by price and personal preferences. This view is backed by the rationality assumption which outlines the aim of the consumers as to maximise their present and future utility (Hammond, 1997). However, several economists sought to give a broader understanding of consumer behaviour by incorporating the role of emotions in the choices that we make under different situations. One such sensitive area that required further explanation was the stark differences observed in the spending behaviour of consumers.

Contrary to the tenets of rationality assumed in economic theories, consumers seemed to take decisions that were not maximising their utility in many real-world purchase situations. One plausible explanation given for this deviation from the rational behaviour was the psychological and hedonic vexations linked with the act of spending money. This area remained largely overlooked in the literature until the work of Zellermayer (1996) on the 'pain of paying' became popular. Zellermeyer's research is the pioneering work in the field explicating the pain or negative emotions associated with spending money. He defined pain of paying as 'the direct and immediate displeasure or pain from the act of making a payment' (Zellermayer, 1996, p. 2). It is the psychological or hedonic exasperation connected with spending money.

Zellermeyer outlined that spending has an emotional aspect that is influenced by both emotional and situational variables. He attributed a positive dimension to the concept of pain of paying as something which helps consumers restrain themselves from splurging. The study also attempts to address the question of why some people are classified as spendthrifts and some as tightwads by associating their behaviour with the pain of paying. The reason why the spendthrifts find it hard to control spending even when a situation is unfavourable is that their perceived pain of paying is much lesser than that of the tightwads who are too frugal albeit the situation is in their best interest (Zellermayer, 1996).

Zellermeyer's work on the pain of paying provided a theoretical support for explaining the differences between spendthrifts and tightwads and also pointed towards the unexplored research area of disinclined frugality. The study also showed that the contextual factors which are considered as seemingly irrelevant from an economic perspective also affect the pain of paying just as the price of the product; hence, it is important for the sellers to address those issues to increase sales. As Prof. Lowenstein rightly remarked, 'People's preferences change quite dramatically, sometimes moment to moment' (Krzyzanowski, 2007). The age-old presumption that human beings always aim at satisfying own preferences remains largely in theory with limited practical application in real-world settings.

An experimental research conducted in this area by Knutson, Rick, Wimmer, Prelec & Loewenstein (2007) showed that the consumers largely consider the immediate pleasure of consumption and the immediate pain of spending rather than choosing between the present good and future alternatives as presumed by theories. The experiment involved 26 adults who were given 20$ each to spend on a range of products. They were allowed to keep the money had they not wished to spend it on any products. After giving the instructions, the participants were asked to lie in a functional MRI scanner and the prices of several products were shown to them. The study examined whether the pain centres in the brain would get stimulated when the participants saw the prices. The results showed that the pain processing centre of the brain, named the insular cortex (insula), was activated when the participants perceived prices as too high. Previous researches have shown that insula is also associated with some emotions such as sadness, disgust, anger and fear (Knutson,

Taylor, Kaufman, Peterson & Glover, 2005; Wager et al., 2013). For the participants who were already classified as tightwads, this area of the brain was activated even when they were shown the prices of products that they could afford. Does being a tightwad or spendthrift have an influence on happiness? A study shows that both spendthrifts and tightwads are apparently less happy than those people who do not experience conflict in mind about buying something new or those who cannot really afford to buy it (Rick, Cryder & Loewenstein, 2008).

Knutson's experiment concluded with the result that for the tightwads, the pain caused by high prices activates the pain processing centre of the brain which then deters spending. Tightwads seem to be spending much lesser than what they would really like to spend in anticipation of the pain that will follow the spending. For them, spending money is painful as their immediate emotions associated with spending outweigh the expected emotions. The concept of immediate and expected emotions is important in explaining these behavioural changes. Expected emotions are those emotions that are felt only when the outcome of a decision actualises, which implies that these emotions are not experienced instantly when a decision is taken. On the other hand, immediate emotions are experienced at the moment when a decision is made (Rick & Loewenstein, 2008). An example for immediate emotions is that the very thought of purchasing junk food would make a consumer feel bad because of the health problems associated with it. In the case of expected emotions, the guilt feeling comes only after consuming it.

According to the consequentialist perspective, a consumer always maximises his/her utility based on the expected emotions rather than the immediate emotions. They consider immediate emotions as an 'epiphenomenal' by-product of a decision and have little influence on the decision itself which is the substance of the intertemporal choice theory propounded Fisher (1930). This deliberate ignorance of the immediate emotions associated with spending was bridged by Zellermeyer and other economists who provided a theoretical foundation for the irrational spending choices of consumers by introducing the concept of pain of paying.

The intensity of pain associated with payments also differs based on many other factors. This implies that not all payments are equally painful. First, as already mentioned, spendthrifts and tightwads experience the pain of paying differently. The pain differs with the nature of the products we buy as well. Buying frivolous products inflicts more pain than buying virtuous products. Buying French fries feels different from buying vegetable salad as the momentary guilt associated with the former is higher than the latter. The pain of paying also varies with the method of payment. Using credit cards for making a payment has been proven to be less agonising than using cash.

## Price Tag Psychology Versus Priceless Products: Should the Price Tag Be on the Product or Away from the Product

If you have been to jewellery shops, you might have noticed that some shops choose not to display the prices of the jewellery pieces in a fully visible manner to the customers. At least for some customers, this might seem quite intriguing and at the same time riveting. Why would the sellers hide the prices from their potential buyers? Does it really matter if the customers see the product first and price later? There are several reasons why shops engage in this price hiding tactic. The most obvious reason is the buyers getting spooked by seeing the exorbitant prices and deciding to walk out of the store. When there is no price tag, the salesperson gets a chance to explain the features of the product first and then price, trying to lure the potential buyer by making the decision to buy more about emotions rather than just a number on the price tag.

There are only a few empirical studies available at present that explain whether the sequence of appearance of products and prices influences the purchase decisions of consumers. The most popular work in this field is the recent research conducted by Karmarkar, Shiv and Knutson (2015). They attempted to examine the changes in consumer decisions when they were exposed to prices before products using advanced functional magnetic resonance imaging (fMRI) techniques. The results of the study reported that when the respondents saw the price first (price primacy) they were mostly thinking if the price is fair enough or whether the product is worth paying the price and so on. The neural correlations in the brain seemed to be more critical about the valuation process when the respondents saw the prices first and they turned out to be very cost conscious. Previous studies conducted in this field also reported that early exposure to price information would make the consumers think of what they will lose in a transaction which in turn increases 'the pain of paying' (Knutson et al., 2007; Rick et al., 2008). On the other hand, when they saw the product first (product primacy), they were considering if they really liked it or if it will satisfy their needs and so on. Under product primacy, the respondents were focusing more on the utility and desirability aspects of the products. Medial prefrontal cortex (MPFC) is the brain area that is associated with decision-making which is responsible for linking the product information with the price. The visual appeal of the product and its desirability are processed by another brain area named as nucleus accumbens (NA). This study has shown that the sequence of appearance of product and price activates these brain areas which affect the valuation of products and influence their purchase decisions.

Karmakar, the lead researcher prior to publishing the results, mentioned in an interview that, 'The pattern of activity in the prefrontal cortex suggested to us that sequence matters: At the very simplest, the neural signals looked different when the price came first versus when the product came first. When the product came first, the decision question seemed to be one of 'Do I like it?' and when the price came first, the question seemed to be 'Is it worth it?' (Nobel, 2014).

Further research conducted by the same team highlighted a significant takeaway for retailers. When selling utilitarian items, price primacy is more likely to be effective provided the seller offers a reasonable discount for the products. An important caveat which sellers need to keep in mind is that they should not try to fool the customers by displaying insignificant offers which will considerably decrease the willingness to pay that is already compromised with price primacy. While selling a luxury product like jewellery, apply product primacy, i.e. display the product first and then highlight the quality, properties and features. This will give an indirect cue to the customers that the product is worth its price (Karmarkar et al., 2015).

The key point here is that the process of consumer valuation before making a purchase decision changes significantly under price primacy and product primacy as visible from the activity patterns in the medial prefrontal cortex of the brain. The sellers should consider the nature of the products they sell before deciding to apply price primacy or product primacy. For utilitarian product, price primacy will increase the purchases, and for expensive and high-quality products, using product primacy is likely to increase the rate of purchases. Now you know why jewellery shops hide the prices!

## 20% Extra to 20% Reduced Price: Which Generates More Positive Experience

Discounts and bonuses are two marketing techniques used by the sellers to ramp up sales. A discount simply means a reduction in price. When a seller advertises that a product is on 50% discount, it means that the price drops while the product remaining the same. If a hand wash worth 8$ is on 50% discount, the buyer gets the hand wash for 4$. On the other hand, a bonus implies an increase in the quantity purchased. If the seller advertises 50% extra for the same hand wash, you still pay 8$, but now walk out of the store with 50% extra hand wash. So, which one creates more positive experience: discounts or bonuses? Although both bonuses and discounts are effective in increasing sales, studies show that bonuses have a slight edge over discounts.

One of the main reasons why people prefer bonuses to discounts is the tendency to frame losses and gains differently. The prospect theory developed by Kahneman and Tversky (1979) propounds that 'Losses hurt more than gains feel good'. This means that people tend to avoid losses more than that of getting an equivalent gain. Most of the time, consumers perceive in a totally different way to an economically equivalent bonus and discount. People perceive discount as a reduced loss in money or gains in money and bonuses as reduced loss in money or gains in product. Why do consumers frame bonuses and discounts differently even though they are economically the same? One study shows that most of the consumers are confused with the numbers and percentages they see in the stores. For example, they have a hard time figuring out the real value difference between a product with 50% discount and a product which offers 33% extra (Kruger & Vargas, 2008). In reality, a 33% bonus is the same as a 50% discount.

Another reason for this difference in perception is the tendency of the consumers to consider percentages as numbers or ignoring the base value. Quantities expressed in percentages are not absolute. A base value is always present which is relative to the value which people see. So, 33% more means base value plus 33% extra, which is the bonus. A study conducted by Chen, Marmorstein, Tsiros and Rao (2012) involving 120 respondents was shown two sets of promotions. In the first set of promotion, 50% extra was given to an 8 oz bottle worth $3.89. In the second set, 35% off was offered to the same 8 oz bottle worth $3.89. The results showed that most of the respondents chose the first option. The respondents fell for 50% extra as it seems to be a greater amount than 35% off. The reason for this behaviour is the ignorance of the base value and the reluctance in converting percentages to numbers. In a study conducted by Munger and Grewal (2001) involving the valuation of modified cars with the same base price, consumers gave more ratings to cars with extra features than cars on discounts. The results showed that the consumers were slightly more likely to buy the cars when extra features were offered at the base price than discounts.

So, if bonuses are more effective, why do stores sell products on discounts? Researches have shown that there are many cases where consumers would prefer discounts to bonuses. This totally depends on the nature of the product sold. Chen et al. (2012) report that consumers are more likely to purchase expensive products on discounts. Bonuses seem to be effective in the case of inexpensive products than high-priced products. The reason might be that while buying expensive products, a discount makes a consumer feel better as it reduces the pain of paying. Another study showed that people would prefer discounts if the product or brand is unfamiliar to them and they want extra if it is familiar. This makes sense as a consumer might be sceptical of getting extra for a product that is unfamiliar but is okay with a price discount as it helps reduce the risk of buying an unfamiliar product (Lowe, 2010). People would also like to buy non-durable or perishable products on discounts than with bonuses. For example, a buy one get one yogurt cup will not be appealing to you if you consume only 1 yogurt cup a week. Because the extra yogurt cup you are getting for free may get spoiled by the time you finish the first one. Putting a 50% discount, in this case, is more likely to make the consumer purchase it (Sinha & Smith, 2000).

However, a recent research conducted by Cai, Bagchi and Gauri (2016) shows that in some cases providing a discount may lower the sales than increase it. This is known as the 'boomerang effect'. It means that offering a low price discount to products that are not really essential to consumers may result in decreasing the likelihood of purchase rather than increasing it. The reason behind this effect is that the low discount may reduce the transactional value of nonessential products and further lowers the likelihood of consumers purchasing them.

Hence as a general rule, it can be concluded that consumers mostly prefer bonuses to discounts as they perceive losses and gains in different ways. However, in certain cases, consumer preferences do change. Depending on the nature and type of products being sold in a store, a seller should use a mix of discounts and bonuses to spice up the sales.

## Virtual Versus Factual: Does the Use of Credit Card Generate More Positive Experience Than Shelling Out Money at the Counter

The rapid increase in the usage of plastic money indicates that the world is moving towards a cashless economy. The World Payment Report (2019) shows that the global non-cash transactions touched 539 billion in 2016–2017 which is the highest record marked in the past two decades. The convenience offered by credit cards lures most people into using it rather than keeping stacks of paper money especially while going out for shopping. In case the card gets stolen, the problem is mostly fixed with a call to the bank. This minimum liability or no liability feature is what makes the credit cards look more attractive in our purses than cash.

Does using credit cards influence our spending behaviour? Well, it turns out that the method of payment matters when it comes to the question of how much we spend. Some purchases can be more painful than others. Although the prices are same, the pain of paying might differ based on the method of payment and people tend to avoid that method for which the perceived pain is higher. A recent study (Raghubir & Srivastava, 2008) shows that as the transparency of a payment outflow increases, the pain of paying associated with the purchase also increases. Paying for any products creates an immediate pain of paying in the minds of the shopper which he/she tries to compensate for the future benefits from consuming it. However, the pain experienced while using credit cards and cash is different. The study also reports that using credit cards curbs the pain of paying as there is a time gap between the time of purchase and the time the bill has to be settled. Furthermore, at the time of settlement, say once every month, the shopper pays for all the products together without thinking too much about the exact money he/she paid for each product.

Research in credit card psychology and behaviour has revealed that some consumers who exhibit greater inclination to spend and make higher value purchases on their credit cards have weaker memories of past credit expenses. They also tend to overvalue their available funds when using credit cards than cash in an otherwise identical purchase situation which is referred to as the 'credit card effect' (Chatterjee & Rose, 2012; Feinberg, 1986).

Previous evidence suggests that the payment method influences the perception of the amount paid. Payment by card may be hedonically advantageous (i.e. reducing the 'pain of payment') but economically disadvantageous (i.e. leading to an underestimation of real cost). Credit cards are also special in terms of timing because the purchase of product precedes the actual depletion of funds. This temporal separation further reduces the pain of paying and has been argued to facilitate more consumer spending (Prelec & Loewenstein, 1998). Credit cards are regarded as a spending stimulus that allows greater ease of spending than cash. Hence, they display certain distinctive purchase attitudes and are more likely to set some individuals on a path towards indebtedness and personal bankruptcy (Pirog & Roberts, 2007; Szmigin & O'Loughlin, 2010).

Prior research has explained the spending effects associated with various payment mechanisms by ascribing them to memory processes (Soman, 2001), separating purchase from the pain of payment (Prelec & Loewenstein, 1998), classical conditioning (Feinberg, 1986) and processing fluency (Mishra, Mishra & Nayakankuppam, 2006).

These arguments of cost salience and pain explain why those paying using credit card tend to spend and anticipate to spend more than cash payers (Raghubir & Srivastava, 2008). Moreover, consumers systematically misjudge past and future credit card bills because it is not easy for them to recall individual credit card expenses (Soman, 2001).

More recent research has added to the picture by showing that the payment method not only affects how much is spent but also what is acquired. In most studies, credit card payments have been shown to increase unintentional spending or hedonic spending when compared to cash payments (Inman, Winer & Ferraro, 2009; Thomas, Desai & Seenivasan, 2011).

When credit cards as a payment method appear to be more accessible, the consumers tend to focus more on a product's benefits rather than the cost. On the contrary, when cash as a payment method is more accessible, consumers are likely to pay more attention to the cost side of the product (broadly defined to include price, delivery time/costs, warranty costs, installation costs, etc.) relative to the benefits accrued.

## Psychology of Ownership: Does 'Buy Now, Pay Later' Help Generate Positive Experience?

Buy now pay later is a service that allows buyers to delay payments or pay in instalments. The convenience and flexibility offered by this method is the feature that attracts buyers. The logic behind this method is very simple. It works based on the presumption that the people will have money in future, so they buy now and pay later. Buy now pay later services are becoming more popular in the online shopping arena. Interest-free equated monthly instalment (EMI) has become a common service offered by almost all big e-commerce companies where a buyer can purchase a product instantly and repay the money in monthly instalments at zero per cent interest rate.

What drives people to own or consume now and pay later is driven by the power of now. People generally are motivated to further the pleasure (ownership/consumption) and farther the pain (payment). Seeing a high price alone for an attractive product may extinguish the very desire to purchase it. Why do we buy things? Because we derive pleasure from consumption. The pleasure of consumption appears to be wholesome and pure without the experience of paying. Anything which can separate the pleasure of consumption from the pain of paying is worth trying. Buy now, pay later is cashing on this phenomenon. It separates the pleasure of consumption from the pain of paying.

We are all surrounded by umpteen numbers of options for indulgence. From a pure economic point of view, a decision is never slanted towards immediate indulgence. The rational economic theory states that a consumer always thinks of the next best option available and compares the pleasure attained from both before making a decision. In the real world, fortunately, or unfortunately, it does not happen that way. Consumers do not spontaneously think of the prices in terms of opportunity costs. Still, many people can control their immediate indulgence, not because of their calculation of the opportunity costs but because of the negative affect which acts as a deterrent. Emotions help to 'shift the relevant future payoffs into the current moment' (Frank, 1988, p. 281). Negative affect may be defined as 'how the brain 'immediatizes' delayed and intangible opportunity costs' (Preston, 2014). The concept of 'pain of paying' can be used to explain the phenomenon of consumers resisting the pull for immediate indulgence (Prelec & Loewenstein, 1998). Knutson et al. (2007) using fMRI techniques proved that 'pain of paying' is real, and it is a major factor that deters spending. This pain may be emphasised in the case of tightwads and spendthrifts, tightwads experiencing too much pain and spendthrifts experiencing too little pain. People who tend to experience a moderate amount of pain of paying and thus tend to spend close to what they would like to spend are termed as 'unconflicted consumers' (Rick et al., 2008).

Most formal economic models do not account for the 'pain of paying'. Fisher (1930), who had proposed that spending is allocated so as to maximise the sum of utilities, with an allowance for time discounting, later on, recognised that it is not psychologically realistic. Consumers derive direct and immediate displeasure or pain from the act of making a payment. This pain is the psychological or hedonic exasperation connected with spending money. Parting with money is associated with utility relinquished. In effect, the consumer is paying twice.

Pain of paying comes with its own advantages. It solves the human problem of attraction towards immediate gratification by bringing the cost of indulgence to the present. It also solves the problem of under-weighing opportunity costs by making a consumer feel the pinch as it is difficult for the human mind to perceive the benefits of forgone consumption. It offers a heuristic cost–benefit analysis than going through the complexities of intertemporal consumption decisions.

A similar dictum to 'buy now, pay later' which has become popular in the recent times is 'pay now, buy later', where the consumer captures the experience of deriving pleasure in the present from anticipating the future. It provides a source of pleasure with purchase, appending the joy of actual consumption. Alongside, it can be a source of torment for some too, as one is not sure whether he/she will be actually able to consume/own the product as the future is always uncertain. A sensible rule for those types of people is that they should not be paying for a delay even when it is sure that a delay can bring in more anticipated pleasure.

A line of inquiry highly relevant to the pain of paying hails from the idea of Thaler, which is termed as 'buffering' (Prelec & Loewenstein, 1996). The researches in the field of buffering mainly study the interactions between payment and consumption, and how they change under a particular time frame. The theory states that while consumption can mitigate the pain that is associated with it, payments can spoil the pleasure of consumption. It observes when people prefer to consume a certain product prior to paying for it, when they prefer an opposite order, and when they prefer some assortment of consumption and payment streams.

Buy now, pay later option alleviates the pain of payment to some extent and brings in the unadulterated pleasure of consumption as well as the pleasure of ownership to the consumer. When the consumer is offered the option of buying now and then and paying later, the nucleus accumbens—aka, the pleasure centre in the brain gets activated which offsets the stimulation of insula, the pain processing centre in the brain. After all, people buy for one of two reasons—to move closer to pleasure or to move further away from pain.

## Power of Free in Generating Positive Experience

Getting something for nothing activates a human response which is different from the normal business transactions—it just feels better. A free product comes with low expectations and neutral perceptions. If that product proves to be useful, expectations are exceeded, and perception elevates from neutral to positive. Getting something for free makes us act a whole lot different than when we pay, even if it is just a few cents. What exactly is going on when the price of something is lowered to zero?

Freebies make the consumer forget for a moment that 'There is no such thing as a free lunch' as everything that is provided free (as a freebie) eventually has to be paid for in one way or another, such as by returning the favour, being indebted to the host or by 'paying' in some other hidden way which is not made explicit at the time. Freebies can take the forms of free samples, free trials, promotional/specialty items and premiums. 'People always enjoy getting something for free' (Scott, 2006, p.26). People associate free offerings with purely 'a benefit without any cost' (Shampanier, Mazar & Ariely, 2007).

Bodur and Grohmann (2005) show that selling a product with freebies can positively influence customers' attitude towards the product, increase the frequency of buying the product or stimulate the probability of purchase from potential consumers. Similarly, providing freebies can strengthen the uniqueness of a product and increase its value (Chu & Keh, 2006). Montaner, De Chernatony and Buil (2011) show that non-monetary promotion not only is an efficient marketing tool but also has proven to be successful in increasing the brand awareness of a company among the public. Offering freebies is more effective when the quality of the product being sold is very high. The study also reveals that the behavioural patterns of consumers change when they get something for free and are more likely to comply with the seller. Free is not just an indicator of price. It is a very powerful emotional trigger that is often so irresistible that it makes people bring home useless stuff just so they get it at no cost.

The emotion that the term 'free' triggers is known as 'zero-price effect'. 'When people are offered something for free, they have this extreme positive reaction that clouds their judgement'. Decisions about free (zero-price) products may seem to be distinct in that people do not simply deduct costs from benefits but instead, they see the benefits linked with free products as higher. Intuitive and anecdotal evidences show that in some sense, people value free things too much. It might not be surprising that the interest towards a product is very high when the price is very low (zero), but the extent of the effect is intuitively too large to be explained by this simple economic argument. Zero price has a special role in consumers' cost–benefit analysis.

A possible psychological mechanism that might be underlying the zero-price effect deals with the norms that might accompany free products. Costly options invoke market exchange norms, whereas free products invoke norms of social exchange (Fiske, 1992; McGraw & Tetlock, 2005), which can create higher value for the product in question. Heyman and Ariely (2004) offer one example in which they demonstrate that people are likely to exert more effort under a social contract (no monetary amounts) than when small or medium monetary amounts are mentioned. Another example of the relationship between social and exchange norms appears in Ariely (2003) in which they examine the behaviour of persons faced with a large box of candies and an offer to receive the candy either for free or for a nominal price (1¢ or 5¢). Not surprisingly, when the cost is zero, many more students take candy than when the price is positive. More interesting, when the price is zero, the majority of the students take one and only one candy, while those who pay to take candy take a much larger amount (effectively creating lower demand as prices decrease). These results suggest that social norms are more likely to emerge when the price is not a part of the exchange, which could increase the valuation for a good.

What makes the idea of 'free' appealing is the notion of reciprocity—one party in a transaction is at the obligation of returning the favour granted to them by complying with the other. Reciprocity is a very strong impulse. Hence when a brand or seller does something for its customers, they feel that they should do something in return.

A second possible psychological mechanism that might explain the overemphasis on free options comes from the findings of Ariely (2003), Hsee, Yu, Zhang and Zhang (2003), and Nunes and Park (2003), which demonstrate that people have difficulty mapping the utility they expect to receive from hedonic consumption into monetary terms.

A third possible psychological mechanism that might account for the zero-price effect pertains to affect, such that options with no downside (no cost) raise a more positive affective response; to the extent that consumers use this affective reaction as a decision-making cue, they opt for the free option (Finucane, Alhakami, Slovic & Johnson, 2000; Gourville and Soman, 2005; Slovic, Finucane, Peters & MacGregor, 2002).

Of late, as consumer base has been turning highly sceptical of the new seller gimmicks, the word 'free' has started losing its charm. Consumers have begun questioning the value and the worth of a product when it is given to them free of cost.

## Conclusion

From the literature, it could be understood that the decision-making process of consumers is influenced by various known and unknown factors. Most of the time, it is more of an unconscious process rather than a conscious effort. The studies explored in this chapter indicate that consumer decisions could be easily manipulated by making minute changes in pricing techniques that remain mostly unnoticed even after consumers being continuously exposed to these practices. This implies that the sellers can take advantage of this ignorance of consumers, or broadly speaking, the cognitive biases in consumer decision-making, and streamline their pricing tactics accordingly to ramp up sales and ultimately profits.

Hence, marketers have to be prudent and sensible while using this sales promotion technique as a marketing tool.

## References

Ariely, D. (2003). Why good CIOs make bad decisions. CIO Magazine.
Bernoulli, D. (1738). *Specimen theoriae novae de mensura sortis. In: Commentarii Academiae Scientiarum Imperialis Petropolitanae* (Sommer Trans.) (1954).
Bodur, H. O., & Grohmann, B. (2005). Consumer responses to gift receipt in business-to-consumer contexts. *Psychology and Marketing, 22*(5), 441–456.
Cai, F., Bagchi, R., & Gauri, D. K. (2016). Boomerang effects of low price discounts: How low price discounts affect purchase propensity. *Journal of Consumer Research, 42*(5), 804–816.
Chatterjee, P., & Rose, R. L. (2012). Do payment mechanisms change the way consumers perceive products? *Journal of Consumer Research, 38*(6), 1129–1139.
Chen, H., Marmorstein, H., Tsiros, M., & Rao, A. R. (2012). When more is less: The impact of base value neglect on consumer preferences for bonus packs over price discounts. *Journal of Marketing, 76*(4), 64–77.
Chu, S., & Keh, H. T. (2006). Brand value creation: Analysis of the Interbrand-Business Week brand value rankings. *Marketing Letters, 17*(4), 323–331.
Feinberg, R. A. (1986). Credit cards as spending facilitating stimuli: A conditioning interpretation. *Journal of Consumer Research, 13*(3), 348–356.
Finucane, M. L., Alhakami, A., Slovic, P., & Johnson, S. M. (2000). The affect heuristic in judgments of risks and benefits. *Journal of Behavioral Decision Making, 13*(1), 1–17.
Fisher, I. (1930). *The theory of interest*. New York: Macmillan.
Fiske, A. P. (1992). The four elementary forms of sociality: Framework for a unified theory of social relations. *Psychological Review, 99*(4), 689.
Frank, R. H. (1988). *Passions within reason: The strategic role of the emotions*. WW Norton & Co.
Gourville, J. T., & Soman, D. (2005). Overchoice and assortment type: When and why variety backfires. *Marketing Science, 24*(3), 382–395.
Hammond, P. J. (1997). Rationality in economics. *Rivista Internazionale Di Scienze Sociali, 105*(3), 247–288.
Heyman, J., & Ariely, D. (2004). Effort for payment: A tale of two markets. *Psychological Science, 15*(11), 787–793.

Hsee, C. K., Yu, F., Zhang, J., & Zhang, Y. (2003). Medium maximization. *Journal of Consumer Research, 30*(1), 1–14.

Inman, J.J., Winer, R.S., & Ferraro, R. (2009). The interplay among category characteristics. *Customer Marketing, 73*(5), 19–29.

Kahneman, D., & Tversky, A. (1979). Prospect theory of decisions under risk. *Econometrica, 47*(2), 1156–67.

Karmarkar, U. R., Shiv, B., & Knutson, B. (2015). Cost conscious? The neural and behavioral impact of price primacy on decision making. *Journal of Marketing Research, 52*(4), 467–481.

Keh, H. T., Chu, S., & Xu, J. (2006). Efficiency, effectiveness and productivity of marketing in services. *European Journal of Operational Research, 170*(1), 265–276.

Knutson, B., Rick, S., Wimmer, G. E., Prelec, D., & Loewenstein, G. (2007). Neural predictors of purchases. *Neuron, 53*(1), 147–156.

Knutson, B., Taylor, J., Kaufman, M., Peterson, R., & Glover, G. (2005). Distributed neural representation of expected value. *Journal of Neuroscience, 25*(19), 4806–4812.

Kruger, J., & Vargas, P. (2008). Consumer confusion of percent differences. *Journal of Consumer Psychology, 18*(1), 49–61.

Krzyzanowski, L. (2007). *The sceience of spending.* https://thetartan.org/2007/1/29/pillbox/shopping. Accessed on March 02, 2020.

Loewenstein, G., Rick, S., & Cohen, J. D. (2008). Neuroeconomics. *Annual Review of Psychology, 59,* 647–672.

Lowe, B. (2010). Consumer perceptions of extra free product promotions and discounts: The moderating role of perceived performance risk. *Journal of Product and Brand Management, 19*(7), 496–503.

McGraw, A. P., & Tetlock, P. E. (2005). Taboo trade-offs, relational framing, and the acceptability of exchanges. *Journal of Consumer Psychology, 15*(1), 2–15.

Mishra, H., Mishra, A., & Nayakankuppam, D. (2006). Money: A bias for the whole. *Journal of Consumer Research, 32*(4), 541–549.

Montaner, T., De Chernatony, L., & Buil, I. (2011). Consumer response to gift promotions. *Journal of Product and Brand Management*.

Munger, J. L., & Grewal, D. (2001). The effects of alternative price promotional methods on consumers' product evaluations and purchase intentions. *Journal of Product and Brand Management, 10*(3), 185–197.

Nobel, C. (2014). *How Our Brain Determines if the Product is Worth the Price.* https://hbswk.hbs.edu/item/how-our-brain-determines-if-the-product-is-worth-the-price. Accessed on January 28, 2020.

Nunes, J. C., & Park, C. W. (2003). Incommensurate resources: Not just more of the same. *Journal of Marketing Research, 40*(1), 26–38.

Pirog, S. F., & Roberts, J. A. (2007). Personality and credit card misuse among college students: The mediating role of impulsiveness. *Journal of Marketing Theory and Practice, 15*(1), 65–77.

Prelec, D., & Loewenstein, G. (1996). The Red and the Black: Mental Accounting of Savings and Debt. Working paper, Massachusetts Institute of Technology, Cambridge, MA, pp. 02138.

Prelec, D., & Loewenstein, G. (1998). The red and the black: Mental accounting of savings and debt. *Marketing Science, 17*(1), 4–28.

Preston, S. D. (2014). *The interdisciplinary science of consumption.* MIT Press.

Raghubir, P., & Srivastava, J. (2008). Monopoly money: The effect of payment coupling and form on spending behavior. *Journal of Experimental Psychology: Applied, 14*(3), 213.

Rick, S., & Loewenstein, G. (2008). The role of emotion in economic behavior. *Handbook of Emotions, 3,* 138–158.

Rick, S. I., Cryder, C. E., & Loewenstein, G. (2008). Tightwads and spendthrifts. *Journal of Consumer Research, 34*(6), 767–782.

Scott, A. (2006). Try Me! Promo, 19(5). Pennsylvania: Chronicle Newspaper Inc.

Shampanier, K., Mazar, N., & Ariely, D. (2007). Zero as a special price: The true value of free products. *Marketing Science, 26*(6), 742–757.

Sinha, I., & Smith, M. F. (2000). Consumers' perceptions of promotional framing price. *Psychology and Marketing, 17,* 257–275.

Slovic, P., Finucane, M., Peters, E., & MacGregor, D. G. (2002). Rational actors or rational fools: Implications of the affect heuristic for behavioral economics. *The Journal of Socio-Economics, 31*(4), 329–342.

Soman, D. (2001). Effects of payment mechanism on spending behavior: The role of rehearsal and immediacy of payments. *Journal of Consumer Research, 27*(4), 460–474.

Szmigin, I., & O'Loughlin, D. (2010). Students and the consumer credit market: Towards a social policy agenda. *Social Policy and Administration, 44*(5), 598–619.

Thomas, M., Desai, K. K., & Seenivasan, S. (2011). How credit card payments increase unhealthy food purchases: Visceral regulation of vices. *Journal of Consumer Research, 38*(1), 126–139.

Tversky, A., & Kahneman, D. (1974). Judgment under uncertainty: Heuristics and biases. *Science, 185*(4157), 1124–1131.

Tversky, A., & Kahneman, D. (1979). Prospect theory: An analysis of decision under risk. *Econometrica, 47*(2), 263–291.

Wager, T. D., Atlas, L. Y., Lindquist, M. A., Roy, M., Woo, C. W., & Kross, E. (2013). An fMRI-based neurologic signature of physical pain. *New England Journal of Medicine, 368*(15), 1388–1397.

World Payments Report. (2019). https://www.europeanpaymentscouncil.eu/sites/default/files/inline-files/World-Payments-Report-2019.pdf. Accessed on January 12, 2020.

Zeki, S., Goodenough, O. R., & Zak, P. J. (2004). Neuroeconomics. *Philosophical Transactions of the Royal Society of London. Series B: Biological Sciences, 359*(1451), 1737–1748.

Zellermayer, O. (1996). The pain of paying. Unpublished dissertation, Department of Social and Decision Sciences, Carnegie Mellon University, Pittsburgh, PA.

# Personality Metatraits, Neurocognitive Networks, and Reasoning Norms for Creative Decision-Making

Paul Hangsan Ahn and Lyn M. Van Swol

**Abstract** Creative decision-making can be viewed as comprising two elements—generative and evaluative thinking (Guilford, 1967; Paulus, Coursey, & Kenworthy, 2019; Sunstein & Hastie, 2015). Generative thinking produces a large number of alternatives. Evaluative thinking eliminates less-promising options (and revises them) for a small number of high-quality solutions. Iterative generative-evaluative thinking is discussed in light of dual personality metatraits, neurocognitive networks, and reasoning modes; these three themes—progressing from creative personal motivation to individual cognition to group norms—provide consistent perspectives for generative-evaluative thinking in terms of plasticity–stability metatraits, default-executive brain networks, and abductive–deductive/inductive modes of reasoning, respectively.

To explain creative decision-making process, this chapter adopts the iterative generative-evaluative framework and identifies corresponding constituents of effective creative decision-making from perspectives on (1) personality metatraits, (2) neurocognitive networks, and (3) reasoning norms. The iterative generative-evaluative framework posits a creative process in which one first generates lots of ideas and then evaluates them.

Creativity, in cognitive terms, involves selectively retrieving ideas from memory and combining them in a novel way (Paulus & Brown, 2007) in order to meet the given creative goal (Nijstad, De Dreu, Rietzschel, & Baas, 2010). The sociocultural definition of creativity is the production of ideas, products, or services that are accepted by a suitably knowledgeable social group as being both useful and new (Amabile, 1988; Brown, 2009; Sawyer, 2011). According to both cognitive and sociocultural definitions, useless or mundane output is not creative; thus, creativity cannot be completely random (Csikszentmihalyi & Sawyer, 2014; Hadamard, 1954). Rather, creativity

---

P. H. Ahn (✉) · L. M. Van Swol
Department of Communication Arts, University of Wisconsin-Madison, 821 University Avenue, Vilas Hall, Madison, WI 53706, USA
e-mail: hahn36@wisc.edu

is motivated by and toward certain personal values (Brown, 2009; Csikszentmihalyi, 1996) and guided primarily by extensive reasoning (Csikszentmihalyi, 1996; Johnson-Laird, 2015) that is socially relevant (Dunne & Martin, 2006), while also supported by relatively random unconscious processing (Hadamard, 1954; Sawyer, 2011). We share the view of many creativity researchers that evaluative (deliberate and top-down) thinking and generative (spontaneous and bottom-up) thinking are both necessary (e.g., Finke, Ward, & Smith, 1992; Guilford, 1957; Paulus, Coursey, & Kenworthy, 2019; Sunstein & Hastie, 2015), with the former playing the crucial role of separating creativity from fantasizing. We first address personality traits, for they are about fundamental motivations to create, i.e., to generate options and criteria to evaluate them by. Then we focus on the neural substrates of creative cognition at the individual level, where the characteristics of generative and evaluative thinking become more explicit and physiological. Finally, we briefly discuss reasoning norms in groups and organizations, an important yet emerging line of creativity research. These three perspectives are far from being exhaustive of the topics studied by researchers regarding creative decision-making. Nonetheless, our choice of these perspectives is intentional, as each—progressing from creative personal motivation to individual cognition to group norms—effectively instantiates generative-evaluative thinking, a recurring theme in this chapter, in ways faithful to these definitions of creativity.

## Terminology

The term creative decision-making has been used to refer either to a multi-stage creative problem-solving process such as iterative divergent-convergent phases (e.g., Nijstad & De Dreu, 2002; Sunstein & Hastie, 2015; West, 2002) or to only the latter stage of this process when options generated during the preceding divergent phase are evaluated and narrowed down, i.e., converged, to the few best ones (e.g., Nijstad & Levine, 2007; Rietzschel, De Dreu, & Nijstad, 2009). Construing creative decision-making as a creative process, we further incorporate the notion that divergent and convergent thinking are cognitive processes that, respectively, diversify options or converge to a few promising candidates, rather than temporal phases within the process. This means both types of thinking are present in both the divergent and convergent "phases," though divergent thinking is emphasized during the divergent phase and convergent thinking is emphasized during the convergent phase. In this chapter, we will use the terms "generative thinking" and "evaluative thinking" to refer to the cognitive processes involved in divergent and convergent thinking, respectively, to differentiate them from the divergent and convergent temporal phases commonly discussed in the literature. Below, we start by briefly introducing generative-evaluative thinking in general.

## *Generative-Evaluative Thinking*

The idea of multi-stage views of creative decision-making traces at least as far back as René Descartes, the seventeenth century creative mathematician, who stated, "Imagination will chiefly be of great use in solving a problem... the results of which need to be coordinated *after* a complete enumeration" (cited in Hadamard, 1954, p. 86). Wallas' (1926) classic four-stage model of insight, i.e., preparation, incubation, illumination, and verification, is perhaps the first modern multi-stage account of creative process (Klein, 2013): Preparation is to consciously and effortfully analyze a problem; incubation is relaxing our mind for some time to let the free working of unconsciousness take over; illumination (as a culmination of a series of unconscious or fringe-conscious associations) is when a sudden but certain insight surfaces; and verification, finally, is testing the validity of the idea (which is then followed by elaboration). Variations of this model range from the twofold divergent-convergent phases (to first create many possible choices and then eliminating and elaborating them; Finke et al., 1992; Guilford, 1957; Paulus et al., 2019; Sunstein & Hastie, 2015) to the five-step IDEAL model (Identification and Definition of the problem, Exploration of possibilities, Application, and Learning from the result; Bransford & Stein, 1984) to the more comprehensive eight-stage model (problem definition, knowledge acquisition, information expansion, incubation, idea generation, combination, selection, and externalization; for an extensive review, see Sawyer, 2011).

There are several reasons why we prefer the simpler generative-evaluative view. First, despite distinctions in detail and emphasis across the multi-step models discussed above, there are a couple of key similarities, i.e., generation of a great number of possibilities and their evaluation to reach the optimal solution. While the ultimate goal in creative thinking is to obtain a small number of highly qualified ideas, generating a large quantity of ideas has been strongly correlated with achieving this goal, as revealed in a meta-analysis (Nijstad et al., 2010). Indeed, decision-making—and creative decision-making as its subset (Newell, Shaw, & Simon, 1962)—presupposes moving "from an initial situation toward a goal situation that satisfies all the conditions for a solution" (Simon, 2001, p. 12, 121). This process involves not only proposing a number of possible paths to reach the goal situation, but also understanding the given problem description and evaluating the relative merits of the alternative paths in keeping with that description (Simon, 2001). Thus, extensive exploration of possible alternatives accompanied by evaluation and elaboration should be key to effective creative decision-making as well, and the generative-evaluative view is the simplest approach, while capturing this essence (Guilford, 1957, 1967; Newell et al. 1962).

Second, rarely is the creative process linear (Sawyer, 2011). For example, incubation in Wallas' (1926) four-stage model has increasingly been understood as unconscious and active formation of new information, e.g., System 1 (heuristic) processing (Abraham, 2018; Kahneman, 2011) or REM sleep (Cai, Mednick, Harrison, Kanady, & Mednick, 2009), rather than necessarily a separate temporal phase with a relaxed mood before the "eureka" moment of a creative insight. Further, stages for problem definition and acquisition of relevant knowledge (e.g., in the IDEAL or the eight-stages models) may either be unnecessary for highly experienced individuals (e.g., see Kelley, 2001; Klein, 2013) or be achieved throughout the creative process, given that problems requiring creativity are often ill-defined ones, for which relevant knowledge is a constantly moving target (Klahr, 2000). Multiple-step approaches thus may be superior in detail but model overfitting is likely to be its liability.

Third, the generative-evaluative framework is generally accepted by researchers (e.g., Guilford, 1957; Nijstad & De Dreu, 2002; Paulus et al., 2019; Sunstein & Hastie, 2015; West, 2002). This framework is useful for research purposes as it exhibits high compatibility with cognitive dual processing models, in that generative thinking is characteristic of association of ordinarily unrelated concepts through (guided) bottom-up, intuitive processing, while evaluative thinking heavily relies on top-down, rule-based processing (Abraham, 2018; Nijstad et al., 2010). These conceptual links can be flexibly applied to the dual pathway to creativity model (Nijstad et al., 2010), dual processing theories (e.g., Kahneman, 2011), or the coupling of the default and control networks (Abraham, 2018) at the cognitive level, or it can be extended to the divergent-convergent phases (e.g., Beaty, Benedek, Silvia, & Schacter, 2016), which we later describe.

Fourth, this twofold model is also widely utilized in professional circles (e.g., Brown, 2009; Osborn, 1953; Kelley, 2001), in which the addition of the temporal distinction is particularly meaningful. Actual generative and evaluative thinking at the individual cognitive level take place together during both the divergent and convergent phases, e.g., by building on or revising more promising ideas during the divergent or convergent phase (Harvey, 2013). An individual can also freely choose to keep generating or evaluating ideas any time during the creative process and be free from process loss incurred by production blocking (when one cannot share ideas while other members in the group are sharing their ideas; Diehl & Stroebe, 1987). When generating or evaluating ideas in groups or teams, the two types of thinking are often temporally distinguished for effective collaboration (Brown, 2009; Kelley, 2001; Paulus et al., 2019). Seeking not just a better but the best answer to, for example, a product design requires generative-evaluative thinking to be iterative at each step of decision-making but also throughout the value chain, which can take months (Brown, 2009; Kelley, 2001). Given the demanding, complex, and often frustrating nature of these real-world problems, guaranteeing the psychological safety of each member is of high importance (Edmondson, 1999). The clear temporal distinction between the divergent and convergent phases can provide a safe environment by reducing evaluation apprehension during the divergent phase (Camacho & Paulus, 1995; Osborn, 1953; Paulus et al., 2019). Using the iterative generative-evaluative view as our analytical lens, we explore how personality traits relate to creative decision-making.

## Plasticity and Stability

Early researchers focused on differences in personality characteristics that might explain variance in creative performance. One of the most well-known models of personality traits is the Big Five, which proposes five major personality traits, openness to experience/intellect, conscientiousness, extraversion, agreeableness, and neuroticism (reversed emotional stability). Each trait is introduced first before discussing the two metatraits—plasticity and stability—that simplify the relationship between creativity and personality traits.

### *Openness to Experience/Intellect*

Openness to experience, alternatively referred to as intellect (Goldberg, 1992), involves the motivation to appreciate unusual ideas and imagination (McCrae & Costa, 1987). Individuals high on openness/intellect are curious, intellectual, exploratory, and independent thinkers who are amenable to unconventional viewpoints (Goldberg, 1992; McCrae & Costa, 1987). McCrae (1987) found that greater self-reported openness/intellect is related to better generative thinking ability. More open/intellectual people are likely to more actively seek out new information and combine previously stored knowledge with it to generate new ideas.

Although openness/intellect has been the trait that has been found to be most consistently and strongly associated with creative achievements and generative thinking ability (Feist, 1998; Fürst, Ghisletta, & Lubart, 2016; Kaufman et al. 2016), it subsumes two distinct subtraits which may affect creativity in different directions. Kaufman et al. (2016) further explored the discriminant validity of openness (engagement with perception, aesthetics, fantasy, and emotions) and intellect (engagement with semantic and abstract information, primarily based on reasoning). Openness encompasses interest in detecting patterns in sensory and artistic experience, which involves implicit learning, while intellect encompasses interest in logical patterns that involve working memory (DeYoung, 2014). Openness was found, as hypothesized, to predict creative achievements in arts, e.g., design, creative writing, whereas intellect predicted achievements in sciences, e.g., scientific discovery and inventions (Kaufman et al., 2016). While openness/intellect motivates individuals to cognitively diverge and generate new ideas, the more relevant domains may differ depending on the subtraits (openness or intellect) in which one is stronger.

### *Conscientiousness*

Conscientiousness is a general appreciation for self-discipline and striving for achievement. Those high in conscientiousness have a strong sense of purpose, are ambitious, work hard toward their goals, and are reliable (Costa & McCrae, 1992; Goldberg, 1992). Given certain creative criteria to focus upon (e.g., usefulness and newness defined in the domain-specific task to evaluate ideas by), people with higher conscientiousness are likely to strive more to achieve these criteria by critically and consistently assessing ideas they have generated. They are likely to be more motivated to persist in the demanding iterative process necessary to innovate (Fürst et al., 2016). Relatedly, concepts similar to conscientiousness such as ambition, precision, and persistence have been found to predict better evaluation and selection of the best alternatives (Fürst et al., 2016). Individuals who are both open and conscientious, by being both generative and evaluative, would be better able to narrow their many alternatives down to few optimal ones. Thus, it is not surprising that in a meta-analysis of scientists and artists, openness/intellect and conscientiousness, out of the five-factor (Big Five) traits, have been found to most strongly correlate with a creative personality (Feist, 1998).

On the other hand, people high in conscientiousness are also obedient to rules and willing to conform to set social norms (Goldberg, 1992). These characteristics of conscientiousness may work against their potential to become creative in social environments where conformity to the status quo and meeting predetermined expectations are emphasized. George and Zhou (2001) showed in a survey of office employees from a petroleum company that higher conscientiousness resulted in low levels of creative behavior if supervisors closely monitored them and coworkers were unsupportive. Their results suggested that high conscientiousness combined with an environment that encourages conformity discourage creative behaviors, and that the relationship between conscientiousness and creativity depends on the social/organizational context (George & Zhou, 2001). High conscientiousness is likely to be a weakness in creativity when generative thinking is not the social norm and when evaluative thinking operates under non-creative criteria.

## *Extraversion*

Extraversion is the general tendency to be positive, enthusiastic, and outgoing in social circumstances, as well as to seek novelty and engage in spontaneous exploratory activity (Depue & Collins, 1999; Fürst et al., 2016). Creative individuals in the popular culture are often portrayed as lone geniuses who are likely to be introverted and able to seclude themselves and concentrate intensely. There is some truth to this portrayal of highly creative individuals because focus and perseverance are instrumental in the demanding iterative generative-evaluative processes required to reach the desired innovative outcome. However, Csikszentmihalyi (1996), who studied exceptionally creative achievers in science and the arts, calls this a myth; the reality is that they behave as if they are both introverts and extroverts, but at different times. Sutton and Hargadon (1996) illustrated that in an innovative design company, each employee usually first exhaustively generates their ideas alone. It is often when they are at a stalemate that they convene a brainstorming team to build on the ideas of others to get "unstuck" and further diverge. The designers are expected to communicate actively with others to generate additional possible alternatives, but such social generative thinking would be difficult for "shrinking violets" (Sutton & Hargadon, 1996, p. 64), or introverts. Furthermore, to generate high-quality ideas, it is necessary in one way or another to constantly expose oneself to specific criteria for creativeness (i.e., usefulness and newness) shared by experts within the same domain to evaluate and direct one's own work (Csikszentmihalyi & Sawyer, 2014). Consistent with these observations, Carne and Kirton (1982), Carson, Peterson and Higgins (2003), and Peterson et al. (2002) found positive correlations between reported levels of generative thinking ability and extraversion. Kaufman et al. (2016) also provided evidence that extraversion predicts creative achievements in the arts (e.g., creative writing, music, visual arts, dance, etc.).

## *Agreeableness*

Agreeableness is the tendency to be good-natured, accepting, and trusting. Those who are highly agreeable at the same time are eager to conform and avoid conflict (McCrae & Costa, 1987), whereas people low in agreeableness tend to exhibit higher levels of arrogance (which is defined by low levels of honesty and humility; Silvia, Kaufman, Reiter-Palmon, & Wigert, 2011). Creativity is related to independent thought and action (Barron & Harrington, 1981), but agreeableness may dampen it, as it leads to conformity (King, Walker, & Broyles, 1996). The more agreeable would want to conform, so their motivation to generate novel connections and effortfully engage in new ideas to evaluate their creative value would be weak. This has been supported by research, as creative individuals have been found to be more autonomous (Perkins, 1993) and less conforming (Guncer & Oral, 1993), and individuals high in agreeableness had fewer creative accomplishments (King et al., 1996).

This understanding of the negative relationship between creativity and agreeableness has recently been updated by Silvia et al. (2011), who split agreeableness into Honesty-Humility (consisting of sincerity, fairness, greed avoidance, and modesty) and Agreeableness (consisting of forgiveness, gentleness, flexibility, and patience) in the HEXACO model (Honesty-Humility, Emotionality, Extraversion, Agreeableness, Conscientiousness, and Openness; Ashton & Lee, 2007). They found a negative relationship between generative thinking and with Honesty-Humility but not with Agreeableness (Silvia et al., 2011). A study by Hunter and Cushenbery (2015) helped interpret this finding by manipulating the supportiveness of the social context: Having a more forceful and independent personality, as is the case with someone who is arrogant (i.e., low in honesty and humility; Silvia et al., 2011), helps combat challenges encountered in the creative process when the environment is not particularly supportive of generating novel ideas.

## *Neuroticism*

Neuroticism, the opposite of which is emotional stability (Mount & Barrick, 1995), represents the orientation to react stressfully to life events (McCrae & Costa, 1987). Neurotic people are more likely to experience negative emotions, e.g., insecurity, anxiety, depression, anger, or shame, while those low in neuroticism (and high in emotional stability) are emotionally stable, hardy, and manage stress well (Reynolds, McClelland, & Furnham, 2014). Though one common stereotype of a highly creative individual is that of a tortured artist, there are arguments and evidence against it (King et al., 1996). Given that the iterative generative and evaluative processes needed to innovate require prolonged commitment, superior coping and heightened well-being should be foundational to creative achievements (Anthony, 1987). Research has found that there is in general a positive correlation between job performance and emotional stability (as well as conscientiousness) (Salgado, 1997). Further, neurotic individuals (who are low on emotional stability) tend to engage less in self-management and are more likely to perceive environmental features as limiting their potential to behave and perform, which in turn negatively correlates with their academic performance (Gerhardt, Rode, & Peterson, 2007). Creativity researchers have further shown that neuroticism is at best uncorrelated with creativity (McCrae & Costa, 1987; Pickering, Smillie, & DeYoung, 2016) or negatively correlated with scores on a creative personality scale (Gough, 1979) and generative thinking ability (Hunter & Cushenbery, 2015).

Each trait relates to creativity in a distinct manner. Openness and extraversion often positively correlate with generative thinking while conscientiousness, agreeableness, and emotional stability (reversed neuroticism) tend to relate to evaluative thinking, although the associations are dependent on the social context. The plasticity-stability approach makes these relationships simpler (Block, 2010) by using two higher-order personality traits, which we introduce next.

## *Plasticity and Stability*

One of the two higher-order factors (Digman, 1997), or metatraits, associated with creativity is plasticity. Plasticity represents the shared variance of openness and extraversion (DeYoung, 2006; Digman, 1997). This construct is rooted in the fact that both openness (curiosity and variety of experience, which are positively related to generative thinking; McCrae, 1987) and extraversion (implying novelty-seeking, which also positively relates to generative thinking; Baas, De Dreu & Nijstad, 2008). This can be characterized by gravitation toward exploratory behavior and exerting effort to seek information and engaging voluntarily with novelty, which is driven by the function of the central dopaminergic (DA) system (DeYoung, 2013; DeYoung, Peterson, & Higgins, 2002; Jauk, 2019). The central DA system, which mediates approach behavior and incentive reward sensitivity, originates in the ventral tegmental area of the midbrain and extends to the limbic system, motor output centers, the anterior cingulate cortex, and the prefrontal cortex (Ashby & Isen, 1999; Panksepp, 1999). Plasticity has been positively associated with reduced latent inhibition (Eysenck, 1995; Peterson, Smith, & Carson, 2002), or the ability to filter stimuli previously experienced as irrelevant out of conscious awareness (Lubow, 1989). This relationship is likely due to higher plasticity (openness and extraversion) allowing a greater inventory of unfiltered stimuli, resulting in a wider range of possible combinations for generating ideas (Peterson et al., 2002). Decreased latent inhibition associated with high plasticity predicted higher lifetime creative achievement and greater originality of responses for high-achieving individuals with IQ over 120 (Carson et al., 2003). Similarly, for a more typical population, decreased latent inhibition was found to be associated with a stronger creative personality (Peterson et al., 2002). On the one hand, being cognitively overwhelmed by the flood of information caused by low latent inhibition is associated with pathologies such as schizophrenia (Serra, Jones, Toone, & Gray, 2001). However, for healthy individuals who can carefully evaluate and select from many options, generating a greater pool of options from greater post-exposure access to the information latent in the stimuli would be helpful to becoming more creative (Peterson et al., 2002).

Stability is the other higher-order personality factor associated with creativity. If plasticity in terms of its motivational emphases can be characterized in terms of divergence, change, novelty, and creativity, stability can be characterized by convergence, order, familiarity, and conformity (Block, 2010). Defined as the shared variance of conscientiousness, agreeableness, and emotional stability (reversed neuroticism) (DeYoung et al., 2002), stability reflects the tendency to be behaviorally and emotionally well-regulated as well as suppressing disruptive impulses and facilitating goal-congruent behavior, which is mediated through the functions of the ascending rostral serotonergic system (DeYoung, 2013; DeYoung et al., 2002). The ascending rostral serotonergic system, which plays an important role in emotional and motivational regulation (Meltzer, 1990), begins primarily in the midbrain and rostral pons and projects upward, to the cerebral cortex, limbic system, and basal ganglia (Tork, 1990). In the popular notion that views creativity as solely generative thinking (Paulus et al., 2019), stability would not be considered essential. However, within the generative-evaluative framework for creativity, plasticity and stability correspond to the former and the latter, respectively. In this vein, DeYoung et al. (2002) showed, using structural equation modeling, that conformity (socially desirable responding) is negatively related to plasticity while positively correlated with stability. Conformity, unless circumstances where creativity is explicitly and effectively encouraged are present (e.g., George & Zhou, 2001), is likely to stifle generative thinking (Nemeth & Staw, 1989), because the status quo as reality will loom greater than possibilities. The implication is similar to our discussion of conscientiousness: Stability will be conducive to creativity if the environment is supportive of generative thinking.

We have seen how generative-evaluative thinking is driven and restricted by motivational forces stemming from personality traits. The duality of plasticity and stability informing generative and evaluative thinking has parallels in how creativity is engaged at the individual cognitive level.

# The Default and Executive Networks

## *Dual Processing Theories*

The dual pathway to creativity model of Nijstad et al. (2010) assumes two non-independent paths to original[1] ideas. One is having the cognitive flexibility to diverge and associate conceptually distant sources of inspiration, as measured by diversity of idea categories. The other is cognitive persistence to effortfully work toward meeting the evaluative criteria, operationalized as the number of ideas within the same idea category. The cognitive flexibility pathway builds on the traditional conception of creativity which proposes that original ideas are generated by combining remote associations (Guilford, 1967; Moraru, Memmert, & Van der Kamp, 2016). In contrast, the persistence path has not usually been associated with creativity (Nijstad et al., 2010; Sawyer, 2011), because it is often assumed that an unsystematic, random process, and defocused attention enables generation of novel connections, which had been considered synonymous with creativity (Paulus et al., 2019; Sawyer, 2011). However, a meta-analysis revealed that flexibility as well as persistence related positively to originality, although the association with flexibility was much stronger than with persistence (Nijstad et al., 2010).

An apparent dilemma is that while flexibility, through unconstrained association, requires more distant semantic associates to be readily considered, persistence—demanding executive control—evaluates and eliminates distractions and irrelevant thoughts from working memory; thus, the two pathways could be inversely related. Nonetheless, results from the meta-analysis (Nijstad et al., 2010) are consistent with the position that we can switch rapidly between flexible and persistent processing on the basis of perceived need (Kahneman, 2011). One can generate novel connections using the flexibility path and then immediately start a more evaluative processing using the persistence path, both during the divergent thinking phase (Harvey, 2013; Paulus et al., 2019); attention can conversely shift from persistence back to a flexible one when it gets hard to generate additional ideas from close semantic categories (Nijstad & Stroebe, 2006).

In sum, using either the flexibility or persistence pathway does not exclude using the other one. Instead, we could benefit from pursuing both paths simultaneously without sacrificing either. We now refer to Abraham (2018) and Beaty et al. (2016) for the parallel emphasis on generative-evaluative thinking based on the functions of the default mode network (DMN) and executive control network (ECN) and their coupling during creative cognition.

**System 1 and System 2 processing**. System 1 (S1) processing draws on similarity-based associations that are intuitive, automatic, and effortless. Contrarily,

---

[1] Idea quality is assessed by combinations of originality and feasibility. However, Nijstad et al. (2010) focused on originality, because empirically originality is a more difficult quality to achieve and, conceptually, originality is considered the hallmark of creative ideas.

System 2 (S2) processing draws upon rule-based logical inference involving deliberation, control, and effort (Kahneman, 2011; Smith & DeCoster, 2000). Creative cognition entails both generative thinking (generation of remote associations) that uses intuitive S1 processing and evaluative thinking focused on given criteria and goals (values to create) to evaluate options with that uses effortful S2 processing (Abraham, 2018). Accordingly, recent studies using functional magnetic resonance imaging (fMRI) demonstrated that brain networks and regions known to be associated with S1 (e.g., the default mode network, medial prefrontal cortex, and posterior cingulate cortex) and S2 (e.g., the executive control network, dorsolateral prefrontal cortex, and anterior cingulate cortex) processing were simultaneously activated during generative (during the alternate uses task and Remote Associates Task; Beaty, Benedek, Kaufman, & Silvia, 2015; Green, Cohen, Raab, Yedibalian, & Gray, 2015; Mayseless, Eran, & Shamay-Tsoory, 2015) and evaluative thinking (during the convergent phase; Ellamil, Dobson, Beeman, & Christoff, 2012; Liu et al. 2015). We will briefly introduce the two networks and revisit these findings.

**DMN and ECN.** The DMN can be divided into three subdivisions: the ventromedial prefrontal cortex (vmPFC), the dorsomedial prefrontal cortex (dmPFC), and the posterior cingulate cortex (PCC), with adjacent precuneus and the lateral parietal cortex (Raichle, 2015). Activity in the default network involves spontaneous, bottom-up, and generative thinking, so its functions are likened to S1 processing (Abraham, 2018). During rest or cognitively undemanding activities, the mind can wander and non-creative or imaginative cognition can engage the DMN. These operations include inferring other people's mental states (i.e., having a theory of mind), episodic future thinking (i.e., mentally simulating a future event), episodic memory (i.e., recollecting specific events in one's past), and self-referential and moral reasoning (i.e., reflecting on one's own thoughts or actions) (Abraham, 2018; Schacter et al. 2012).

The ECN, which largely corresponds to the fronto-cingulo-parietal network, encompasses a broader set of regions in comparison to the DMN: the anterior prefrontal cortex, the dorsolateral prefrontal cortex (dlPFC), the anterior cingulate cortex (ACC), the parietal cortex, and cerebellar structures (Blasi et al. 2006). Regions in the ECN have been associated with deliberate, top-down, and evaluative thinking, similar to S2 processing (Abraham, 2018). Examples of executive network operations include task-set switching (i.e., consciously shifting attention between different tasks), externally directed attention (e.g., to meet objective task demands), relational integration (i.e., coordinating information elements into new relations and structures such as metaphor processing or conceptual expansion), and working memory (Beaty et al., 2016; Seeley et al. 2007).

At rest or during many non-creative tasks, these two brain networks tend to exhibit an antagonistic relationship. For example, the DMN is activated and the ECN deactivated while mind-wandering or resting; and ECN activity typically increases and the DMN deactivates during non-creative behavioral tasks that demand working memory (Anticevic et al., 2012; Beaty et al., 2016), as the ECN presumably suppresses task-irrelevant thoughts associated with the DMN to minimize distraction (Anticevic et al., 2012).

In other conditions, however, the two networks have been found to cooperate. Specifically, during episodic future planning when participants imagine detailed sequences and strategies for their future, DMN activity was coupled with ECN activity (Beaty et al., 2016). Such goal-directed self-generative thinking and generative thinking during the divergent phase of creative process are much alike. During the divergent phase, one generates candidate ideas, and during the following convergent phase these ideas are evaluated. Although we could assume that generative momentum based on similarities (S1) that elicit the DMN would be necessary during the divergent phase while more evaluative processing based on logical rules (S2) recruiting the ECN might seem rather optional, in practice both types of thinking are present during both phases (Harvey, 2013), and there is the need to leverage both during the divergent phase to generate more novel associations (Beaty et al., 2016). Beaty et al. (2016) compiled fMRI studies supporting this positive relationship between the two networks during generative (and evaluative) thinking.

**Coupling during generative thinking.** Researchers have extensively used the alternate uses task (AUT) to measure strength of generative thinking. During the AUT, participants are given a common everyday object (e.g., a brick) and asked to generate as many unconventional uses for it (e.g., grind and make red pigment out of it, instead of using it to build a house) as they can for a limited length of time. Beaty et al. (2015) found that during AUT, participants' PCC (a region in the DMN) was coupled with the dlPFC (a region in the control network). Another AUT study (Mayseless et al., 2015) reported that participants' response quality, as assessed by independent raters, increased with greater functional coupling between the left angular gyrus (part of the DMN) and the ventral ACC (part of the ECN involved in cognitive control in evaluating social valence).

The Remote Associates Test (RAT) is another paradigm for measuring creativity in which the experimenter provides participants with nouns and asks them to generate a correct noun that is relationally remote to varying degrees from the given noun. An example of a close association is giving "blue" as the correct answer when given "print," "berry," and "bird," while "game," when given "piece," "mind," and "dating" would be an example of a more distant association. Green et al. (2015), using a variation of RAT that further stressed generative thinking, also found that the more semantically distant responses were from nouns given as prompts, the greater the coupling of the medial PFC, a hub of the DMN, with the ventral ACC in the ECN.

Further, Pinho et al. (2014) recruited professional pianists and asked them to improvise under either one of the two cognitively contrasting strategies, i.e., playing a specific set of keys (evaluative condition) or expressing a certain emotion content (generative condition). Pianists who improvised following the evaluative condition (playing fixed keys) showed brain activity in the dlPFC (ECN) functionally associated with cognitive motor control areas, i.e., the dorsal pre-motor and presupplementary motor area, instead of the default network. In contrast, the right dlPFC (ECN) activity of those who played to express specific emotions (generative condition) exhibited increased functional connectivity with regions in the DMN such as the medial PFC, the PCC, and the bilateral inferior parietal lobule (IPL). These studies support the dual processing view of generative thinking, as the DMN and ECN dynamically interact with each other (Abraham, 2018; Beaty et al., 2016). Unlike cognition during which the relationship is negative or uncorrelated (e.g., working memory tasks or playing using fixed keys), in generative thinking the two networks appear to be cooperative.

**Coupling during evaluative thinking.** Other fMRI studies that divided the task into the divergent and convergent phases found coupling also during the (evaluative) convergent phase. Professional poets were asked to first generate a novel poem during the divergent phase and then modify it in a subsequent convergent phase (Liu et al., 2015). During the divergent phase, activity in DMN regions, including the medial PFC, increased, while activity in the ECN regions, including dlPFC and intraparietal sulcus, decreased. In the convergent phase, during which the poets evaluated and revised their poetry, the correlation (i.e., coupling) between the activity of the DMN and ECN increased. In another study, visual art students were first asked to generate ideas for a book cover during the divergent phase by sketching on a drawing tablet built into the fMRI scanner, and then to evaluate their sketches in a subsequent convergent phase (Ellamil et al., 2012). The divergent phase was characterized by the preferential recruitment of the medial temporal lobe, a subsystem of the DMN. By contrast, regions in the DMN (e.g., medial PFC and PCC) and the ECN (e.g., dlPFC and ACC) were jointly recruited during the convergent phase.

We can consider a couple of possible explanations for the predominant function of the DMN during the divergent phase separated from the convergent phase. For one, when it is known to participants that they will have time to evaluate and revise many or most of their ideas, idea generation might primarily be supported by the self-generative, intuitive, and spontaneous characteristics of the DMN, as the task demand becomes relatively relaxed and simple (e.g., "Freely generate any new poetic lines"). Another reason self-generation of ideas could depend heavily on the DMN is that the primary function of this network is to retrieve episodic memory (Schacter, Addis, & Buckner, 2007). The reliance on past memory to generate new ideas seems so natural that a study was devoted to separating recalling past memory and creating a new idea only to find that the more episodic-memory-dependent the participant was, the more creative the ideas generated (Benedek et al., 2014). Also, there is sound ecological validity to the position that there is a positive correlation between memory and creativity, since in the real world, for anyone to be recognized as creative the person must first absorb a vast amount of existing knowledge in a specific domain (Csikszentmihalyi, 1996).

In contrast, idea evaluation and revision during the convergent phase that is separate from the divergent phase benefited from both brain networks. During idea evaluation, top-down task demands for high-quality ideas should be accounted for by goal-directed, deliberate, and strategic executive control. The reason idea evaluation also engages the default mode is likely to be the goal of a creativity task—to generate something not only appropriate but also novel, so that separating the two, generation and evaluation, is inherently impossible (Abraham, 2018). In other words, it is "creative evaluation" (Ellamil et al., 2012, p. 2). In the example of a pianist improvising based on specific emotions (Pinho, de Manzano, Fransson, Eriksson, & Ullén, 2014), creative and spontaneous melodic sequences would be generated while at the same time aesthetically undesirable melodies or melodies incongruent with the evaluative goal of expressing specific emotions are consciously avoided. Without the (evaluative) deliberate avoidance of certain melodic sequences during (generative) musical improvisation, the play would probably not be of high quality.

One last question could be the directionality of this relationship. Vartanian et al. (2018) used dynamic causal modeling to examine whether the relationship between the two networks is unidirectional or bidirectional during AUT. They found the inferior frontal gyrus (IFG), a key region in the ECN, exerted unidirectional control over the inferior parietal lobule and middle temporal gyrus. This suggests unidirectionality from the ECN to the DMN is the likely cause of the ECN-DMN coupling in creative cognition.

In brief, the brain network associated with S2 processing, the ECN, coactivates with the DMN, which is associated with S1 processing. A better balance between the two disparate and often inversely related networks leads to better generative and evaluative processing (Abraham, 2018; Beaty et al, 2015). We now turn from individual processing modes to how generative-evaluative thinking is engaged at the team or organizational level.

## Abductive-Deductive/Inductive Reasoning

### Modes of Reasoning

Creative reasoning is underlain by induction, deduction, and abduction (Johnson-Laird, 2015). Induction is based on particular observations (e.g., facts or problems) and looks to acquire general knowledge (e.g., principles or patterns). A classic example of inductive reasoning is "Socrates died," "Plato died," and "Aristotle died," thus "all humans die." Deduction moves from premises (often general knowledge) to a particular case, such as "All humans are mortal," "Socrates is a human," thus "Socrates is a mortal." Abduction moves from a particular problem to a particular solution or explanation. For example, "Socrates wants a beard," thus "Socrates should not shave." Abductive creative reasoning is thinking up something that is likely to serve the goal (referred to as "innovative abduction," which is to be distinguished from explanatory abduction; for explanation, see Dorst, 2011; Kroll & Koskela, 2015; Roozenburg, 1993).

A creative person skillfully switches among these three modes of reasoning (Dunne & Martin, 2006). Taking an example from Roozenburg and Eekels (1995), imagine creating a consumer product such as a dishwasher for sailing boats, which must function properly using seawater when there are rough swells. One could think abductively by going from the particular problem of having to control yaws and pitches due to the waves to thinking of a particular solution, e.g., suspending only the inside of the machine using mechanisms of a stabilized gyroscope. This rational person would then deduce that, because something suspended like the gyroscope stays properly in the position, once the dishwasher has been suspended, it will stay in the proper position. Still confronting the issue of working with seawater, the designer would think inductively that because a kind of stainless steel could take seawater and there was a detergent that dissolved well in seawater, they would also work on board in the dishwasher.

## *Modes of Reasoning and the Phases of Thinking*

Creative reasoning is complex and involves all three modes of reasoning, as shown in the example above. Nevertheless, they are employed to different degrees during different phases of the creative process (Roozenburg, 1993). Each mode of reasoning has distinct characteristics that serve the purpose of being involved in either generative or evaluative thinking. Abduction, which is thinking up how something new would meet a creative goal (Roozenburg, 1993), derives a conclusion that is likely to be the case in the future (Peirce, 1932; Roozenburg, 1993). However, there is no guarantee as to the truth of the conclusion (Martin, 2009). Abductive reasoning would encourage people to share and not self-censor their preliminary guesses because, with abduction, a conclusion only "might be" true (Dunne & Martin, 2006, p. 513), and this is acceptable only as long as the proposed solution is at least in theory capable of satisfying the need (Roozenburg, 1993). When abductive reasoning is the norm and a legitimate thinking tool, people would be encouraged to view their social context as welcoming and not punishing anyone for suggesting something new that only "might be" a good idea (Martin, 2009). Those who allow for this would become more "open and optimistic" (Martin, 2009, p. 154) while attempting to come up with bold ideas. This makes abduction a good fit for the generation of ideas, and it is the predominant mode of reasoning during the divergent phase (e.g., see Dunne & Martin, 2006).

Deduction and induction reach a conclusion that is more certain to work in the future since they depend on knowledge derived from the past. The conclusions are unlikely to be very innovative (Martin, 2009) because, in order to be creative, one should think of not only something useful, but also something that is new (Amabile, 1988). Deduction requires one to reach a conclusion (i.e., an idea) from generalized knowledge distilled from past experiences, and induction looks to reach a conclusion likewise from a variety of past experiences (McBurney, 1936). Because they do not emphasize newness, deduction and induction may not be the optimum reasoning mode for the divergent phase. Furthermore, sound deductive and inductive reasoning accepts only "what should be" or "what is" the case (Dunne & Martin, 2006, p. 513), since the conclusion must be based on actual past events. This implies that the level of certainty of the conclusion is likely expected to be higher for deduction and induction than for abduction. For this reason, deduction and induction may be useful in evaluating ideas that were already generated, but they are inherently limited in their ability to generate—or fully appreciate the potential of—ideas for solving a problem in a previously unknown (i.e., creative) way (Martin, 2009).

## *The Modes of Reasoning and Creativity for Teams and Organizations*

Notwithstanding the limitations of deduction and induction, these two modes of reasoning are in general highly valued to the exclusion of abduction in traditional organizations, including business corporations, non-profit companies, higher education institutions, the government, etc. (Brown, 2009; Kelley, 2001; Dunne & Martin, 2006; Martin, 2009). In a social context, a person has to communicate and persuade other people of their ideas according to the prevalent social norms. Because idea generation is often a collective and communicative process (Csikszentmihalyi & Sawyer, 2014), group or organizational norms that emphasize deduction and induction are likely to work against the generation and communication of new and creative ideas within a group (Martin, 2009). Despite the fact that abduction is the predominant mode of reasoning during generative thinking (Roozenburg, 1993), the importance of abductive reasoning in traditional organizations is underemphasized during the divergent phase (Dunne & Martin, 2006). Often, it is the members' experience that they are given tasks that require generative thinking while they are also required to observe organizational norms such as applying deduction and induction in all stages of creative process, which does not allow them to be generative of ideas (Brown, 2009; Dunne & Martin, 2006; Kelley, 2001; Martin, 2009). This leads right into the innovation dilemma (Martin, 2009). Using the words of March (1991), the inability to allow sufficient room for "exploring" ideas that are abductively derived and risky, but potentially innovative, while being unable to stop "exploiting" old and reliably profitable solutions that are deductively and inductively supported narrows the window of possible innovative solutions available for an organization (Dunne & Martin, 2006). While appreciating the role of abductive reasoning and elevating it as a social norm for the divergent phase would alleviate the dilemma, requiring individuals in groups, teams, and organizations to communicate and justify their ideas in deductive and inductive ways during the divergent phase might "drive out the pursuit of valid answers to new questions" (Martin, 2009, p. 43).

Few empirical studies have been done on the modes of reasoning and creativity, though these few have produced some intriguing results. Dong et al. (2015) found that participants using deductive/inductive reasoning during the convergent phase evaluated ideas more harshly than when using abductive reasoning. This provides evidence that a greater use of abductive reasoning during the convergent phase is associated with a greater tolerance for trying out risky ideas (Dong, Lovallo, & Mounarath, 2015). Cramer-Petersen, Christensen, and Ahmed-Kristensen (2019) found in a content analysis of conversations among creative professionals that during the divergent phase, abduction and deduction were heavily involved while the frequency of deductive reasoning surpassed abduction. They also found that inductive reasoning was minimally used. This result may appear to contradict our account of the place of deduction during the divergent phase. However, there are different types of deductive inferences, e.g., propositional, relational, syllogistic, and multiply quantified (Johnson-Laird, 1995). A propositional deduction uses the word "if," e.g., "If A, then B; A; thus B," and in practice this may be indistinguishable from abduction, at least as observed by coders during the content analysis. Abduction is characterized by hypothesizing (Roozenburg, 1993), the use of imagination (Johnson-Laird, 2009), or guessing (Fann, 2012). These instances involve an "if–then" reasoning structure, which is similar to some examples of deductive reasoning in Cramer-Petersen et al. (2019). This might help explain how their findings are in accordance with Roozenburg (1993), who maintains that abductive reasoning is predominant during the divergent phase.

To our knowledge, no creativity studies have yet directly manipulated the modes of reasoning during the divergent phase. More research could elucidate the role of different modes of reasoning during the divergent or convergent phase.

## Conclusion

The creative process is integral to decision-making at both the individual and group level. Utilizing generative-evaluative thinking as the guiding analytical lens to explicate the creative decision-making process, a selective review of the creativity literature was performed. The duality of plasticity-stability metatraits, flexibility-persistence cognitive pathways, System 1-System 2 processing, default-executive brain networks, abductive-inductive/deductive reasoning norms documented by previous research concretizes how generative-evaluative thinking aimed toward a preset creative goal is engaged from different perspectives and levels. The two essential forces in play are the top-down, deliberate, rule-based, and regulatory one to converge and evaluate alternatives on the one hand and the bottom-up, spontaneous, intuitive, and exploratory one to diverge and generate possibilities on the other. How these interact and shape creative decision-making is differentially motivated by individual personality traits, processed either sequentially or simultaneously at the individual cognitive level, and imposed on individuals as socially relevant reasoning norms.

It is encouraging to be able to draw a similar pattern across different strands of evidence. Thus far, relatively few studies have been conducted on reasoning norms and their relationships with creative decision-making processes. We expect it will be interesting and fruitful to further explore how individual creative momentum is reinforced or deflected depending on the reasoning norm, because reasoning modes such as deduction, induction, or abduction are one of the most fundamental forces shaping creative reasoning (Habermas, 1978), and thus are likely to have a highly generalizable effect on individual, team, and organizational creativity across different domains (Joullie, 2016).

# References

Abraham, A. (2018). The forest versus the trees: Creativity, cognition and imagination. In R. E. Jung & O. Vartanian (Eds.), *The Cambridge handbook of the neuroscience of creativity* (pp. 195–210). Cambridge, UK: Cambridge University Press. https://doi.org/10.1017/9781316556238.012

Amabile, T. M. (1988). A model of creativity and innovation in organizations. *Research in Organizational Behavior, 10*(1), 123–167.

Anthony, E. (1987). Risk, vulnerability, and resilience: An overview. In E. Anthony & B. Cohler (Eds.), *The invulnerable child* (pp. 3–48). New York: Guilford Press.

Anticevic, A., Cole, M. W., Murray, J. D., Corlett, P. R., Wang, X. J., & Krystal, J. H. (2012). The role of default network deactivation in cognition and disease. *Trends in Cognitive Sciences, 16*(12), 584–592. https://doi.org/10.1016/j.tics.2012.10.008.

Ashby, F. G., & Isen, A. M. (1999). A neuropsychological theory of positive affect and its influence on cognition. *Psychological Review, 106*(3), 529. https://doi.org/10.1037/0033-295X.106.3.529.

Ashton, M. C., & Lee, K. (2007). Empirical, theoretical, and practical advantages of the HEXACO model of personality structure. *Personality and Social Psychology Review, 11,* 150–166. https://doi.org/10.1177/1088868306294907.

Baas, M., De Dreu, C. K. W., & Nijstad, B. A. (2008). A meta-analysis of 25 years of mood-creativity research: Hedonic tone, activation, or regulatory focus? *Psychological Bulletin, 134,* 779–806. https://doi.org/10.1037/a0012815.

Barron, F., & Harrington, D. M. (1981). Creativity, intelligence, and personality. *Annual Review of Psychology, 32,* 439–476. https://doi.org/10.1146/annurev.ps.32.020181.002255.

Beaty, R. E., Benedek, M., Kaufman, S. B., & Silvia, P. J. (2015). Default and executive network coupling supports creative idea production. *Scientific Reports, 5*(1), 1–14. https://doi.org/10.1038/srep10964.

Beaty, R. E., Benedek, M., Silvia, P. J., & Schacter, D. L. (2016). Creative cognition and brain network dynamics. *Trends in Cognitive Sciences, 20*(2), 87–95. https://doi.org/10.1016/j.tics.2015.10.004.

Benedek, M., Jauk, E., Fink, A., Koschutnig, K., Reishofer, G., Ebner, F., & Neubauer, A. C. (2014). To create or to recall? Neural mechanisms underlying the generation of creative new ideas. *NeuroImage, 88,* 125–133. https://doi.org/10.1016/j.neuroimage.2013.11.021.

Blasi, G., Goldberg, T. E., Weickert, T., Das, S., Kohn, P., Zoltick, B., Bertolino, A., Callicott, J.H., Weinberger, D.R., & Mattay, V.S. (2006). Brain regions underlying response inhibition and interference monitoring and suppression. *European Journal of Neuroscience, 23*(6), 1658–1664. https://doi.org/10.1111/j.1460-9568.2006.04680.x.

Block, J. (2010). The five-factor framing of personality and beyond: Some ruminations. *Psychological Inquiry, 21,* 2–25. https://doi.org/10.1080/10478401003596626.

Bransford, J. D., & Stein, B. S. (1984). *The IDEAL problem solver: A guide for improving thinking, learning, and creativity.* New York: W.H. Freeman.

Brown, T. (2009). *Change by design.* New York, NY: HarperCollins.

Cai, D. J., Mednick, S. A., Harrison, E. M., Kanady, J. C., & Mednick, S. C. (2009). REM, not incubation, improves creativity by priming associative networks. *Proceedings of the National Academy of Sciences, USA, 106,* 10130. https://doi.org/10.1073/pnas.0900271106.

Camacho, L. M., & Paulus, P. B. (1995). The role of social anxiousness in group brainstorming. *Journal of Personality and Social Psychology, 68*(6), 1071–1080. https://doi.org/10.1037/0022-3514.68.6.1071.

Carne, G. C., & Kirton, M. J. (1982). Styles of creativity: Test-score correlations between Kirton Adaption-Innovation Inventory and Myers-Briggs Type Indicator. *Psychological Reports, 50,* 31–36. https://doi.org/10.2466/pr0.1982.50.1.31.

Carson, S., Peterson, J. B., & Higgins, D. (2003). Decreased latent inhibition is associated with increased creative achievement in high-functioning individuals. *Journal of Personality and Social Psychology, 85,* 499–506. https://doi.org/10.1037/0022-3514.85.3.499.

Costa, P. T., & McCrae, R. R. (1992). Reply to Eysenck. *Personality and Individual Differences, 13*(8), 861–865. https://doi.org/10.1016/0191-8869(92)90002-7.

Cramer-Petersen, C. L., Christensen, B. T., & Ahmed-Kristensen, S. (2019). Empirically analyzing design reasoning patterns: Abductive-deductive reasoning patterns dominate design idea generation. *Design Studies, 60,* 39–70. https://doi.org/10.1016/j.destud.2018.10.001.

Csikszentmihalyi, M. (1996). *Creativity: Flow and the psychology of discovery and invention.* New York. NY: HarperCollins.

Csikszentmihalyi, M., & Sawyer, K. (2014). Shifting the Focus from Individual to Organizational Creativity. In M. Csikszentmihalyi (Ed.), *The Systems Model of Creativity* (pp. 67–71). Dordrecht, Netherlands: Springer. https://doi.org/10.1007/978-94-017-9085-7_6

Depue, R. A., & Collins, P. F. (1999). Neurobiology of the structure of personality: Dopamine, facilitation of incentive motivation, and extraversion. *Behavioral and Brain Sciences, 22,* 491–569. https://doi.org/10.1017/s0140525x99002046.

DeYoung, C. G. (2006). Higher-order factors of the Big Five in a multi-informant sample. *Journal of Personality and Social Psychology, 91,* 1138–1151. https://doi.org/10.1037/0022-3514.91.6.1138.

DeYoung, C. G. (2013). The neuromodulator of exploration: A unifying theory of the role of dopamine in personality. *Frontiers in Human Neuroscience, 7,* 1–26. https://doi.org/10.3389/fnhum.2013.00762.

DeYoung, C. G. (2014). Openness/intellect: A dimension of personality reflecting cognitive exploration. In M. L. Cooper & R. J. Larsen (Eds.), *APA handbook of personality and social psychology: Personality processes and individual differences* (Vol. 4, pp. 369–399). Washington, DC: American Psychological Association. https://doi.org/10.1037/14343-017

DeYoung, C. G., Peterson, J. B., & Higgins, D. M. (2002). Higher-order factors of the Big Five predict conformity: Are there neuroses of health? *Personality and Individual Differences, 33,* 533–552. https://doi.org/10.1016/s0191-8869(01)00171-4.

Diehl, M., & Stroebe, W. (1987). Productivity loss in brainstorming groups: Toward the solution of a riddle. *Journal of Personality and Social Psychology, 53*(3), 497–509. https://doi.org/10.1037/0022-3514.53.3.497.

Digman, J. M. (1997). Higher-order factors of the Big Five. *Journal of Personality and Social Psychology, 73,* 1246–1256. https://doi.org/10.1037/0022-3514.73.6.1246.

Dong, A., Lovallo, D., & Mounarath, R. (2015). The effect of abductive reasoning on concept selection decisions. *Design Studies, 37,* 37–58. https://doi.org/10.1016/j.destud.2014.12.004.

Dorst, K. (2011). The core of 'design thinking' and its application. *Design Studies, 32*(6), 521–532. https://doi.org/10.1016/j.destud.2011.07.006.

Dunne, D., & Martin, R. (2006). Design thinking and how it will change management education: An interview and discussion. *Academy of Management Learning & Education, 5*(4), 512–523. https://doi.org/10.5465/amle.2006.23473212.

Edmondson, A. (1999). Psychological safety and learning behavior in work teams. *Administrative Science Quarterly, 44*(2), 350–383. https://doi.org/10.2307/2666999.

Ellamil, M., Dobson, C., Beeman, M., & Christoff, K. (2012). Evaluative and generative modes of thought during the creative process. *Neuroimage, 59*(2), 1783–1794. https://doi.org/10.1016/j.neuroimage.2011.08.008.

Eysenck, H. J. (1995). *Genius: The natural history of creativity.* Cambridge, UK: Cambridge University Press.

Fann, K. T. (2012). *Peirce's theory of abduction.* New York, NY: Springer.

Feist, G. J. (1998). A meta-analysis of personality in scientific and artistic creativity. *Personality and Social Psychology Review, 2,* 290–309. https://doi.org/10.1207/s15327957pspr0204_5.

Finke, R. A., Ward, T. B., & Smith, S. M. (1992). *Creative cognition: Theory, research and applications.* Cambridge, MA: MIT Press.

Fürst, G., Ghisletta, P., & Lubart, T. (2016). Toward an integrative model of creativity and personality: Theoretical suggestions and preliminary empirical testing. *The Journal of Creative Behavior, 50*(2), 87–108. https://doi.org/10.1002/jocb.71.

George, J. M., & Zhou, J. (2001). When openness to experience and conscientiousness are related to creative behavior: An interactional approach. *Journal of Applied Psychology, 86,* 513–524. https://doi.org/10.1037/0021-9010.86.3.513.

Gerhardt, M. W., Rode, J. C., & Peterson, S. J. (2007). Exploring mechanisms in the personality-performance relationship: Mediating roles of self-management and situational constraints. *Personality and Individual Differences, 43,* 1344–1355. https://doi.org/10.1016/j.paid.2007.04.001.

Goldberg, L. R. (1992). The development of markers for the Big-Five factor structure. *Psychological Assessment, 4,* 26–42. https://dx.doi.org/ https://doi.org/10.1037/1040-3590.4.1.26

Gough, H. G. (1979). A creative personality scale for the Adjective Check List. *Journal of Personality and Social Psychology, 37,* 1398–1405. https://doi.org/10.1037/0022-3514.37.8.1398.

Green, A. E., Cohen, M. S., Raab, H. A., Yedibalian, C. G., & Gray, J. R. (2015). Frontopolar activity and connectivity support dynamic conscious augmentation of creative state. *Human Brain Mapping, 36*(3), 923–934. https://doi.org/10.1002/hbm.22676.

Guilford, J. P. (1957). Creative abilities in the arts. *Psychological Review, 64*(2), 110–118. https://doi.org/10.1037/h0048280.

Guilford, J. P. (1967). *The nature of human intelligence.* New York, NY: McGraw-Hill.

Guncer, B., & Oral, G. (1993). Relationship between creativity and nonconformity to school discipline as perceived by teachers of Turkish elementary school children, by controlling for their grade and sex. *Journal of Instructional Psychology, 20,* 208–214.

Habermas, J. (1978). *Knowledge and human interests* (2nd ed.) (J. Shapiro, Trans.). London: Heinemann. https://doi.org/10.2307/2149243.

Hadamard, J. (1954). *An essay on the psychology of invention in the mathematical field.* Courier Corporation.

Harvey, S. (2013). A different perspective: The multiple effects of deep level diversity on group creativity. *Journal of Experimental Social Psychology, 49*(5), 822–832. https://doi.org/10.1016/j.jesp.2013.04.004.

Hunter, S. T., & Cushenbery, L. (2015). Is being a jerk necessary for originality? Examining the role of disagreeableness in the sharing and utilization of original ideas. *Journal of Business and Psychology, 30,* 621–639. https://doi.org/10.1007/s10869-014-9386-1.

Jauk, E. (2019). A bio-psycho-behavioral model of creativity. *Current Opinion in Behavioral Sciences, 27,* 1–6. https://doi.org/10.1016/j.cobeha.2018.08.012.

Johnson-Laird, P. N. (1995). Mental models, deductive reasoning, and the brain. In M. S. Gazzaniga (Ed.), *The cognitive neurosciences* (pp. 999–1008). Cambridge: MIT Press.

Johnson-Laird, P. N. (2009). *How we reason.* Oxford, UK: Oxford University Press.

Johnson-Laird, P. N. (2015). Problem solving and reasoning, Psychology of. In J. D. Wright (Ed.), *International encyclopedia of the social & behavioral sciences* (2$^{nd}$ ed., Vol. 19, pp. 61–67). Amsterdam: Elsevier. https://doi.org/10.1016/b978-0-08-097086-8.43076-x.

Joullié, J. E. (2016). The philosophical foundations of management thought. *Academy of Management Learning & Education, 15*(1), 157–179. https://doi.org/10.5465/amle.2012.0393.

Kahneman, D. (2011). *Thinking, fast and slow.* New York, NY: Farrar, Straus and Giroux.

Kaufman, S. B., Quilty, L. C., Grazioplene, R. G., Hirsh, J. B., Gray, J. R., Peterson, J. B., & DeYoung, C. G. (2016). Openness to experience and intellect differentially predict creative achievement in the arts and sciences. *Journal of Personality, 84*(2), 248–258. https://doi.org/10.1111/jopy.12156.

Kelley, T. (2001). *The art of innovation: lessons in creativity from IDEO, America's leading design firm.* New York: Doubleday.

King, L. A., Walker, L. M., & Broyles, S. J. (1996). Creativity and the five-factor model. *Journal of Research in Personality, 30,* 189–203. https://doi.org/10.1006/jrpe.1996.0013.

Klahr, D. (2000). *Exploring science: The cognition and development of discovery processes.* Cambridge, MA: MIT Press.

Klein, G. (2013). *Seeing what others don't: The remarkable ways we gain insights*. New York, NY: Public Affairs.
Kroll, E., & Koskela, L. (2015). On abduction in design. In J. S. Gero (Ed.), *Design computing and cognition '14* (pp. 327–344). London, UK: Springer. https://doi.org/10.1007/978-3-319-14956-1_19.
Liu, S., Erkkinen, M. G., Healey, M. L., Xu, Y., Swett, K. E., Chow, H. M., & Braun, A. R. (2015). Brain activity and connectivity during poetry composition: Toward a multidimensional model of the creative process. *Human Brain Mapping, 36*(9), 3351–3372. https://doi.org/10.1002/hbm.22849.
Lubow, R. E. (1989). *Latent inhibition and conditioned attention theory*. Cambridge, UK: Cambridge University Press.
March, J. G. (1991). Exploration and exploitation in organizational learning. *Organization Science, 2*(1), 71–87. https://doi.org/10.1287/orsc.2.1.71.
Martin, R. (2009). *The design of business: Why design thinking is the next competitive advantage*. Boston, MA: Harvard Business Press.
Mayseless, N., Eran, A., & Shamay-Tsoory, S. G. (2015). Generating original ideas: The neural underpinning of originality. *Neuroimage, 116*, 232–239. https://doi.org/10.1016/j.neuroimage.2015.05.030.
McBurney, J. H. (1936). The place of the enthymeme in rhetorical theory. *Communication Monographs, 3*(1), 49–74. https://doi.org/10.1080/03637753609374841.
McCrae, R. R. (1987). Creativity, divergent thinking, and openness to experience. *Journal of Personality and Social Psychology, 52*(6), 1258–1265. https://doi.org/10.1037/0022-3514.52.6.1258.
McCrae, R. R., & Costa, P. T. (1987). Validation of the five-factor model of personality across instruments and observers. *Journal of Personality and Social Psychology, 52*(1), 81–90. https://doi.org/10.1037/0022-3514.52.1.81.
Meltzer, H. Y. (1990). Role of serotonin in depression. *Annals of the New York Academy of Sciences, 600*, 400–486. https://doi.org/10.1111/j.1749-6632.1990.tb16904.x.
Moraru, A., Memmert, D., & Van der Kamp, J. (2016). Motor creativity: The roles of attention breadth and working memory in a divergent doing task. *Journal of Cognitive Psychology, 28*, 856–867. https://doi.org/10.1080/20445911.2016.1201084.
Mount, M. K., & Barrick, M. R. (1995). The Big Five personality dimensions: Implications for research and practice in human resources management. In K. M. Rowland & G. Ferris (Vol. Eds.), *Research in personnel and human resources management. Research in personnel and human resources management* (Vol. 13, pp. 153–200). Greenwich, CT: JAI. https://doi.org/10.1080/20445911.2016.1201084.
Nemeth, C. J., & Staw, B. M. (1989). The tradeoffs of social control and innovation in small groups and organizations. In L. Berkowitz (Ed.). *Advances in experimental social psychology* (Vol. 22, pp. 75–210). New York: Academic Press. https://doi.org/10.1016/S0065-2601(08)60308-1.
Newell, A., Shaw, J., & Simon, H. (1962). The processes of creative thinking. In H. Gruber, G. Terrell & M. Wertheimer (Eds.), *Contemporary approaches to creative thinking* (pp. 63–119). New York: Atherton Press. https://doi.org/10.1037/13117-003
Nijstad, B. A., & De Dreu, C. K. W. (2002). Creativity and group innovation. *Applied Psychology: An International Review, 51*, 401–407. https://doi.org/10.1111/1464-0597.00984.
Nijstad, B. A., De Dreu, C. K. W., Rietzschel, E. F., & Baas, M. (2010). The dual pathway to creativity model: Creative ideation as a function of flexibility and persistence. *European Review of Social Psychology, 21*(1), 34–77. https://doi.org/10.1080/10463281003765323.
Nijstad, B. A., & Levine, J. M (2007). Group creativity and the stages of creative group problem solving. In M. Hewstone, H. A. W. Schut, J. B. F. de Wit, K. van den Bos, & M. S. Stroebe (Eds.), *The scope of social psychology: Theory and applications* (pp. 159–172). New York: Psychology Press. https://doi.org/10.4324/9780203965245

Nijstad, B. A., & Stroebe, W. (2006). How the group affects the mind: A cognitive model of idea generation in groups. *Personality and Social Psychology Review, 10,* 186–213. https://doi.org/10.1207/s15327957pspr1003_1.

Osborn, A. F. (1953). *Applied imagination: Principles and procedures of creative thinking* (1st ed.). New York, NY: Scribner.

Panksepp, J. (1999). *Affective neuroscience.* New York: Oxford University Press.

Paulus, P. B., & Brown, V. R. (2007). Toward more creative and innovative group idea generation: A cognitive-social-motivational perspective of brainstorming. *Social and Personality Psychology Compass, 1*(1), 248–265. https://doi.org/10.1111/j.1751-9004.2007.00006.x.

Paulus, P. B., Coursey, L. E., & Kenworthy, J. B. (2019). Divergent and convergent collaborative creativity. In I. Lebuda & V. P. Glăveanu (Eds.), *The Palgrave Handbook of Social Creativity Research* (pp. 245–262). Cham, Switzerland: Palgrave Macmillan.

Peirce, C. S. (1931–1958). *Collected papers* (Vols. 1–8, C. Hartshorne, P. Weiss, & A. Burks, Eds.). Cambridge, MA: Harvard University Press.

Perkins, R. M. (1993). Personality variables and implications for critical thinking. *College Student Journal, 27*(1), 106–111.

Peterson, J. B., Smith, K. W., & Carson, S. (2002). Openness and extraversion are associated with reduced latent inhibition: Replication and commentary. *Personality and Individual Differences, 33,* 1137–1147. https://doi.org/10.1016/S0191-8869(02)00004-1.

Pickering, A. D., Smillie, L. D., & DeYoung, C. G. (2016). Neurotic individuals are not creative thinkers. *Trends in Cognitive Science, 20*(1), 1–2. https://doi.org/10.1016/j.tics.2015.10.001.

Pinho, A. L., de Manzano, Ö., Fransson, P., Eriksson, H., & Ullén, F. (2014). Connecting to create: Expertise in musical improvisation is associated with increased functional connectivity between premotor and prefrontal areas. *Journal of Neuroscience, 34*(18), 6156–6163. https://doi.org/10.1523/JNEUROSCI.4769-13.2014.

Raichle, M. E. (2015). The brain's default mode network. *Annual Review of Neuroscience, 38,* 433–447. https://doi.org/10.1146/annurev-neuro-071013-014030.

Reynolds, J., McClelland, A., & Furnham, A. (2014). An investigation of cognitive test performance across conditions of silence, background noise and music as a function of neuroticism. *Anxiety, Stress, and Coping, 27*(4), 410–421. https://doi.org/10.1080/10615806.2013.864388.

Rietzschel, E. F., De Dreu, C. K., & Nijstad, B. A. (2009). What are we talking about, when we talk about creativity? Group creativity as a multifaceted, multistage phenomenon. *Research on Managing Groups and Teams, 12,* 1–28. https://doi.org/10.1108/S1534-0856(2009)0000012004.

Roozenburg, N. F. (1993). On the pattern of reasoning in innovative design. *Design Studies, 14*(1), 4–18. https://doi.org/10.1016/S0142-694X(05)80002-X.

Roozenburg, N. F., & Eekels, J. (1995). *Product design: Fundamentals and methods* (Vol. 2). Chichester, UK: Wiley.

Salgado, J. F. (1997). The five-factor model of personality and job performance in the European community. *Journal of Applied Psychology, 82,* 30–43. https://doi.org/10.1037/0021-9010.82.1.30.

Sawyer, R. K. (2011). *Explaining creativity: The science of human innovation.* Oxford, UK: Oxford University Press.

Schacter, D. L., Addis, D. R., & Buckner, R. L. (2007). Remembering the past to imagine the future: The prospective brain. *Nature Reviews Neuroscience, 8*(9), 657–661. https://doi.org/10.1038/nrn2213.

Schacter, D. L., Addis, D. R., Hassabis, D., Martin, V. C., Spreng, R. N., & Szpunar, K. K. (2012). The future of memory: Remembering, imagining, and the brain. *Neuron, 76*(4), 677–694. https://doi.org/10.1016/j.neuron.2012.11.001.

Seeley, W. W., Menon, V., Schatzberg, A. F., Keller, J., Glover, G. H., Kenna, H., Reiss, A. L., & Greicius, M. D. (2007). Dissociable intrinsic connectivity networks for salience processing and executive control. *Journal of Neuroscience, 27*(9), 2349–2356. https://doi.org/10.1523/JNEUROSCI.5587-06.2007.

Serra, A. M., Jones, S. H., Toone, B., & Gray, J. A. (2001). Impaired associative learning in chronic schizophrenics and their first-degree relatives: A study of latent inhibition and the Kamin blocking effect. *Schizophrenia Research, 48*(2), 273–289. https://doi.org/10.1016/S0920-9964(00)00141-9.

Silvia, P. J., Kaufman, J. C., Reiter-Palmon, R., & Wigert, B. (2011). Cantankerous creativity: Honesty-humility, agreeableness, and the HEXACO structure of creative achievement. *Personality and Individual Differences, 51,* 687–689. https://doi.org/10.1016/j.paid.2011.06.011.

Simon, H. A. (2001). Problem Solving and Reasoning, Psychology of. In N. J. Smelser & P. B. Baltes (Eds.), *International Encyclopedia of the Social and Behavioral Sciences* (pp. 12120–12123). Oxford, UK: Elsevier Science. https://doi.org/10.1016/b0-08-043076-7/00543-x.

Smith, E. R., & DeCoster, J. (2000). Dual-process models in social and cognitive psychology: Conceptual integration and links to underlying memory systems. *Personality and Social Psychology Review, 4*(2), 108–131. https://doi.org/10.1207/S15327957PSPR0402_01.

Sunstein, C. R., & Hastie, R. (2015). *Wiser: Getting beyond groupthink to make groups smarter.* Harvard Business Press.

Sutton, R. I., & Hargadon, A. (1996). Brainstorming groups in context: Effectiveness in a product design firm. *Administrative Science Quarterly, 41*(4), 685–718. https://doi.org/10.2307/2393872.

Tork, I. (1990). Anatomy of the serotonergic system. *Annals of the New York Academy of Science, 600,* 9–35. https://doi.org/10.1111/j.1749-6632.1990.tb16870.x.

Vartanian, O., Beatty, E. L., Smith, I., Blackler, K., Lam, Q., & Forbes, S. (2018). One-way traffic: The inferior frontal gyrus controls brain activation in the middle temporal gyrus and inferior parietal lobule during divergent thinking. *Neuropsychologia, 118,* 68–78. https://doi.org/10.1016/j.neuropsychologia.2018.02.024.

Wallas, G. (1926). *The art of thought.* New York: Harcourt-Brace.

West, M. A. (2002). Sparkling fountains or stagnant ponds: An integrative model of creativity and innovation implementation in work groups. *Applied Psychology, 51*(3), 355–387. https://doi.org/10.1111/1464-0597.00951.

# Recommender Systems Beyond E-Commerce: Presence and Future

**Alexander Felfernig, Thi Ngoc Trang Tran, and Viet-Man Le**

**Abstract** Recommender systems are supporting users in the identification of items that fulfill their wishes and needs and are also helping to foster consumer happiness. These systems have been successfully applied in different application domains—examples thereof are the recommendation of movies, books, digital cameras, points of interest, financial services, and software requirements. The major objectives of this chapter are to provide an overview of recommendation approaches including criteria when to use which algorithm, to show different applications of recommendation algorithms going beyond standard e-commerce scenarios and to discuss issues for future research.

## Introduction

Recommender systems help to find items that fulfill the specific needs of their users. They have been established as an own research discipline in the 1990s (Goldberg, Nichols, Oki, & Terry, 1992). Recommender systems guide users in a personalized way to interesting or useful objects in a large space of possible objects or produce such objects as output (Burke, 2000; Felfernig, Boratto, Stettinger, & Tkalcic, 2018a; Konstan & Riedl, 2012). Examples of successful deployments of recommendation technologies in industrial contexts are *amazon.com* (Linden, Smith, & York, 2003)and *netflix.com* (Tuzhilin & Koren, 2008).

Recommender systems are regarded as a basic means to support decision processes related to *simple items* such as movies (*which movie to watch?*) (Koren, Bell, &

---

A. Felfernig (✉) · T. N. T. Tran · V.-M. Le
Institute of Software Technology, Applied Artificial Intelligence, TU Graz, Inffeldgasse 16b/2, A-8010 Graz, Austria
e-mail: alexander.felfernig@ist.tugraz.at; afelfern@ist.tugraz.at

T. N. T. Tran
e-mail: ttrang@ist.tugraz.at

V.-M. Le
e-mail: vietman.le@ist.tugraz.at

Volinsky, 2009) and music (*which song to play?*) (Janssen, Broek, & Westerink, 2011) as well as to *complex items* such as financial services (*which pension product to choose?*) (Felfernig, Friedrich, Jannach, & Zanker, 2006) and software requirements (*which software requirements to implement next?*) (Felfernig et al., 2013). Since the 1990s, these systems have evolved toward core technologies of many online sales platforms where a significant share of items is sold due to the provision of recommendation services (Linden, Smith, & York, 2003).

There exist different recommendation approaches which can be classified as follows. (1) *Content-based filtering* (Pazzani & Billsus, 1997) is an information filtering approach that exploits user-liked items (described in terms of keywords or categories) for recommending new similar items to the same user. Content-based recommendation is applied, for example, by *amazon.com* for the recommendation of similar items (e.g., *since a user A purchased a music album X of interpreter Y, a new album of Y is recommended to A*).

(2) *Collaborative filtering* is based on the concept of *word-of-mouth promotion* where items purchased by users with a similar purchase behavior (the nearest neighbors) are recommended to the current user (Konstan et al., 1997; Koren, Bell, & Volinsky, 2009). For example, if user *Donald* has consumed movies similar to the ones that have been consumed by *Dagobert* then *netflix.com* or a similar platform would recommend movies to *Donald* which have been consumed by *Dagobert* but not by *Donald*. In contrast to content-based filtering, collaborative filtering does not need detailed information about item properties.

(3) When dealing with complex items such as financial services and software-related artifacts (e.g., release plans), *knowledge-based recommendation techniques* (Burke, 2000; Felfernig et al., 2006; Felfernig & Burke, 2008) are applied. These techniques allow to take into account semantic item properties and constraints/rules (e.g., *a specific financial service is not available for customers with an age greater than X*, or *a software requirement X must be implemented before software requirement Y*). In contrast, approaches such as collaborative filtering and content-based filtering do not allow the inclusion of constraints which makes them less applicable especially in complex item domains. A disadvantage of knowledge-based recommenders is the existence of knowledge acquisition bottlenecks, i.e., additional efforts related to the communication between knowledge engineers in charge of defining the recommendation knowledge and domain experts in charge of explaining the product/item domain to knowledge engineers (Felfernig, Friedrich, Jannach, & Zanker, 2006).

(4) In contrast to the previously discussed approaches, *group recommender systems* (Felfernig et al., 2018a) are focusing on the determination of recommendations for groups of users. An example thereof is the recommendation of *software release plans* or a *requirements prioritization* where a group of stakeholders is in charge of deciding about the sequence in which a set of requirements should be implemented (Felfernig et al., 2018d). Scenarios where recommendations are determined for groups rather than for individuals are many-fold and range from the recommendation of a *restaurant for dinner* (Tran, Atas, Felfernig, & Stettinger, 2018) to the recommendation of a *skiing resort* for a group of tourists (McCarthy et al., 2006). Although recommendation focuses on user groups in this context, the underlying

algorithms used in group recommendation are in most of the cases the ones that are used for single user recommendation. An often used approach in group recommender systems is to combine recommendations or ratings determined for single users by using aggregation functions such as *least misery* or *average voting* (Felfernig et al., 2018a).

The overview presented in this chapter is based on an in-depth analysis of research results published in workshops, conferences, and journals related to recommender systems research (see Table 1). In this context, we provide an overview of different algorithmic approaches (Felfernig et al., 2018d; Jannach, Zanker, Felfernig, & Friedrich, 2010; Schafer, Konstan, Riedl, 2011) but focus more on new/upcoming application scenarios that represent new ways of exploiting the potentials of recommendation technologies (Ducheneaut, Patridge, Huang, Price, & Roberts, 2009; Konstan & Riedl, 2012). An overview of these application scenarios is given in Table 2. The work presented in this chapter is an extension (new references, applications, and examples) of the recommender systems overview presented in Felfernig, Jeran, Ninaus, Reinfrank, and Reiterer (2013).

The major contributions of this chapter are the following. *First*, we provide an overview of basic recommendation approaches and discuss their advantages and limitations. *Second*, we introduce various application scenarios that go beyond e-commerce (the mainstream application area of recommender systems). *Third*, in order to stimulate further research in the field, we discuss a couple of open research issues.

The remainder of this chapter is organized as follows. Sections 14.2–14.7 provide an overview of different application domains of recommendation technologies—related discussions are accompanied by different working examples. In Sect. 14.8,

**Table 1** Overview of journals and conferences (including workshops) used as the basis for our literature analysis (papers on recommender system applications 2005–2020)

| Journals and conferences |
| --- |
| International Journal of Electronic Commerce (IJEC) |
| Journal of User Modeling and User-Adapted Interaction (UMUAI) |
| AI Magazine |
| IEEE Intelligent Systems |
| Communications of the ACM |
| Expert Systems with Applications |
| ACM Recommender Systems (ACM RecSys) |
| ACM User Modeling, Adaptation, and Personalization (ACM UMAP) |
| ACM Symposium on Applied Computing (ACM SAC) |
| ACM International Conference on Intelligent User Interfaces (ACM IUI) |
| International Conference on Artificial Intelligence (IJCAI) |
| AAAI Conference on Artificial Intelligence (AAAI) |

**Table 2** Overview of identified recommender applications which are not in the e-commerce mainstream (*KBR* Knowledge-based Recommendation, *CF* Collaborative Filtering, *CBR* Content-based Recommendation, *ML* Machine Learning, *GR* Group Recommendation, *PR* Probability-based Recommendation, *DM* Data Mining)

| Domain | Recommended Items | | Recommendation Approach |
|---|---|---|---|
| Open Innovation | Innovation teams | KBR | Brocco and Groh (2009) |
| | Business plans | KBR | Jannach and Joergensen (2007) |
| | Ideas | DM | Thorleuchter, Van Den Poel, and Prinzie (2010) |
| | New software requirements | ML | Stanik and Maalej (2019) |
| | New service features | GR | Felfernig, Gruber, Brandner, Blazek, and Stettinger (2018b) |
| | Citizen requirements | ML | Felfernig, Stettinger, Wundara, and Stanik (2019) |
| | New software functions | CF | Li, Matejka, Grossmann, and Fitzmaurice (2015) |
| Software Engineering | Effort estimation methods | KBR | Peischl, Zanker, Nica, and Schmid (2010) |
| | Problem solving approaches | KBR | Burke and Ramezani (2010) |
| | API call completions | CF | McCarey, Cinneide, and Kushmerick (2005) |
| | Example code fragments | CBR | Holmes, Walker, and Murphy (2006) |
| | Contextualized artifacts | KBR | Kersten and Murphy (2010) |
| | Software defects | ML | Misirli, Bener, and Kale (2011) |
| | Requirements prioritizations | GR | Felfernig, Jeran, Ninaus, Reinfrank, and Reiterer (2011) |
| | Software requirements | CF | Mobasher and Cleland-Huang (2011) |
| | Product & service requirements | ML | Fucci et al. (2018) |
| | Requirements dependencies | ML | Samer et al. (2019) |
| Knowledge Engineering | Related constraints | KBR | Felfernig, Jeran, Ninaus, Reinfrank, and Reiterer (2012) |
| | Explanations | KBR | Felfernig, Schubert, and Zehentner (2011) |
| | Constraint rewritings | KBR | Felfernig, Mandl, Pum, and Schubert (2010) |
| | Database queries | CF | Chatzopoulou, Eirinaki, and Poyzotis (2009) |
| | | CBR | Fakhraee and Fotouhi (2011) |
| Knowledge-based Configuration | Relevant features | CF | Felfernig and Burke (2008) |
| | Requirements repairs | CBR | Felfernig et al. (2009) |
| | Group-based configurations | GR | Felfernig, Atas, Tran, and Stettinger (2016) |
| | Group diagnoses | GR | Polat Erdeniz, Felfernig, and Atas (2019a) |
| | Search heuristics | CF | Polat-Erdeniz, Felfernig, Atas, Samer (2019b) |

(continued)

**Table 2** (continued)

| Domain | Recommended Items | | Recommendation Approach |
|---|---|---|---|
| Persuasive Technologies | Game task complexities | CF | Berkovsky, Freyne, Coombe, and Bhandari (2010) |
| | Development practices | KBR | Pribik and Felfernig (2012) |
| | Exam configurations | GR | Atas, Felfernig, Stettinger, and Tran (2017) |
| | Explanations | GR | Tran, Atas, Le, Samer, and Stettinger (2019) |
| Smart Homes | Equipment configurations | KBR | Leitner, Fercher, Felfernig, and Hitz (2012) |
| | Smart home control actions | CF | LeMay, Haas, and Gunter (2009) |
| | IoT workflows | CF | Felfernig et al. (2018c) |
| People | Criminals | PR | Tayebi, Jamali, Ester, Glaesser, and Frank (2011) |
| | Reviewers | CF | Kapoor et al. (2007) |
| | Physicians | CF | Hoens, Blanton, and Chawla (2010) |
| | Experts | CBR | Felfernig et al. (2013) |
| | Datings | CF | Wobcke et al. (2015) |
| Points of Interest (POIs) | Tourism services | CF | Huang, Chang, and Sandnes (2010) |
| | Passenger hot spots | PR | Yuan, Zheng, Zhang, and Xie (2012) |
| | POIs for groups | GR | Felfernig et al. (2018b) |
| Help Services | Schools | CBR | Wilson et al. (2009) |
| | Counteracting forgetting | CBR | Stettinger et al. (2020) |
| | Financial services | KBR | Fano and Kurth (2003) |
| | Lifestyles | KBR | Hammer, Kim, and André (2010) |
| | Recipes | CBR | Pinxteren, Gelijnse, and Kamsteeg (2011) |
| | Healthy eating | CBR | Tran, Atas, Felfernig, and Stettinger (2018) |
| | Running practice | ML | Smyth (2018) |

criteria for the selection of recommendation algorithms are discussed. The chapter is concluded with a discussion of issues for future research (Sect. 14.9).

## Recommender Systems in Open Innovation

Integrating customer knowledge into product and service innovation (also denoted as *Open Innovation* (Chesbrough, 2003) is important for successful new product and service development. In the line of open innovation, innovations are also triggered by consumers. Platforms such as *sourceforge.net* or *ideastorm.com* follow this trend of integrating consumers into innovation processes. These platforms base their service on the knowledge and preferences of consumers with the goal to come up with ideas for new product and service features.

The informality of collected consumer preferences and ideas outstrips an innovation manager's capabilities to have a structured and up-to-date overview. Recommender systems can help out in various ways, for example, by automatically identifying ideas, by categorizing user opinions about items, by identifying power innovators, and by (semi-) automatically filtering out the most promising ideas (*idea mining* (Brocco and Groh 2009; Jannach & Bundgaard-Joergensen, 2007; Thorleuchter, VanDen Poel, & Prinzie, 2010). Brocco and Groh (2009) introduce an approach to recommend teams in the context of open innovation processes. Their approach includes different model parameters such as *psychological variables* (e.g., personality type), *demographic variables*, *functional variables* (e.g., expertise), and *social network variables* (e.g., role in different project-related social networks). Jannach and Bundgaard-Joergensen (2007) present a knowledge-based advisory approach for start-ups that focuses on important aspects to be taken into account when developing business plans. Thorleuchter et al. (2010) propose a text mining approach which is based on the following assumption: *If there is a large textual overlap between idea description and problem description, there is not much new to be found in the idea description.* Thus, content-based recommendation can be used to identify relevant ideas. Stanik and Maalej (2019) introduce an approach to the identification of new and relevant software requirements. This analysis is based on a machine learning approach that uses text analysis of TWITTER[1] channels and labeled data for learning a classification model (e.g., *is a textual description (a tweet) a requirement?*). The same approach has also been evaluated in public administration scenarios where TWITTER channels have been analyzed, for example, with regard to new problem reports from citizens (Felfernig, Stettinger, Wundara, & Stanik, 2019). Finally, Li et al. (2015) introduce an approach to recommend new functions to users of the AUTOCAD[2] environment. The underlying idea is that users should be able to easily detect new functions of relevance and that the software provider has the chance to analyze the user feedback to even better understand the relevance of features.

An important task in open innovation scenarios is to figure out the most relevant features of a product. For example, a start-up software company tries to figure out the features that should be included in a minimum viable product (MVP), i.e., a first fully operable version of the software that should be offered to customers. Figure 1 depicts a screenshot of the EVENTHELPR[3] system (Felfernig et al., 2018b) which supports the definition of such decision tasks. EVENTHELPR is a platform for managing different kinds of decision tasks where a potentially large group of persons is in charge of selecting a subset of a set of candidate items (alternatives). Related decision processes are configurable, for example, one can define to which extent information can be exchanged between different group members, to which extent it is allowed to evaluate individual alternatives, and if group members are allowed to add new alternatives within the scope of a group decision process.

---

[1] twitter.com.
[2] autodesk.com.
[3] eventhelpr.com.

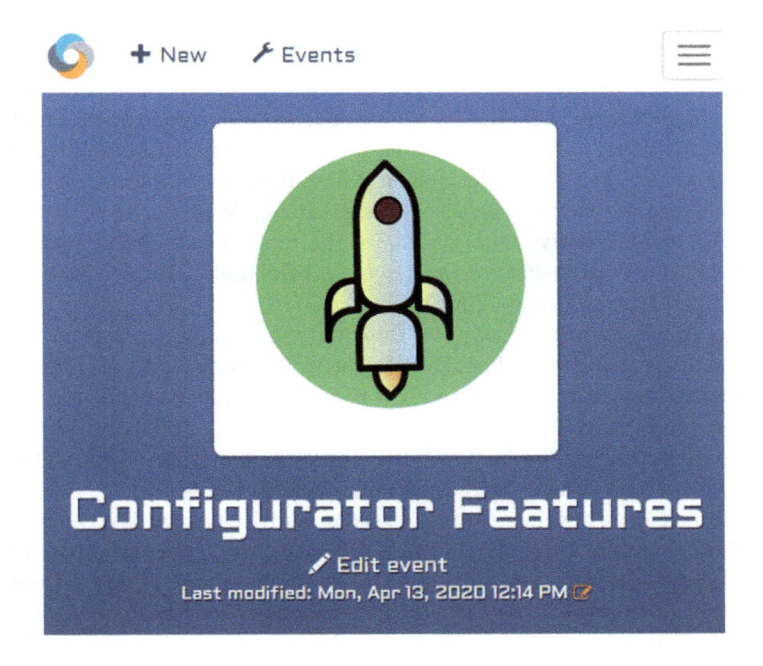

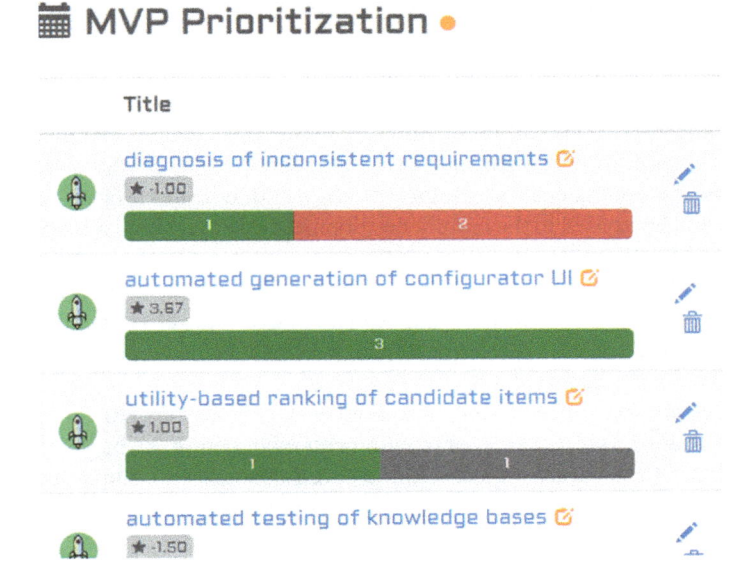

**Fig. 1** Prioritizing the features of a minimum viable configurator product (MVP) in EVENTHELPR (Felfernig et al., 2018b). The star-value with gray background represents the *item relevance* (the higher the better) derived from arguments for and against specific alternatives

## Recommender Systems in Software Engineering

In software engineering scenarios, recommender systems support stakeholders by reducing the information overload (Robillard, Walker, & Zimmermann, 2010). Stakeholder support can be provided throughout different phases of a software process—examples thereof are *method recommendation* (Burke & Ramezani, 2010; Peischl et al., 2010), *requirements recommendation* (Felfernig et al., 2011; Mobasher & Huang, 2011), and *code artifact recommendation* (Cubranic, Murphy, Janice Singer, & Booth, 2005; McCarey et al., 2005).

*Software Method Recommendation.* Depending on the project setting, different software methods are relevant. For example, the waterfall method should be selected in the context of risk-free projects but is not recommended to be applied in medium- or high-risk projects. Method recommendation approaches have been developed in the context of algorithmic problem solving (Burke & Ramezani, 2010) and the recommendation of appropriate effort estimation methods (Peischl et al., 2010). These approaches have been developed on the basis of knowledge-based recommendation (Burke, 2000; Felfernig, Friedrich, Jannach, & Zanker, 2006) since recommendations are based on well-defined rules that are the basis for method selection. Examples of related rules are *waterfall-type software development processes are incompatible with high-risk project settings*, *software engineering research projects should be organized as iterative development process*, and *effort estimation by analogy should be applied if more than one similar project exists*.

*Code Recommendation.* High change rates in software team configurations require the provision of intelligent techniques supporting the understanding, locating, and quality assurance of software (McCarey et al., 2005). Code recommendation is applied in the context of call completion scenarios, i.e., which API methods are relevant in a specific development context (McCarey et al., 2005). Furthermore, code fragment recommendation (Holmes et al., 2006) indicates which sequence of methods could be needed in the current development context. Finally, recommender systems can be applied to tailor displayed software artefacts to the current development context (Kersten & Murphy, 2010) and to predict software defects on the basis of classification methods (Misirli et al., 2011). An overview of code recommendation approaches in software development can be found, for example, in Happel and Maalej (2008), Robillard et al. (2010).

*Requirements Recommendation.* Requirements engineering (RE) is a critical phase in software development: Poor requirements engineering can be a major reason for the failure of a software project (Hofmann & Lehner, 2001). Important tasks in RE processes are *elicitation & definition*, *quality assurance*, *negotiation*, and *release planning* (2007). All these tasks can be supported on the basis of recommendation technologies. Examples thereof are the recommendation of similar requirements (Mobasher & Cleland-Huang, 2011), the group-based prioritization of requirements (Felfernig et al., 2011), the classification of requirements in the context of bidding processes (Fucci et al., 2018), and the recommendation of (hidden) dependencies between requirements (Samer et al., 2019).

Recommender Systems Beyond E-Commerce: Presence and Future 211

**Table 3** Example of a content-based filtering recommendation problem: recommendation of similar requirements based on *category* and/or *keyword* information

| Requirement | Category | Person days | Description |
|---|---|---|---|
| $req_1$ | Database | 155 | Store component configuration in DB |
| $req_2$ | User interface | 70 | User interface with online help available |
| $req_3$ | Database | 265 | Separate tier for DB independence |
| $req_4$ | User interface | 45 | User interface with corporate identity |

*Example: Content-based Recommendation of Similar Requirements.* The following simple example sketches the application of *content-based filtering* (Pazzani & Billsus, 1997) to the recommendation of similar requirements. Similar requirements should be recommended, for example, in the context of reuse scenarios where requirements defined in already finished projects or requirements that have been defined by other stakeholders in the same software project should be recommended to avoid the reinvention of the wheel. Table 3 depicts an enlisting of software requirements that have been defined within the scope of a software project. Individual requirements $req_i$ can be characterized by the attributes *category*, the estimated *person days* needed for implementation, and a textual *description* of the requirement.

If we assume that the stakeholder currently interacting with the requirements engineering system, already took a look at requirement $req_3$ (assigned to the *database* category), a content-based recommender system would recommend the requirement $req_1$ (if this one has not been analyzed by the stakeholder up to now). If categories are not available to describe requirements on a semantic level, keywords extracted from the textual descriptions of requirements can be exploited (Roy & Mooney, 2000). An approach to determine the similarity between requirements is sketched with Formula 1. For example, sim($req_3$, $req_1$) = 0.17, if we assume $keywords(req_1)$ = {store, component, configuration, DB} and keywords($req_3$) = {tier, DB, independence}.

$$\text{sim}(req_1, req_2) = \frac{|keywords(req_1) \cap keywords(req_2)|}{|keywords(req_1) \cup keywords(req_2)|} \quad (1)$$

Figure 2 shows a screenshot of an ECLIPSE[4] user interface that supports the recommendation of tasks (bugs) that should be solved by individual contributors of an open source community. The underlying recommendation approach (Felfernig et al., 2018d) is hybrid (content-based and knowledge-based) and takes into account the *similarity between keywords of task descriptions already completed by a contributor* and *new task descriptions* and further properties such as *global relevance* (how many

---

[4] www.eclipse.org.

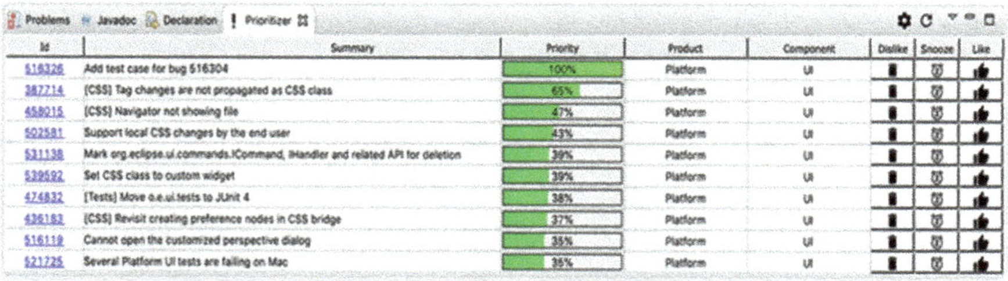

**Fig. 2** ECLIPSE plugin for open source prioritization (Felfernig et al., 2018d) based on a hybrid recommendation approach. Tasks (bugs) to be completed are ranked according to their estimated relevance for an open source developer

**Table 4** Example of a group recommendation task: recommendation of requirements prioritizations to a group of stakeholders (used heuristics = average voting)

| Requirement | Stakeholder$_1$ | Stakeholder$_2$ | Stakeholder$_3$ | Stakeholder$_4$ | *Recommendation* |
|---|---|---|---|---|---|
| req$_1$ | 1 | 1 | 1 | 2 | 1.25 (4) |
| req$_2$ | 3 | 3 | 2 | 3 | 2.75 (3) |
| req$_3$ | 5 | 4 | 4 | 5 | 4.5 (2) |
| req$_4$ | 5 | 5 | 4 | 5 | 4.75 (1) |

contributors took a look at a task) and how many *comments* are associated with this task.

*Example: Group-based Recommendation of Requirements Prioritizations.* Group recommender systems exploit heuristics (Felfernig et al., 2018a; Masthoff, 2011) that aggregate the preferences of individual group members. Requirements prioritization is the task of defining which requirements should be implemented within the scope of the next release. This can be regarded as a group decision process. In group recommender systems, different aggregation functions (heuristics) can be applied to aggregate the preferences of individual stakeholders. For example, *majority voting* selects the alternative with the highest majority level. Furthermore, *average voting* recommends the item with the highest average value (see $req_4$ in Table 4). For an introduction to group recommender systems and related aggregation functions, we refer to Masthoff (2011), Felfernig et al. (2018a).

Figure 3 depicts a screenshot of the OPENREQ[5] user interface that supports an argumentation-based prioritization of requirements. Each requirement gets evaluated by different stakeholders (members of the software team). Thereafter, a weighted average function is used to determine a ranking of the proposed alternatives. If the diversity of individual item evaluations is too high, a discussion is triggered in order to restore consistency in the evaluations and to establish consensus among stakeholders.

---

[5]openreq.eu.

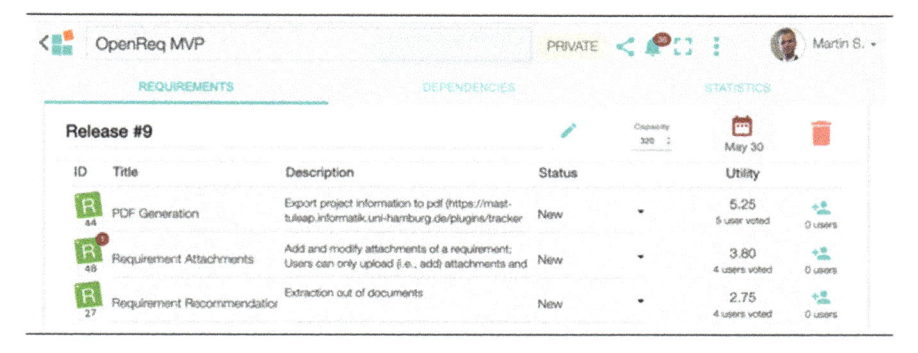

**Fig. 3** Requirements prioritization user interface of OPENREQ (Felfernig et al., 2018a; Fucci et al., 2018). A group recommender ranks requirements according to their utility for a group of stakeholders

## Recommender Systems in Knowledge Engineering

Successfully developing knowledge-based systems strongly depends on the efficiency of the underlying development processes. Knowledge base development suffers from frequent changes in organizational structures and project teams. In such project contexts, recommender systems can help knowledge engineers to better cope with the complexity of knowledge bases and related artefacts (Chatzopoulou et al., 2009; Felfernig et al., 2009). These systems can improve the understandability of knowledge bases by suggesting knowledge chunks that should experience an increased testing and debugging (Felfernig et al., 2009; Felfernig, Schubert, & Zehentner, 2012). Repair and refactoring actions can be proposed if faulty and ill-formed parts of a knowledge base have been identified (Felfernig et al., 2010). Recommenders can also help to improve database accessibility by recommending queries that lead to relevant results (Chatzopoulou et al., 2009).

*Knowledge Base Understanding.* Efficient knowledge base development and maintenance requires an easy but still expressive kind of knowledge representation. Recommender systems can help to make knowledge bases more accessible by supporting knowledge engineers, for example, by recommending *constraints to be investigated* when analyzing a knowledge base or by recommending *similar constraints*, for example, which are referring to common variables (Felfernig, Schubert, & Zehentner, 2012).

*Knowledge Base Testing & Debugging.* Changes in the product knowledge have to be taken into account. The integration of such changes is a time-critical task (Felfernig, Friedrich, Jannach, & Zanker, 2006), and recommender systems can help to improve the efficiency of testing and debugging processes. This can be achieved on the basis of recommending diagnosis & repair actions that help to focus on the relevant parts of a knowledge base (Felfernig et al., 2009, 2011). Popular approaches to the identification of such *diagnoses* are model-based diagnosis introduced by Reiter (1987) and different variants thereof (Felfernig et al., 2011). A diagnosis

**Table 5** Example of a collaborative recommendation problem. The entry $ij$ with value 1 denotes that fact that knowledge engineer$_i$ has inspected constraint$_j$

|  | constraint$_1$ | constraint$_2$ | constraint$_3$ | constraint$_4$ |
|---|---|---|---|---|
| knowledge engineer$_1$ | 1 | 0 | 1 | 1 |
| knowledge engineer$_2$ | 1 | 0 | 1 | ? |
| knowledge engineer$_3$ | 1 | 1 | 0 | 1 |

(hitting set) can be regarded as a (minimal) set of elements of a knowledge base that have to be adapted or deleted in order to restore consistency.

*Knowledge Base Refactoring.* Knowledge base maintainability is extremely important for efficient knowledge base development and maintenance (Barker, O'Connor, Bachant, & Soloway, 1989). Low quality can trigger enormous maintenance overheads. In knowledge engineering scenarios, recommenders are applied to recommending refactorings of knowledge base structures in a semantics-preserving fashion. These refactorings can be regarded as constraint rewritings (Felfernig et al., 2010) coming along with simplifications in terms of reduced redundancy and increased understandability.

*Recommender Systems & Databases.* Chatzopoulou et al. (2009) introduce collaborative filtering concepts that support users when exploring databases (*complex SQL queries*). Queries can be recommended on the basis an analysis of the navigation behavior of the nearest neighbors (users with a similar navigation behavior). Fakhraee et al. (2011) propose an approach to the recommendation of database queries derived from database attributes that contain the keywords part of the initial user query. In contrast to Chatzopoulou et al. (2009), Fakhraee and Fotouhi (2011) focus on *keyword-based database queries*.

*Example: Collaborative Recommendation of Constraints.* In the context of knowledge base analysis, recommender systems can recommend constraints that could be relevant in a certain navigation context. Table 5 depicts a scenario where the constraints {constraint$_{1..4}$} have partially been analyzed by knowledge engineers {knowledge engineer$_{1..3}$}. The first knowledge engineer has already investigated constraint$_1$, constraint$_3$, and constraint$_4$ but was not interested in taking a look at constraint$_2$. Collaborative filtering (CF) analyzes the ratings (the rating = 1 if a knowledge engineer has already investigated a constraint and it is 0 if the knowledge engineer was not interested in analyzing the constraint).

User-based collaborative filtering (Konstan et al. 1997) identifies the k-nearest neighbors (knowledge engineers with a similar knowledge navigation preferences) and determines a prediction for the rating of a constraint the knowledge engineer did not take a look at up to now (but did not specify disinterest). In Table 5, knowledge engineer$_1$ is the 1-nearest neighbor of knowledge engineer$_2$ since knowledge engineer$_1$ has the same knowledge base analysis behavior as knowledge engineer$_2$. At the same time, knowledge engineer$_2$ did not analyze (change) constraint$_4$. In our example, CF would recommend constraint$_4$ to knowledge engineer$_2$ since it has been investigated by knowledge engineer$_1$.

## Recommender Systems in Knowledge-Based Configuration

Configuration is regarded as a design activity where a target product is composed from a set of predefined components in such a way that it is consistent with a set of predefined constraints (Sabin & Weigel, 1998). Similar to knowledge-based recommenders (Burke, 2000; Felfernig et al., 2006), configuration systems (configurators) support users in finding solutions (Falkner, Felfernig, & Haag, 2011). Configurators and recommender systems differ in their way of item knowledge representation: Configurators are based on an implicit knowledge representation in terms of variables and constraints (and/or rules), whereas their recommender counterpart is based on an item table semantically described by product properties. Configuration knowledge bases are used in scenarios where the amount of (potential) alternatives does not allow an explicit representation due to a combinatorial explosion (Falkner et al., 2011).

Recommender systems can be exploited to increase the *usability of a configurator*. For example, filtering out product features of relevance (Falkner et al., 2011; Garcia-Molina, Koutrika, & Parameswaran, 2011) and also proposing feature values and thus providing support in situations where knowledge about some product properties is not available (Falkner et al., 2011; Polat-Erdeniz, Felfernig, Atas, & Samer, 2019b), and by determining personalized explanations (Felfernig et al. 2009; Polat Erdeniz, Felfernig, & Atas, 2019a). In the context of group-based configuration scenarios (configurations are determined for groups of users), group decision heuristics are applied to personalize configurations and explanations for groups (Felfernig et al., 2016; Tran et al., 2019).

*Selecting Features.* Users are often not interested in all features offered by a configurator. For example, some users might be interested in high-speed photography features of a digital camera, whereas others are primarily interested in a low price. Recommenders can pre-select features of relevance, i.e., features the user is interested to specify. An approach to the implementation of such a recommendation functionality is presented in Felfernig and Burke (2008).

*Determining Feature Values.* Users prefer not to specify features they are not interested in or about which they do not have the needed technical knowledge (Falkner et al., 2011; Polat-Erdeniz, Felfernig, Atas, & Samer, 2019b). Recommender systems can automatically recommend feature values and thus reduce the burden for users. Feature value recommendation can be implemented, for example, on the basis of collaborative recommendation (Falkner, Felfernig, & Haag, 2011; Konstan et al. 1997). Note that feature value recommendation can trigger decision biases (e.g., through the presentation of default values). As a consequence, there is the danger of manipulation which has to be taken into account (Mandl, Felfernig, Tiihonen, & Isak, 2011).

**Table 6** Example of collaborative feature recommendation. The table entries denote orders in which users specified feature values

|  | $feature_1$ | $feature_2$ | $feature_3$ | $feature_4$ |
|---|---|---|---|---|
| $session_1$ | 1 | 4 | 2 | 3 |
| $session_2$ | 1 | 2 | 3 | 4 |
| $session_3$ | 1 | 3 | 4 | 2 |
| $session_4$ | 1 | 3 | 2 | 4 |
| $session_5$ | 1 | 2 | ? | ? |

*Plausible Repairs for Inconsistent Requirements.* In situations where no solution can be found for the defined customer preferences, configurators can propose repair alternatives which help to restore consistency. Often, there exist many alternatives which make the selection of repair alternatives a recommendation task (Felfernig et al., 2009; Polat Erdeniz, Felfernig, & Atas, 2019a).

*Example: Collaborative Feature Recommendation.* Table 6 depicts an interaction log indicating in which session which features have been selected in which order—in $session_1$, $feature_1$ was selected first, then $feature_4$ and $feature_2$, and finally $feature_3$. In order to determine relevant features for the current user in $session_5$, collaborative filtering can be applied. Assuming that the user in $session_5$ has already specified $features_{1,2}$, the most similar session (assuming a 1-nearest neighbor approach) is $session_2$. In this case, $feature_3$ would be recommended due to the selection behavior of the nearest neighbor. A more detailed discussion of feature selection techniques can be found in Falkner et al. (2011).

Figure 4 shows a screenshot of a workflow configuration user interface that supports the recommendation of IoT workflow nodes. Collaborative filtering is applied to predict additional nodes of interest (Felfernig et al. 2018c). The collaborative approach does not guarantee consistency in every case since collaborative recommendation does not allow the inclusion of constraints.

## Recommender Systems & Persuasion

Persuasive technologies (Fogg, 2003) aim to trigger changes in a user's attitudes and behavior with the concepts of human computer interaction (HCI). Their impact can be significantly increased by recommendation technologies. Thus, one-size-fits-all approaches can be replaced with personalized environments capable of providing personalized persuasive messages (Berkovsky, Freyne, & Oinas-Kukkonen, 2012). Examples of recommendation technologies in persuasive systems are the enforcement of physical activity during computer games (Berkovsky, Freyne, Coombe, & Bhandari, 2010), the encouraging of software developers to focus on software quality (Pribik & Felfernig, 2012), and the generation of diverse recommendations that help

Recommender Systems Beyond E-Commerce: Presence and Future 217

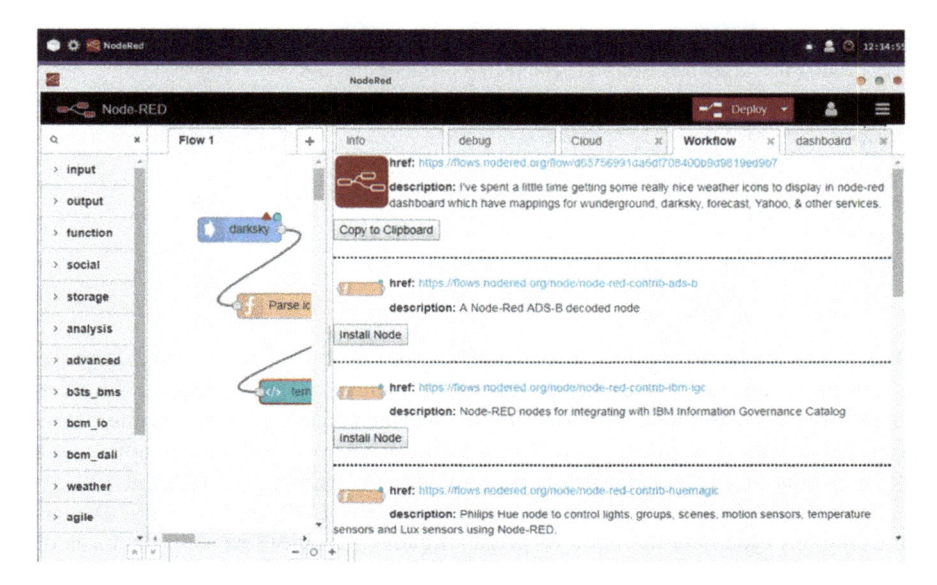

**Fig. 4** Workflow configuration in Internet Of Things (IoT) scenarios (Felfernig et al. 2018c). A collaborative filtering-based approach determines the nearest neighbors (similar workflows) and recommends workflow nodes used by similar neighbors but have not been included in the current workflow

to foster communication in group decision scenarios (Atas, Felfernig, Stettinger, & Tran, 2017).

*Persuasive Games.* Persuasive games focus on pushing physical activities on the basis of reward mechanisms. Berkovsky et al. (2010) introduce a collaborative filtering approach (Konstan et al., 1997) that helps to estimate the personal difficulty of game playing. This recommendation (estimation) is exploited to adapt difficulty levels of game sessions.

*Persuasive Software Environments.* Software development teams are often under a high time pressure with resulting negative impacts on software quality. *Software quality* is strongly correlated with the aspects of *understandability* and *maintainability*. Recommender systems can inform programmers about critical code segments and also recommend actions to be performed in order to increase the quality of software artefacts. For example, Pribik and Felfernig (2012) propose an ECLIPSE[6] plugin that is based on the idea of providing additional feedback to programmers to encourage them to invest more time in quality-related development activities.

*Persuasion in Group Decisions.* In group decisions, communication among group members has a major impact on decision quality (Felfernig et al., 2018a). The more decision-relevant knowledge is exchanged among group members, the higher is the probability, that the optimal decision will be taken. Atas et al. (2017) introduce a group recommendation approach where communication among group members is

---

[6]eclipse.org.

**Table 7** Collaborative filtering-based determination of personal game difficulty levels. A table entry denotes the time a user needed to complete a certain game task

|  | $task_1$ | $task_2$ | $task_3$ | $task_4$ |
|---|---|---|---|---|
| $session_1$ | 1 | 2 | 2 | 2 |
| $session_2$ | 4 | 6 | 5 | 7 |
| $session_3$ | 4 | 5 | 5 | 6 |
| $session_4$ | 1 | 1 | 1 | 2 |
| $session_5$ | 4 | 6 | 4 | ? |

fostered on the basis of diverse recommendations. Even in the case that a group agrees on a specific alternative, diverse recommendations can trigger a situation where other alternatives are still taken into account and possibly a different alternative is finally chosen. The impact of different explanation styles on the perception of a recommendation (e.g., in terms of decision quality) is analyzed in Tran et al. (2019).

*Example: Collaborative Estimation of Game Level Difficulties.* Table 7 depicts an example of the determination of personal game difficulty levels on the basis of collaborative filtering. Let us assume that the current user in $session_5$ has already completed the tasks 1..3; by determining the nearest neighbor (1-NN) of $session_5$, we can estimate the duration of the open task: $session_2$ is the NN of $session_5$, and the user in $session_2$ needed seven time units for completing $task_4$. This approach can be exploited to adapt the game level complexity to better push physical activity (Berkovsky et al., 2010).

## Further Recommender Applications

Our discussion of recommendation technologies focused on scenarios that go beyond typical e-commerce settings (Linden, Smith, & York, 2003; Schafer, Konstan, Riedl, 2011). In this section, we provide an overview of further applications.

*Recommender Systems in Smart Homes.* Smart home technologies follow the goal of improving the quality of life inside the home. Leitner, Fercher, Felfernig, A., and Hitz (2012) show the applicability of knowledge-based recommendation technologies in ambient assisted living (AAL) where users are guided through a preference construction process in the design phase of a smart home. Recommender systems can be applied for smart home design as well as to control smart home equipment. During the "runtime" of a smart home, recommenders can generate activity- and context-dependent recommendations such as activating air-conditioning and informing relatives about potential dangerous situations, for example, the status of elderly relatives living remotely is unclear. In the line of Leitner et al. (2012), LeMay, Haas, and Gunter (2009) show how to apply collaborative recommendation to support the control of complex smart home installations.

*People Recommender.* Platforms such as *facebook.com* or *linkedin.com* include recommender applications that exploit the information from underlying social networks (Golbeck, 2009). Following a similar approach, some recommender applications exploit social networks for identifying crime suspects. In such a scenario, social networks represent relationships between criminals. *CrimeWalker* (Tayebi, Jamali, Ester, Glaesser, & Frank, 2011) is based on a random-walk method that "recommends" crime suspects. In the line of *CrimeWalker*, *TechLens* focuses on person recommendation, for example, persons who could act as reviewers for a scientific conference. *TechLens* is based on collaborative filtering. With the goal of improving the perceived quality of and satisfaction with personal physicians, Hoens, Blanton, and Chawla (2010) propose a collaborative filtering approach to physician recommendation. Wobcke et al. (2015) introduce a collaborative filtering-based people-2-people recommender system. These systems differ from standard recommendation approaches since a recommendation that is of interest for one person could be ignored or rejected by another person. Finally, Samer et al. (2018) introduce a content-based recommendation approach which is based on the identification of stakeholders who have the expertise to take over quality assurance tasks related to a specific software requirement.

*POI Recommenders.* Personalized services become ubiquitous. Examples thereof are museums and art galleries which already provide personalized service access to their customers. In such scenarios, users often do not define their preferences explicitly, and recommendations are determined on the basis of navigation behavior analysis. Huang, Chang, and Sandnes (2010) introduce a handheld museum guide which is based on collaborative filtering. RFID exploits radio-frequency electromagnetic fields to transfer tag data for the purpose of object tracking (Thiesse & Michahelles, 2009). The EVENTHELPR system (Felfernig et al. 2018b) supports group decision processes related to the visit of travel locations. In such scenarios, groups are supported by argumentation-based prioritization approaches. Finally, Yuan, Zheng, Zhang, and Xie (2012) introduce a passenger hot spot recommender system in the taxi domain which helps to reduce idle times.

*Help Agents.* Help agents support users who are non-experts in a domain. Such agents can be implemented, for example, on the basis of recommendation technologies (Terveen & Hill, 2001). An example of a *help agent* based on recommendation technologies is *SmartChoice*. It uses content-based recommendation for supporting low-income families in the choice of a public school for their children. Such a recommendation service appears to be extremely useful since (1) parents do not dispose of detailed knowledge about the different school types and (2) sub-optimal decisions could have extremely negative consequences. *Personal Choice Point* (Fano & Kurth 2003) is a knowledge-based recommendation and visualization environment for the financial services domain where users can directly experience the impact of their financial decisions. A lifestyle recommender for supporting diabetes patients is introduced in Hammer, Kim, and André (2010). The introduced *MED-StyleR* system helps to improve care provision, enhancing a patient's life quality and also to lower costs of public health. Another lifestyle recommender is introduced in Pinxteren, Gelijnse, and Kamsteeg (2011)—this recommender focuses on determining health-supporting recipes in a personalized fashion. An overview of recommendation agents in the healthy eating domain is provided by Tran et al. (2018). Smyth (2018) reports insights regarding pacing strategies in marathon runs—with a special focus on non-professionals. Finally, Stettinger et al. (2020) introduce a knowledge-based recommendation environment that focuses on the recommendation of test questions in test-enhanced learning scenarios. A screenshot of the KNOWLEDGECHECKR[7] environment is depicted in Fig. 5.

## Selection of Algorithms

In this chapter, we have discussed four basic recommendation approaches, namely collaborative filtering (CF), content-based filtering (CBF), knowledge-based recommendation (KBR), and group recommender systems (GR). In the following, we will discuss evaluation criteria related to these approaches.

*Setup costs.* Setup costs for collaborative filtering are low since only basic item information (e.g., the item name) and information about user × item preferences is needed. Similarly, content-based recommendation setup costs are low, since only item information and information about past item consumptions of a user are needed to determine a recommendation. In contrast, knowledge-based recommenders have higher setup costs due to the fact that a semantic model in terms of a knowledge base needs to be defined. Especially, knowledge-based recommender systems are based on conversational recommendation, i.e., users explicitly define their preferences within the scope of a dialog. Conversational recommendation is often associated with significantly higher costs related to the development of the recommender user interface. The setup costs of group recommenders depend on the underlying recommendation approach since group recommenders can exploit all types of single user recommenders as a basis. Collaborative filtering and content-based filtering are typically integrated into user interfaces for single-shot recommendation, i.e., no dialog is provided to support decision making.

*Adaptation costs.* Collaborative and content-based filtering have low adaptation costs since new ratings provided by users and new item consumptions are automatically taken into account. Knowledge-based recommendation has higher adaptation costs since new items and new selection criteria (constraints) have to be integrated manually. The adaptation costs of group recommenders depend on the chosen basic recommendation approach.

*Achievement of Serendipity Effects.* The term *serendipity* is associated with a situation where a user of a recommender system receives recommendations, he/she never would have expected (a.k.o. something that is completely surprising in the positive sense). Collaborative filtering style recommender systems are the first choice when it comes to the triggering of serendipity effects. Content-based recommendation does not provide much help in this context, since primarily items similar to

---

[7]knowledgecheckr.com.

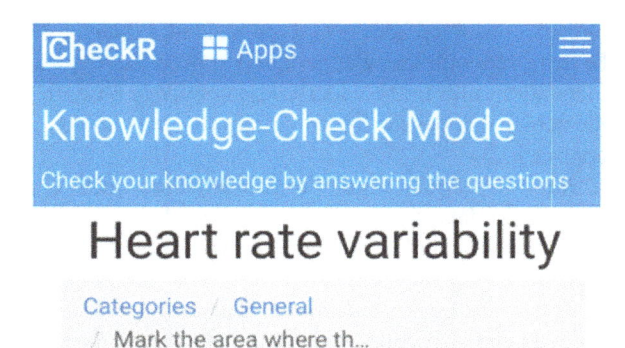

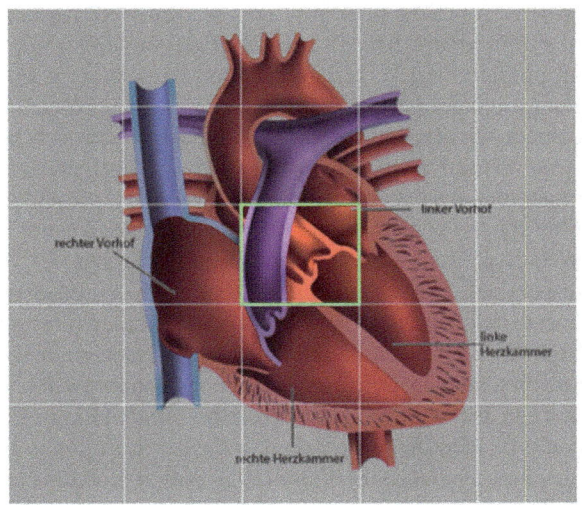

**Fig. 5** Question answering user interface of KNOWLEDGECHECKR (Stettinger et al. 2020). The system offers different question types—in the provided example, the task (question) is to find relevant areas in a shown picture. Question recommendation is implemented as a combination of content-based and knowledge-based recommendation

already consumed ones get recommended. Finally, the degree of serendipity that can be achieved with knowledge-based approaches depends on the definition of recommendation rules in the underlying knowledge base. In group recommender systems, the achievable degree of serendipity depends on the chosen basic recommendation approach.

*Ramp-Up Issues.* Ramp-up problems can occur if a recommender system relies on specific data that might not be available, for example, user × item ratings are not available in collaborative filtering and content-based filtering if a user interacts with the system the first time (*new user* problem). Collaborative filtering suffers from the *new item problem*, i.e., if a new item is added, corresponding ratings are needed before a recommendation can be determined. The new item problem is not an issue in the context of content-based recommendation since new items are primarily recommended depending on the content-wise similarity with items already consumed by the user. New user problems are typically solved by asking these users for feedback regarding a proposed set of items or asking them to specify their personal item preferences. New item problems are resolved, for example, by reusing the ratings of similar items. Knowledge-based recommender systems do not suffer from ramp-up problems since recommendation knowledge is encoded in an explicit fashion. For group recommender systems, a ramp-up problem only exists if collaborative or content-based recommendation has been chosen as basic recommendation approach.

*Assuring Transparency.* The degree of transparency specifies to which extent recommendations can be explained to users of a recommender system. In collaborative filtering and content-based filtering, explanations are based on similarities between users (e.g., users who purchased $X$, also purchased $Y$) or similarities between items, respectively (e.g., $X$ is similar to $Y$ which you already purchased in the past). In contrast to collaborative filtering and content-based filtering, explanations in knowledge-based recommendation are more fine-grained since they refer to individual aspects (attributes) of an item. Furthermore, explanations in knowledge-based recommendation can deal with situations where no solution for given set of customer requirements could be identified (answers to *why not?* questions). Explanations in group recommenders depend on the chosen aggregation function but also in which way the chosen function is used in the explanation. For example, if $AVG$ has been chosen as aggregation function, an explanation could simply indicate that a recommendation has received the highest average value. Alternatively, the explanation could go further and take into account individual ratings of group members, for example: *this alternative has the highest average rating, and none of the group members is unsatisfied.*

*Item Types and Recommendation Approaches.* High-involvement items are often complex and expensive and selected after a careful consideration since sub-optimal decisions have a high negative impact. Examples of such items are cars, financial services, digital equipment, and apartments. In contrast, *low-involvement items* are simple and less expensive. Examples thereof are movies, books, and news articles. Collaborative and content-based filtering are used when recommending low-involvement items where ratings appear with a higher frequency. Knowledge-based approaches are the first choice when it comes to the recommendation of high-involvement items since ratings related to such items have a rather low frequency and user preferences can change over longer time periods without the recommender system noticing.

## Research Issues

The major goal of most existing recommender systems is to achieve the defined business goals. Rarely, the viewpoint of customers is taken into account in the first place (Martin, Donaldson, Ashenfelter, Torrens, & Hangartner, 2011). For example, *amazon.com* recommenders inform users about new books of interest; however, a more user-centered recommender would take into account (if available) information about books that have already been purchased by friends (Martin et al., 2011). In the sense of this example, recommendation technologies have to evolve toward more user-centered technologies—the corresponding issues for future research will be discussed in the following.

*Focusing on the User.* There are many other similar scenarios, for example, consumer packaged goods (CPG) are already offered using recommenders (Dias, Locher, Li, El-Deredy & Lisboa, 2008). Digital camera recommenders are "new technology" focused but do not take into account the current photography equipment of the user. An approach which is in the line of the idea of a stronger focus on the quality of user support is the RADAR personal assistant (Faulring, Mohnkern, Steinfeld, & Myers, 2009) that supports multi-task coordination of personal emails.

*Sharing Recommendation Knowledge.* A major reason for the mentioned low level of customer orientation is the lack of recommendation knowledge. The global availability of CPG goods information seems to be theoretically possible but is definitely in the need of a cloud and mobile computing infrastructure. More customer-centered recommender systems will follow the idea of personal agents providing an integrated and multi-domain recommendation service (Chung et al., 2007; Martin et al., 2011). Following the approach of *ambient intelligence* (Ramos, Augusto, & Shapiro, 2008), such recommenders will be based on global object information (Ramiez-Gonzales, Munoz-Merino, & Delgado, 2010; Thiesse & Michahelles, 2009) and support users in different contexts in a cross-domain fashion.

*Unobtrusive Preference Identification.* Preference knowledge is a key for high-quality recommendations. In this context, technologies are needed which support preference elicitation in an unobtrusive fashion (Foster & Oberlander, 2010; Lee, Park, & Park, 2008; Winoto & Tang, 2010). *Three major modalities* are the *detection of facial expressions*, the *interpretation of recorded speech*, and the *analysis of physiological signals*. An approach to preference extraction from eye tracking patterns is introduced by Xu, Jiang, and Lau (2008). Attention times are used as an additional data-source to improve the quality of a content-based filtering recommender. Janssen, Broek, and Westerink (2011) introduce an approach to preference elicitation from physiological signals—in this context, information about skin temperature.

*Psychological Aspects.* Building high-quality recommender applications requires a deep understanding of human decision making. To achieve this goal, existing psychological theories of human decision making and their impact on the construction of recommender systems have to be taken into account. Lin, Shen, Chen, Zhu, and Xiao (2019) show how models of non-compensatory decision making can be integrated into collaborative recommendation approaches (more precisely, matrix factorization (Koren et al., 2009). Cosley, Lam, Albert, Konstan, and Riedl(2003) show that the item rating style has an impact on a users' rating behavior. Adomavicius, Bockstedt, Curley, and Zhang (2011) show the existence of anchoring effects in collaborative filtering recommender systems. They confirm the results of Cosley et al. (2003) and show in which way rating drifts can have an impact on the rating behavior. There are a couple of negative effects of decision biases that can occur within the scope of decision processes (Jameson et al. 2015). The existence of such biases has been shown in a couple of recommender systems-related publications. Examples thereof are the existence of anchoring effects in group decision scenarios (Stettinger, Felfernig, Leitner, & Reiterer, 2015), primacy/recency effects (Stettinger et al., 2015), and decoy effects (Teppan & Felfernig, 2012). For a detailed overview on the occurrence of decision biases in recommender systems, we refer to Jameson et al. (2015). Future recommenders should exploit the information provided by the mentioned preference elicitation methods (Janssen et al. 2011).

*Further Insights and Trends.* For more than two decades, mainstream recommender systems research primarily focused in one way or another on the optimization of prediction quality. In addition to this mainstream of algorithmic approaches, the aspect of user experience and intelligent user interfaces has become an increasingly important aspect (Konstan & Riedl, 2012). An insight in this context is that prediction quality measures do not take into account user experience aspects which can have a major impact on the applicability of a recommender system. For example, different explanation interfaces have different impacts on user acceptance (Herlocker, Konstan, & Riedl, 2000). Focusing research on algorithmic aspects is still quite attractive for recommender systems researchers (and also practitioners); however, user interface aspects become increasingly important.

Recommender systems are in the majority of the cases regarded as systems supporting single users in their decision processes. A topic of upcoming relevance is group recommender systems (Felfernig et al., 2018a) that allow the determination of recommendations for groups. A specific aspect extremely relevant for both types of systems is the set of goals and stakeholders that are taken into account when developing a recommender system. In this context, multi-stakeholder recommendation (Abdollahpouri, Burke, & Mobasher, 2017) helps to take into account the preferences of different stakeholders who are not necessarily the users of the recommender system—this can be considered as a major difference compared to group recommender systems (Felfernig, Boratto, Stettinger, & Tkalcic, 2018a). An example of the application of multi-stakeholder approaches is *e-learning* (Burke & Abdollahpouri, 2016) where, for example, the preferences of students, instructors, and even external partners have to be considered.

From an algorithmic point of view, deep learning approaches (Batmaz, Yurekli, Bilge, & Kaleli, 2019) (which are similar to matrix factorization (Koren et al., 2009) have the ability to include the aspect of time into the recommendation model. This is important since in many cases there are seasonal aspects that have to be taken into account. For example, in the movie domain, christmas movies have the highest popularity in December, and similar effects can be observed with christmas songs.

## Conclusions

In this chapter, we give a short introduction to basic recommendation approaches and compare their applicability, advantages, and disadvantages in a structured fashion. Furthermore, we provide an overview of recommender systems applications that go beyond typical e-commerce scenarios. This overview is not complete but the result of an in-depth literature analysis of research in recommender systems. In order to stimulate further related research, we provide an overview of relevant topics for future research.

## References

Abdollahpouri, H., Burke, R., & Mobasher, B. (2017). Recommender systems as multistakeholder environments. *25th Conference on User Modeling, Adaptation, and Personalization* (pp. 347–348). Bratislava, Slovakia.

Adomavicius, G., Bockstedt, J., Curley, S., & Zhang, J. (2011). Recommender systems, consumer preferences, and anchoring effects. In: *RecSys 2011 Workshop on Human Decision Making in Recommender Systems* (pp. 35–42).

Atas, M., Felfernig, A., Stettinger, M., & Tran T. (2017). Beyond item recommendation: Using recommendations to stimulate information exchange in group decisions. In: *9th International Conference on Social Informatics (SocInfo'17)* (pp. 368–377). Oxford, UK.

Barker, V., O'Connor, D., Bachant, J., & Soloway, E. (1989). Expert systems for configuration at digital: XCON and beyond. *Communications of the ACM, 32*(3), 298–318.

Batmaz, Z., Yurekli, A., Bilge, A., & Kaleli, C. (2019). A review on deep learning for recommender systems: challenges and remedies. *Artificial Intelligence Review, 52*, 1–37.

Berkovsky, S., Freyne, J., Coombe, M., & Bhandari, D. (2010). Recommender algorithms in activity motivating games. *ACM Conference on Recommender Systems (RecSys'09)* (pp. 175–182).

Berkovsky, S., Freyne, J., & Oinas-Kukkonen, H. (2012). Influencing individually: Fusing personalization and persuasion. *ACM Transactions on Interactive Intelligent Systems, 2*(2), 1–8.

Brocco M., & Groh, G. (2009). Team recommendation in open innovation networks. In *ACM Conference on Recommender Systems (RecSys'09)* (pp. 365–368). NY, USA.

Burke, R., & Abdollahpouri, H. (2016). Educational recommendation with multiple stakeholders. *IEEE/WIC/ACM International Conference on Web Intelligence, Workshops* (pp. 62–63). NE, USA: Omaha.

Burke, R., & Ramezani M. (2010). Matching recommendation technologies and domains. *Recommender systems handbook* (pp. 367–386).

Burke, R. (2000). Knowledge-based recommender systems. *Encyclopedia of Library and Information Systems, 69*(32), 180–200.

Chatzopoulou, G., Eirinaki, M., & Poyzotis, N. (2009). Query recommendations for interactive database exploration. In *21st Internationl Conference on Scientific and Statistical Database Management* (pp. 3–18).

Chesbrough, H. (2003). *Open innovation: The new imperative for creating and profiting from technology.* Boston, MA: Harvard Business School Press.

Chung, R., Sundaram, D., & Srinivasan, A. (2007). Integrated personal recommender systems. *9th ACM International Conference on Electronic Commerce* (pp. 65–74). MN, USA: Minneapolis.

Cosley, D., Lam, S., Albert, I., Konstan, J., & Riedl, J. (2003). Is seeing believing—how recommender system interfaces affect users' opinions. In *CHI03* (pp. 585–592).

Cubranic, D., Murphy, G., Singer, J., & Booth, K. (2005). Hipikat: A project memory for software development. *IEEE Transactions of Software Engineering, 31*(6), 446–465.

Dias, M., Locher, D., Li, M., El-Deredy, W., & Lisboa, P. (2008). The value of personalized recommender systems to e-business. In *2nd ACM Conference on Recommender Systems (RecSys'08)* (pp. 291–294). Lausanne, Switzerland.

Ducheneaut, N., Patridge, K., Huang, Q., Price, B., & Roberts, M. (2009). Collaborative filtering is not enough? Experiments with a mixed-model recommender for leisure activities. *17th International Conference User Modeling, Adaptation, and Personalization (UMAP 2009)* (pp. 295–306). Italy: Trento.

Fakhraee, S., & Fotouhi, F. (2011). TupleRecommender: A recommender system for relational databases. *22nd International Workshop on Database and Expert Systems Applications (DEXA)* (pp. 549–553). France: Toulouse.

Falkner, A., Felfernig, A., & Haag, A. (2011). Recommendation technologies for configurable products. *AI Magazine, 32*(3), 99–108.

Fano, A., & Kurth, S. (2003). Personal choice point: helping users visualize what it means to buy a BMW. *8th International Conference on Intelligent User Interfaces (IUI 2003)* (pp. 46–52). Miami, FL, USA.

Faulring, A., Mohnkern, K., Steinfeld, A., & Myers, B. (2009). The design and evaluation of user interfaces for the RADAR learning personal assistant. *AI Magazine, 30*(4), 74–84.

Felfernig, A., & Burke, R. (2008). Constraint-based recommender systems: technologies and research issues. *10th ACM International Conference on Electronic Commerce (ICEC'08)* (pp. 17–26). Innsbruck, Austria.

Felfernig, A., Atas, M., Tran, I., Stettinger, M. (2016). Towards group-based configuration. In *International Workshop on Configuration 2016 (ConfWS'16)* (pp. 69–72). Toulouse, France.

Felfernig, A., Boratto, L., Stettinger, M., & Tkalcic, M. (2018a). *Group Recommender Systems.* Springer.

Felfernig, A., Friedrich, G., Schubert, M., Mandl, M., Mairitsch, M., & Teppan, E. (2009). Plausible repairs for inconsistent requirements. In *IJCAI'09*, (pp. 791–796). Pasadena, CA.

Felfernig, A., Gruber, I., Brandner, G., Blazek, P., & Stettinger, M. (2018b). Customizing events with EventHelpR. In *8th International Conference on Mass Customization and Personalization (MCP-CE 2018)* (pp. 88–91). Novi Sad, Serbia.

Felfernig, A., Jeran, M., Ninaus, G., Reinfrank, F., Reiterer, S. (2013). Toward the next generation of recommender systems: applications and research challenges. In *Multimedia services in intelligent environments, smart innovation, systems and technologies* (pp. 81–98). Springer.

Felfernig, A., Mandl, M., Pum, A., & Schubert, M. (2010). Empirical knowledge engineering: Cognitive aspects in the development of constraint-based recommenders. *23rd International Conference on Industrial, Engineering and Other Applications of Applied Intelligent Systems (IEA/AIE 2010)* (pp. 631–640). Cordoba, Spain.

Felfernig, A., Ninaus, G., Grabner, H., Reinfrank, E., Weninger, L., Pagano, D., & Maalej, W. (2013). An overview of recommender systems in requirements engineering. In *Managing Requirements Knowledge* (pp. 315–332). Springer.

Felfernig, A., Reinfrank, F., & Ninaus, G. (2012). Resolving anomalies in feature models. *20th International Symposium on Methodologies for Intelligent Systems* (pp. 1–10). Macau, China

Felfernig, A., Schubert, M., Zehentner, C. (2011). An efficient diagnosis algorithm for inconsistent constraint sets. *Artificial Intelligence for Engineering Design, Analysis, and Manufacturing (AIEDAM), 25*(2), 175–184.

Felfernig, A., Stettinger, M., Atas, M., Samer, R., Nerlich, J., Scholz, S., Tiihonen, J., Raatikainen, M. (2018d). Towards utility-based prioritization of requirements in open source environments. In *26th IEEE Conference on Requirements Engineering* (pp. 406–411). Banff, Canada: ACM.

Felfernig, A., Stettinger, M., Wundara, M., & Stanik, C. (2019). Ai in public administration. In *Handbuch e-Government* (pp. 491–504). Springer

Felfernig, A., Friedrich, G., Jannach, D., & Zanker, M. (2006). An integrated environment for the development of knowledge-based recommender applications. *International Journal of Electronic Commerce (IJEC), 11*(2), 11–34.

Felfernig, A., Polat-Erdeniz, S., Uran, C., Reiterer, S., Atas, M., Tran, T., et al. (2018). An overview of recommender systems in the Internet of Things. *Journal of Intelligent Information Systems (JIIS), 52,* 285–309.

Felfernig, A., Zehentner, C., Ninaus, G., Grabner, H., Maalej, W., Pagano, D., et al. (2011). Group decision support for requirements negotiation. *Springer Lecture Notes in Computer Science, 7138,* 1–12.

Fogg, B. J. (2003). *Persuasive technology—Using computers to change what we think and do.* Morgan Kaufmann Publishers.

Foster, M., & Oberlander, J. (2010). User preferences can drive facial expressions: Evaluating an embodied conversational agent in a recommender dialog system. *User Modeling and User-Adapted Interaction (UMUAI), 20*(4), 341–381.

Fucci, D., C. Palomares, X. Franch, D. Costal, M. Raatikainen, M. Stettinger, Z. Kurtanović, T. Kojo, L. Koenig, A. Falkner, G. Schenner, F. Brasca, T. Männistö, A. Felfernig, and W. Maalej. Needs and challenges for a platform to support large-scale requirements engineering: a multiple-case study. In *12th ACM/IEEE International Symposium on Empirical Software Engineering and Measurement (ESEM'18),* pages 1–10, Oulu Finland, 2018.

Garcia-Molina, H., Koutrika, G., & Parameswaran, A. (2011). Information seeking: convergence of search, recommendations, and advertising. *Communications of the ACM, 54*(11), 121–130.

Golbeck, J. (2009). *Computing with social trust.* Springer.

Goldberg, D., Nichols, D., Oki, B., & Terry, D. (1992). Using collaborative filtering to weave an information Tapestry. *Communications of the ACM, 35*(12), 61–70.

Hammer, S., Kim, J., André, E. (2010). MED-StyleR: METABO diabetes-lifestyle recommender. In *4th ACM Conference on Recommender Systems* (pp. 285–288). Barcelona, Spain.

Happel, H., & Maalej, W. (2008). Potentials and challenges of recommendation systems for software engineering. In *International Workshop on Recommendation Systems for Software Engineering* (pp. 11–15), Atlanta, GA, USA.

Herlocker, J., Konstan, J., Riedl, J. (2000). Explaining collaborative filtering recommendations. In *ACM Conference on Computer-Supported Cooperative Work* (pp. 241–250). Philadelphia, PA, USA.

Hoens, T., Blanton, M., Chawla N. (2010). Reliable medical recommendation systems with patient privacy. *1st ACM International Health Informatics Symposium (IHI 2010)* (pp. 173–182). Arlington, Virginia, USA.

Hofmann, H., & Lehner, F. (2001). Requirements engineering as a success factor in software projects. *IEEE Software, 18*(4), 58–66.

Holmes, R., Walker, R., & Murphy, G. (2006). Approximate structural context matching: An approach to recommend relevant examples. *IEEE Transactions on Software Engineering, 32*(12), 952–970.

Huang, Y., Chang, Y., Sandnes, F. (2010). Experiences with RFID-based interactive learning in museums. *International Journal of Autonomous and Adaptive Communication Systems, 3*(1), 59–74.

Jameson, A., Willemsen, M., Felfernig, A., de Gemmis, M. Lops, P., Semeraro, G., & Chen, L. (2015). Human decision making and recommender systems. In *Recommender Systems Handbook* (pp. 619–655). Springer.

Jannach, D., Bundgaard-Joergensen, U. (2007). SAT: A Web-based interactive advisor for ivestor-ready business plans. In International Conference on e-Business (ICE-B 2007) (pp. 99–106).

Jannach, D., Zanker, M., Felfernig, A., & Friedrich, G. (2010). *Recommender systems—An introduction.* Cambridge University Press

Janssen, J., Broek, E., & Westerink, J. (2011). Tune in to your emotions: a robust personalized affective music player. *User Modeling and User-Adapted Interaction (UMUAI), 22*(3), 255–279.

Kapoor, N., Chen, J., Butler, J., Fouty, G., Stemper, J., Riedl, J., & Konstan. J. (2007). Techlens: a researcher's desktop. In *1st Conference on Recommender Systems* (pp. 183–184). Minneapolis, Minnesota, USA.

Kersten, M., Murphy, G. (2010). Using task context to improve programmer productivity. In *14th ACM SIGSOFT Intl. Symposium on Foundations of Software Engineering* (pp. 1–11).

Konstan, J., Miller, B., Maltz, D., Herlocker, J., Gordon, L., & Riedl, J. (1997). GroupLens: applying collaborative filtering to Usenet news. *Communications of the ACM, 40*(3), 77–87.

Konstan, J., & Riedl, J. (2012). Recommender systems: From algorithms to user experience. *User Modeling and User-Adapted Interaction (UMUAI), 22*(1), 101–123.

Konstan, J., & Riedl, J. (2012). Recommender systems: From algorithms to user experience. *User Modeling and User-Adapted Interaction (UMUAI), 22*(1–2), 101–123.

Koren, Y., Bell, R., & Volinsky, C. (2009). Matrix factorization techniques for recommender systems. *IEEE Computer, 42*(8), 30–37.

Lee, T., Park, Y., & Park, Y. T. (2008). A time-based approach to effective recommender systems using implicit feedback. *Expert Systems with Applications, 34*(4), 3055–3062.

Leitner, G., Fercher, A., Felfernig, A., & Hitz, M. (2012). Reducing the entry threshold of AAL systems: Preliminary results from Casa Vecchia. *13th Intlernational Conference on Computers Helping People with Special Needs* (pp. 709–715). Linz, Austria.

LeMay, M., Haas, J., & Gunter, C. (2009). Collaborative recommender systems for building automation. *Hawaii International Conference on System Sciences* (pp. 1–10). Waikoloa, Hawaii, USA.

Li, W., Matejka, J., Grossmann, T., & Fitzmaurice, G. (2015). Deploying community commands: A software command recommender system case study. *AI Magazine, 36*(3), 19–34.

Lin, C., Shen, X., Chen, S., Zhu, M., & Xiao, Y. (2019). Non-compensatory psychological models for recommender systems. In: *33rd AAAI Conference on Artificial Intelligence (AAAI-19)* (pp. 4304–4311). Honolulu, Hawaii, USA

Linden, G., Smith, B. & York, J. (2003). Amazon.com Recommendations—Item-to-item collaborative filtering. *IEEE Internet Computing, 7*(1), 76–80.

Mandl, M., Felfernig, A., Tiihonen, J., & Isak, K. (2011). Status quo bias in configuration systems. *24th Intlernational Conference on Industrial Engineering and Other Applications of Applied Intelligent Systems (IEA/AIE 2011)* (pp. 105–114). Syracuse, NY, USA.

Martin, F., Donaldson, J., Ashenfelter, A., Torrens, M., & Hangartner, R. (2011). The big promise of recommender systems. *AI Magazine, 32*(3), 19–27.

Masthoff, J. (2011). Group recommender systems: Combining individual models. *Recommender Systems Handbook* (pp. 677–702).

McCarey, F., Cinneide, M., & Kushmerick, N. (2005). Rascal—A recommender agent for agile reuse. *Artificial Intelligence Review, 24*(3–4), 253–273.

McCarthy, K., Salamo, M., Coyle, L., McGinty, L., Smyth, B. & Nixon, P. (2006). Group recommender systems: a critiquing based approach. In *International Conference on Intelligent User Interfaces (IUI'06)* (pp. 267–269), Sydney, Australia.

Misirli, A., Bener, A., & Kale, R. (2011). AI-based software defect predictors: applications and benefits in a case study. *AI Magazine, 32*(2), 57–68.

Mobasher, B., & Cleland-Huang, J. (2011). Recommender systems in requirements engineering. *AI Magazine, 32*(3), 81–89.

Pazzani, M., & Billsus, D. (1997). Learning and revising user profiles: The identification of interesting web sites. *Machine learning, 27*, 313–331.

Peischl, B., Zanker, M., Nica, M., & Schmid, W. (2010). Constraint-based recommendation for software project effort estimation. *Journal of Emerging Technologies in Web Intelligence, 2*(4), 282–290.

Pinxteren, Y., Gelijnse, G., & Kamsteeg, P. (2011). Deriving a recipe similarity measure for recommending healthful meals. *16th International Conference on Intelligent User Interfaces* (pp. 105–114). Palo Alto, CA, USA.

Polat Erdeniz, S., Felfernig, A., & Atas, M. (2019a). Learned constraint ordering for consistency based direct diagnosis. In *International Conference on Industrial, Engineering and Other Applications of Applied Intelligent Systems* (pp. 347–359). Graz, Austria.

Polat-Erdeniz, S., Felfernig, A. Atas, M., & Samer, R. (2019b). Matrix Factorization based heuristics for constraint-based recommenders. In *34th ACM/SIGAPP Symposium on Applied Computing (ACM/SAC'19)* (pp. 1655–1662). ACM: Limassol, Cyprus.

Pribik, I., & Felfernig, A. (2012). Towards persuasive technology for software development environments: an empirical study. In *Persuasive Technology Conference (Persuasive 2012)* (pp. 227–238).

Ramiez-Gonzales, G., Munoz-Merino, P., & Delgado, K. (2010). A collaborative recommender system based on space-time similarities. *IEEE Pervasive Computing, 9*(3), 81–87.

Ramos, C., Augusto, J., & Shapiro, D. (2008). Ambient intelligence—The next step for artificial intelligence. *IEEE Intelligent Systems, 23*(2), 15–18.

Reiter, R. (1987). A theory of diagnosis from first principles. *AI Journal, 23*(1), 57–95.

Robillard, M., Walker, R., & Zimmermann, T. (2010). Recommendation systems for software engineering. *IEEE Software, 27*(4), 80–86.

Roy, L., & Mooney, R. (2004). Content-based book recommending using learning for text categorization. *User Modeling and User-Adapted Interaction, 14*(1), 37–85.

Sabin, D., & Weigel, R. (1998). Product configuration frameworks—A survey. *IEEE Intelligent Systems, 14*(4), 42–49.

Samer, R., Atas, M., Felfernig, A., Stettinger, M., Falkner, A. & Schenner, G. (2018). Group decision support for requirements management processes. In *20th Workshop on Configuration* (pp. 19–24). Graz, Austria.

Samer, R., Stettinger, M., Atas, M., Felfernig, A., Ruhe, G., & Deshpande, G. (2019). New approaches to the identification of dependencies between requirements. In *31st International Conference on Tools with Artificial Intelligence (ICTAI'19)* (pp. 1265–1270). Portland, OR, USA: IEEE.

Schafer, J., Konstan, J., & Riedl, J. (2011). E-commerce recommendation applications. *Journal of Data Mining and Knowledge Discovery, 5*(1–2), 115–153.

Smyth, B. (2018). Fast starters and slow finishers: A large-scale data analysis of pacing at the beginning and end of the marathon for recreational runners. *Journal of Sports Analytics, 4*(3), 229–242.

Sommerville, I. (2007). *Software Engineering*. Pearson.

Stanik, C., & Maalej, W. (2019). Requirements intelligence with OpenReq analytics. In *27th International Requirements Engineering Conference (RE'19)* (pp. 482–483). Jeju Island, South Korea: IEEE.

Stettinger, M., Felfernig, A., Leitner, G., & Reiterer, S. (2015). Counteracting anchoring effects in group decision making. *23rd Conference on User Modeling, Adaptation, and Personalization (UMAP'15)* (pp. 118–130). Dublin, Ireland.

Stettinger, M., Felfernig, A., Leitner, G., Reiterer, S., Jeran, M. (2015). Counteracting serial position effects in the CHOICLA group decision support environment. In *20th ACM Conference on Intelligent User Interfaces (IUI2015)* (pp. 148–157). Atlanta, Georgia, USA.

Stettinger, M., Felfernig, A., Pribik, I., Tran, I., Samer, R., et al. (2020). KnowledgeCheckR: Intelligent techniques for counteracting forgetting. In *24th European Conference on AI*, Santiago de Compostela, Spain.

Tayebi, M., Jamali, M., Ester, M., Glaesser, U., & Frank, R. (2011). Crimewalker: A recommender model for suspect investigation. In *ACM Conference on Recommender Systems (RecSys'11)* (pp. 173–180). Chicago, IL, USA.

Teppan, E., & Felfernig, A. (2012). Minimization of decoy effects in recommender result sets. *Web Intelligence and Agent Systems, 1*(4), 385–395.

Terveen, L., & Hill, W. Beyond Recommender systems: helping people help each other. In *HCI in the New Millennium* (pp. 487–509). Addison-Wesley.

Thiesse, F., & Michahelles, F. (2009). Building the Internet of Things using RFID. *IEEE Internet Computing, 13*(3), 48–55.

Thorleuchter, D., VanDenPoel, D., & Prinzie, A. (2010). Mining ideas from textual information. *Expert Systems with Applications, 37*(10), 7182–7188.

Tran, T., Atas, M., Le, V., Samer, R., & Stettinger, M. (2019). Towards social choice-based explanations in group recommender systems. *27th ACM Conference on User Modeling, Adaptation and Personalization* (pp. 13–21). Larnaca, Cyprus.

Tran, T., Atas, M., Felfernig, A., & Stettinger, M. (2018). An overview of recommender systems in the healthy food domain. *Journal of Intelligent Information Systems (JIIS), 50*, 61–70.

Tuzhilin, A., Koren, Y. (2008). *2nd KDD Workshop on Large-Scale Recommender Systems and the Netflix Price Competition* (pp. 1–340).

Wilson, D., Leland, S., Godwin, K., Baxter, A., Levy, A., Smart, J., et al. (2009). SmartChoice: An online recommender system to support low-income families in public school choice. *AI Magazine, 30*(2), 46–58.

Winoto, P., & Tang, T. (2010). The role of user mood in movie recommendations. *Expert Systems with Applications, 37*(8), 6086–6092.

Wobcke, W., Krzywicki, A., Kim, Y., Cai, X., Bain, M., Compton, P., et al. (2015). A deployed people-to-people recommender system in online dating. *AI Magazine, 36*(3), 5–18.

Xu, S., Jiang, H., & Lau, F. (2008). Personalized online document, image and video recommendation via commodity eye-tracking. In *ACM Conference on Recommender Systems (RecSys'08)* (pp. 83–90).

Yuan, N., Zheng, Y., Zhang, L., & Xie, X. (2012). T-finder: A recommender system for finding passengers and vacant taxis. *IEEE Transactions on Knowledge and Data Engineering (TKDE)*, 1–14.

# An Overview of How VR/AR Applications Assist Specialists in Developing Better Consumer Behavior and Can Revolutionize Our Life

**Rocsana Bucea-Manea-Țoniș, Elena Gurgu, Oliva Maria Dourado Martins, and Violeta Elena Simion**

**Abstract** Augmented reality (AR) and virtual reality (VR) have become the presence of daily life. We draw on outstanding research to substantiate that VR/AR are scientific tools that can be integrated in many fields of research, in learning, in marketing campaigns and product design, in psychology, medicine, economy, etc., meeting the principles of circular economy at the same time. Due to the fulminant evolution of communication and interconnected modern devices, VR/AR are more and more present in every business, in every research field. VR/AR facilitate the implementation of simulations and experiments under a safe environment, avoiding possible damages and spending money on expensive technologies. Our study, using references on VR and AR applied in current activities and different fields of business, shows what the current state of knowledge in the field is and opens interesting perspectives regarding the impact of VR and AR on the future market and research, building an interdisciplinary bridge between technology and learning, psychology, medicine, economy (marketing, tourism, and industry). The context in which these technologies will be developed must be seen not only from the perspective of a particular field, but especially as a concrete impact on humans.

**Keywords** VR · AR · Real life · Psychology · Medicine · Economy · Consumer behavior

R. Bucea-Manea-Țoniș (✉)
Doctoral School, University of Physical Education and Sports, Bucharest, Romania
e-mail: rocsense39@yahoo.com

E. Gurgu
Department of Economic Science, Spiru Haret University, Bucharest, Romania

O. M. D. Martins
Department of Business Science, Institute Polytechnic of Tomar, Tomar, Portugal

V. E. Simion
Department of Veterinary Medicine, Spiru Haret University, Bucharest, Romania

© The Author(s), under exclusive license to Springer Nature Singapore Pte Ltd. 2021
T. Dutta and M. K. Mandal (eds.), *Consumer Happiness: Multiple Perspectives*,
Studies in Rhythm Engineering,
https://doi.org/10.1007/978-981-33-6374-8_12

## Introduction

Advances in technology, such as the Internet of things (IoT), augmented reality (AR), virtual reality (VR), mixed reality (MR), or artificial intelligence (AI), bring significant value to many fields today (education, economics, medicine, architecture, etc.) but also improve our daily lives. These technologies are already attractive to most people, from vocational training to personal development as a way to improve reality.

VR is a computer-generated experience taking place within a simulated environment that provides auditory, visual and sensory feedback, simulating a real-world event/place and offering the possibility of experimenting fantastical environments or finding innovative future applications in different fields. AR overlaps the physical environment of a person with digital images usually generated by a mobile device. On the other hand, VR is an immersive reality technology that involves wearing a headset that creates a 360° simulation that places the user in a digital environment. AR offers the ability to view three-dimensional images due to a live camera feed into a headset or through a smartphone or tablet device (Bucea-Manea-Țoniș, Pistol, & Gurgu, 2019). VR uses headsets or multi-projected environments, and the person using virtual reality equipment is able to interact with virtual features or items. Furthermore, VR is frequently immersed in entertainment, scientific research, social networks, artistic creation, and psychotherapy. In art therapy, VR is used for three-dimensional painting, an immersive creative experience, dynamic scaling, and embodied expression, due to its characteristics, such as presence, immersivity, point of view, and perspective (Hacmun, Regev, & Salomon, 2018).

On the other hand, AR was defined as augmenting natural feedback to the operator with simulated cues, being a form of virtual reality where the participant's head-mounted display is transparent, allowing a clear view of the real world (Milgram, Takemura, Utsumi, & Kishino, 1995). Now, MR represents an AR+ that places objects or creatures just like AR, but with an ability to engage in the real, physical world around them and vice versa. Most likely, they will evolve in mobile AR and location-based VR. These technologies offer new methods and tools for environment protection, increasing the online traffic, reducing the footprint of carbon dioxide, air and soil pollution. 3D game engines are used for environmental research allowing system multimodal interaction, gesture capturing, and voice input. The participants (Kefi, Hoang, Richard, & Verhulst, 2018) prefer voice command for best performance. Taking into account the amount in which they will be produced in the next few years, it is necessary to evaluate their impact over the environment, in the context of different usage patterns. Virtual technologies should be an integral part of the circular economy. Furthermore, the relationship between economy/industry and education has been a pillar of creativity regarding evolutionary technologies as shown in the model of innovation and creativity exchange between universities and the business field (Englund, Olofsson, & Price, 2017). The interaction with the business field is crucial, because it comes with the technology (e.g., VR, AR, halo effects) that offers students and specialists a certain environment and tools for exploration and re-creation of the world. Modern students are coming to classes in order experience interactive and experimental learning through bending reality, not just for listening. The interaction with the business field is crucial, because it comes with the technology (VR/AR, halo effects) that offers students and specialists the environment and tools for exploration and re-creation of the world. These technologies have been adopted even in real-life activities (Akçayır & Akçayır, 2017; Alexander, Westhoven, & Conradi, 2017).

Virtual technologies are clear, strong tendencies for the communication and education of the future. The advantages are obvious and positive, both for the individual and for society as a whole. Despite some studies showing a possible negative impact on humans (Garrett et al., 2018; Klempous, Kluwak, Idzikowski, Nowobilski, & Zamojski, 2017; Plechawska-Wojcik & Semeniuk, 2018; Ribeiro, Martins, & Garcia, 2019; Stepanova, Quesnel, & Riecke, 2018; Zaidi, Duthie, Carr, & Abd El Maksoud, 2018), these technologies are part of our lives and will continue to develop and become ever more present. In the age of technology, imagination, innovation, inspiration is the only limits of human growth. This overview shows the extent to which virtual technologies can help specialists in improving but also in revolutionizing their lives, in the context of a better consumer behavior.

## VR and AR Immersion in Real Life

The study is limited to the research results regarding the application of VR technology in the field of science and academics, which is their trend, and if they can be widely applied in education. This is a comparatively short period for academic research, meant to show the increase in the importance of VR/AR in recent years and not an exhaustive study. Also, besides VR/AR applied in current activities and different fields of education and business, we show if these technologies can affect human health.

VR and AR are becoming more and more popular, offering unlimited possibilities to improve computer applications in education, using this technology (Bucea-Manea-Țoniş et al., 2019):

- VR is associated with the feeling of being transported effectively to a new event or place, and this feature is called presence.
- A pleasant and intense presence is associated with the user's tactile response when immersed in VR, this phenomenon being called haptic.
- Users can more easily focus on a VR experience if they use an head-mounted display (HMD).
- For an intense and interesting experience, the manager must add interactivity in VR.

- The VR experience can be enhanced with 360° videos, even without wearing a VR headset, allowing customers to see everything around using a mouse, trackpad or moving their physical phone.
- Adding the fourth dimension to a VR adventure is another way to enhance the experience—4D VR.
- Adding the stereoscopic effect, capturing two photographs/videos at slightly different angles and looking deeply if viewed together.
- Creating a 360° experience using many cameras.
- Moving pictures and videos at the same time as users' positions in the head or eyes.
- Maximizing the "field of view," extending as far as possible the VR experience, to get closer to the 360° view.
- Confuse the interval between a VR stimulus and a response.
- It creates confusion when a person receives visual cues of a movement, without being physically registered.

VR/AR are frequently used for collaboration and for sharing knowledge between different scientists, helping them to discover and develop new brain comprehensive models with clinical dimension empowered by computer science, engineering, and allied sciences (Cipresso, Giglioli, Raya, & Riva, 2018). The therapist can explore a wide variety of controlled stimuli and measure the participants' responses based on which they may suggest a therapeutic solution. VR offers the illusion of a parallel life for the participants, of a different body, experiencing new activities in different places than the ones where they really are: white people embodied in dark-skinned virtual bodies diminish the racial bias; offender males embodied in women bodies (victims) changed perspectives on domestic violence; adults embodied in a child body play more, etc. (Barberia, Oliva, Bourdin, & Slater, 2018; Bergstrom, Papiotis, Saldanha, Azevedo, & Slater, 2017; Hasler, Spanlang, & Slater, 2017; Seinfeld et al., 2018; Skarbez, Neyret, Brooks, Slater, & Whitton, 2017). Furthermore, the interaction between neuroscientists and engineers led to the development of hands with contactless devices (i.e., without gloves), the leap motion device, tactile and haptic device, the motion tracking system, and the concept of "sonoception." These newly developed devices allow almost natural interaction with virtual environments (VE) and an increased sense of presence (Azevedo et al., 2017; Di Lernia, Cipresso, Pedroli, & Riva, 2018a; Di Lernia et al., 2018b; Riva, Wiederhold, & Mantovani, 2018). VE technology is used to expose individuals to live information in an experiential learning game, in which participants use water for different tasks. Participants received exaggerated feedback to intensify the negative consequences of water consumption and/or environmental damage that emphasized personal affective responses (Hsu, Tseng, & Kang, 2018). Furthermore, VR is frequently immersed in entertainment, scientific research, social networks, artistic creation, and psychotherapy. In art therapy, VR is used for 3D painting, a deeply engaging creative experience, dynamic scaling, and embodied expression, due to its characteristics, such as presence, immersivity, point of view, and perspective (Hacmun et al., 2018). Also, the 3D games engines are used for environmental research allowing system multimodal interaction, gesture capturing, and voice input. For best performance (Kefi et al., 2018; Schmidt, Beck, Glaser, & Schmidt, 2017), participants prefer voice command.

Immersive virtual environment technology is now used in Taiwan to expose individuals to live information in an experiential learning game, in which participants were situated in a virtual bathroom and asked to repeatedly use a 600-ml bottle to fill a specific water tank for washing tasks. Participants received exaggerated feedback to intensify the negative consequences of water consumption and/or environmental damage that emphasized personal affective responses. The game has caused significant changes in cognition and behavior intention for Taiwan' residents, helping the water conservation. These experiences can be reproduced in other countries characterized of water shortages (Hsu et al., 2018).

These technologies bring interactivity, dynamics, learning by doing effect, testing in a virtual safe environment, 3D visualization, experimenting new sensations (e.g., flying imponderability), in a safe mode. Most of all the subjects of VR experiences declared that they were having fun when testing it and had the feeling of mindfulness, freedom, and higher curiosity.

## *VR and AR—A Bright Future in Psychology and Medicine*

In psychology, VR tools help the specialists analyze where experiment subjects consider that their bodies are located (e.g., in one or more specific part(s) of their body), due to the perception and self-consciousness. The VR headset showed that subjects mainly located themselves in the upper face and upper torso (Van der Veer, Alsmith, Longo, Wong, & Mohler, 2018). VR is also used for analyzing crowd behavior in order to avoid collision risk in pedestrian traffic and mass events (Meerhoff, Bruneau, Vu, Olivier, & Pettre, 2018).

Moreover, VR technology and social robotics were designed to help children with autism disorder. The system integrates a virtual teacher instructing sight words and included a humanoid robot emulating a peer. The results were very good: 100% of the words were acquired, maintained, and generalized by participants. It improved the operator's reaction time, awareness, hit-rate, and performance (Oyekan et al., 2019; Saadatzi, Pennington, Welch, & Graham, 2018; Tang, Webb, & Thrower, 2019). Commercially, VR and video games are used as active distraction methods to mitigate subjective discomfort, increasing pain tolerance (Boylan, Kirwan, & Rooney, 2018; Erra, 2019). A VR scenario was created in order to asses and treat people experiencing paranoia in social situations (Riches et al., 2018).

Very similar with psychology, in medicine, the VR therapy gained notoriety in clinical conditions such as anxiety disorders (most of the studies), eating disorders, addictions, phobias, stroke rehabilitation, posttraumatic stress disorder, pediatric cancer-related treatment, hemodynamic in cerebral aneurysms, and for pain management, autism, depression, psychosis, schizophrenia although there are negative aspects, too.

As Tsai et al. (2018) demonstrated, the VR/AR therapies are very efficient in claustrophobia: In VR environment, the participant's heart rates variability (HRV) is significantly better. In the therapy of mental illness, VR/AR are also recognized as effective intervention and care therapy (Wiederhold, Miller, & Wiederhold, 2018). Thus, technologies such as VR/AR and brain–computer interfaces (BCI) can be successfully used as therapy for auditory verbal hallucination (AVH) in patients with schizophrenia (Fernandez-Caballero et al., 2017). Oculus rift head-mounted-display is a potentially powerful tool for a wide array of basic research and clinical applications (Chessa, Maiello, Borsari, & Bex, 2019; Liu et al., 2018). Other studies in hemodynamics had successful results in cerebral aneurysms magnetic resonance imaging (4D pcMRI) immersed in AR environment (Karmonik et al., 2018). Introducing serial section immunological original 3D model viewing and annotation in VR was a major breakthrough in quality control (Steiniger, Wilhelmi, Berthold, Guthe, & Lobachev, 2018). Nowadays, VR offers viable software for testing the effect of acute static stretching (ASS) on the lower limb reaction time. ASS decreases the risk of falling and injuries in situations requiring a rapid reaction (Ameer & Muaidi, 2018; Schuster-Amft et al., 2018). Additionally, physiotherapy is one of the forms of healthcare for which VR/AR have many advantages, especially in terms of monitoring and analysis of muscle activity and patient movements. Postolache (2017) indicates a set of technologies that can detect and monitor from a distance the rehabilitation of upper and lower limbs: sensors based on microwave Doppler radar and infrared technologies (e.g., infrared LED, laser, and thermograph). Likewise, the vestibular otolithic disorders can be rehabilitated through Curator subjective visual vertical (SVV) assessments that allow the ability to perceive verticality. It can be used in clinical contexts, such as at the bedside or in head and body positions (Martelli, Xia, Prado, & Agrawal, 2019; Chiarovano, McGarvie, Szmulewicz, & MacDougall, 2018). VR was also the proposed solution for rehabilitation of lateropulsion patients (Tsatsis et al., 2017), and in surgical medical practice, and professional education (Yu, Lee, & Luo, 2018a, 2018b). There are studies that debate hand grasps identification in VR due to signals acquired by FMG (force myography), a muscular hand gesture recognition method, and also by leap motion (Jiang, Xiao, & Menon, 2018; Van Dam & Stephens, 2018). These technologies support therapies and should become a current part of present and future medicine. Although there is now a tendency to increase the number of publications regarding the impact of virtual technologies in medicine, studies show that they are rather used in investigations—computer-assisted surgery, three-dimensional imaging, and computed X-ray tomography, and less in therapy, showing a clear trend in technologies assisting the actual treatment of patients, compared to technologies in a training environment (Eckert, Volmerg, Friedrich, & Christoph, 2019).

In veterinary medicine, VR and AR can be used for teaching anatomy (Seo et al., 2018), and surgical planning (Xu et al., 2018). As well, one of the VR technologies with a good applicability in veterinary medical training is the haptic simulator, a combination of haptic feedback to palm and fingers, and a visual digital simulation of the anatomical structures, while AR technologies are helpful to clinicians in viewing complex data from 3D scanning (Son & Park, 2018).

## VR/AR Will Undoubtedly Have a Major Impact on the Economy

The efficiency of these technologies should not be seen only as an impact on the individual; in short, individuals live in a society and carry out an activity. In the economy, VR technologies bring benefit in the management of urban facilities. In this regard, Google Tango software tracks the life cycle of the underground infrastructure and gets exact location and orientation in just a few minutes without using additional markers or hardware (Gregorio, Ortega, & Feito, 2018). Similarly, in domestic and industrial robotics, as well as autonomous vehicles, VR facilitates simultaneous localization and mapping (SLAM). The latter allows selecting and configuring the appropriate algorithm, hardware, and compilation path to meet performance, precision, and power goals (Saeedi et al., 2018). Likewise, VR is very useful in electro-energetic industrial plants in order to track virtualization at work through an AR system for mobile devices, as a tool for protecting health and safety, as well as to secure performance of tasks in a technological process (Tatic, 2018), which shows the impact of these technologies both in industry and medicine at the same time.

In construction, architecture for example, these technologies are also beginning to find their place. VR brings key advantages for indoor simulations like room acoustics. During processing a dynamic scene, it is easier to obtain a visual rendering, but much more difficult to get high-quality audio rendering, because sound waves propagate substantially slower than light. Through VR has been introduced a hierarchical state-based data structure with time history, which fulfills the requirements for outdoor auralizations and for indoor simulations (Wefers & Vorlaender, 2018). VR technology is also used with success in archaeoacoustics for reconstruction of the historical sound inside heritage buildings, in order to characterize the sound of the past inside that building by using virtual sound reconstruction. A test had been done in Islamic temples—West the Aljama Mosque of Cordoba, where the original state of the Mosque has been reconstructed in the different spatial configurations throughout its history from the eighth to the tenth century (Suarez, Alonso, & Sendra, 2018). In addition, the interactive system based on VR provides users with a real-world experience in the development of open architecture products. The system allows users to review a product design by operating and evaluating the product in VR, by recording user operations and sending feedback to designers to improve the product (Song, Chen, Peng, Zhang, & Gu, 2018). The maintenance of complex products can save substantial costs and reduce incidents and accidents throughout the life cycle of the products and therefore should be fully considered in the early design stages.

Nowadays, innovative products are the result of an open communication between client and producers. Thus, VR/AR are an extraordinary tool for product maintainability design through immersive and non-immersive simulations. The method of virtual maintenance integrates human motion data from simulations, diminishing the problems regarding cumbersome operation, time-consuming labor, and inadequate precision that arise in complicated and repeated operations (Geng et al., 2018). A manual in AR compliance to Industry 4.0 principles was designed (Gattullo et al.,

2019) as innovative productivity increases accordingly with the principles of circular economy. A multi-user design review experience in which all designers, engineers, and end-users actively cooperate within the interactive VR with their own head-mounted display, seems more suited to detect relevant errors than standard systems characterized by mixed usage of assets (Rigutti et al., 2018). The maintenance of complex products can save substantial costs and reduce incidents and accidents throughout the life cycle of the product and should be fully considered in the early design stages. In the virtual environment, the user can have intuitive feelings and can interact with virtual objects, which offers unlimited possibilities of immersive simulations for collecting and analyzing the state of maintenance. A case study that applies immersive maintainability verification and evaluation system (IMVES) in an aero-motor project is presented to demonstrate the system's effectiveness and feasibility (Guo et al., 2018).

In marketing, VR experience has to be accepted as a lasting change to consumer behavior (Verhulst, Normand, Lombart, Sugimoto, & Moreau, 2018). Moreover, organizations have to be in permanent contact with customers to evaluate the function and performance of the product (Gârdan et al., 2018). VR and AR dedicated to future personalized learning systems.

## VR and AR Dedicated to Future Personalized Learning Systems

Nowadays, students must be seen by universities as consumer of knowledge, because knowledge brings value added to economic system and will have a returning positive impact on the universities that offer the most performant specialist to the market.

Millennials cannot be caught up in the old educational system. They learn by experimenting, by creating their own 3D virtual reality, by bending reality. Some studies analyze how children collaborated to create their own stories in a 3D VR environment (Lin & Hsu, 2017; Schmidt et al., 2017; Yeh, Lan, & Lin, 2018). Various studies reveal the fields in which students consider VR/AR a useful tool for sharing and comprehending knowledge and creates for those who have experimented a strong sense of perception, involving the students in the activity, and motivating them in the future: *Mathematics and Natural Sciences* for simulation and testing, *Social sciences* to enrich the students environment with a simulated reality, *Economic sciences* for simulations, *Technical sciences* for developing models/blueprints, *Humanistic sciences* to achieve specific environment in accordance with the taught theme, *Arts and architecture* for models/blueprints, *Medicine,* for the study of anatomy and dissections, for simulating laser operations (Detyna & Kadiri, 2019; Kharitonova, Kharitonova, & Pulyaeva, 2019; Zhihan, Xiaoming, & Wenbin, 2017). Virtual reality technologies bring benefits in the teaching process and can be a useful tool in creating innovative and educational experiences, and helping students to improve their skills and knowledge (Ardiny & Khanmirza, 2018; Mikolajczyk, 2019).

According to Mekacher (2019), the virtual environment allows us to develop the ability to interact and experiment, in a similar way to the real world. Through visualization, the AR-Glass offers additional virtual information that, inserted in a context and with objects associated with cognitive and psycho-physiological objectives, and pedagogical strategies oriented to the teaching objectives, allows the application of these technologies in different areas of education. VR is often used in learning for ludic activities (e.g., gamification), comprehension, and conceptualization. VR/AR engagement in learning is associated with a high degree of motivation, creativity, and innovation in students. From a neurological perspective, AR doubles the visual attention in the brain. AR experiences have a very high rate (70%) of encoding into memory (Fombona, Pascual-Sevillano, & Gonzalez-Videgaray, 2017). Additionally, e-learning comes to support the balance between lucrative, learning, and social life (Sousa & Rocha, 2019). VR/AR facilitate better comprehension of studied material, due to 3D visualization, interactiveness, clarity, and learning by doing effect (Sousa & Rocha, 2019). Professional career is enhanced by the knowledge acquired through the e-learning method, exercised during the spare time, offering a work-life balance (Encalada & Sequera, 2017; Mumtaz et al., 2017; Yip, Wong, Yick, Chan, & Wong, 2019). Although the use of virtual platforms requires digital education for teachers, studies have shown that they are able to accomplish the tasks with 37–64% less user operations, and up to 72% less machine operations (Horvath, 2018). Fernandez (2017) proposes a six-step methodology to aid adoption of these technologies in education: "training teachers; developing conceptual prototypes; teamwork involving the teacher, a technical programmer, and an educational architect; and producing the experience, which then provides results in the subsequent two phases, wherein teachers are trained to apply augmented and virtual reality solutions within their teaching methodology using an available subject-specific experience and then finally implementing the use of the experience in a regular subject with students." This proposal can be solution for including VR technologies in education.

Other studies converge in the same direction: a study conducted by the Open University of Catalonia shows that collaborative skills are developed by gamification, MR, and social media, having a much more positive impact than other Information and Communication Technology (ICT) tools (media, wikis, open educational resources, cloud for sharing files with peers, etc.) (Martinez-Cerda, Torrent-Sellens, & Gonzalez, 2018). Similar studies conducted in the Middle East University show that VR allows students to discover and explore their own knowledge, making learning more interesting and fun (Alfalah, 2018; Voinescu & David, 2019). Other authors identified and taxonomized the elements and the factors that affect learner engagement in virtual worlds, when hybrid virtual learning models are used. This method mitigated the drawbacks of each educational approach and broadened the network of interactions (Christopoulos, Conrad, & Shukla, 2018). Modern students prefer collaborative learning and to be placed in contextual environments, through VR and halo effects, while taking social interactions. Thus, they are "projected in time, in the era of study" (Alonso, Prieto, García, & Corchado, 2018). It is naturally somewhat for these generations to look for and to find in their future work/activities such applications.

Applications such as Nearpod VR or AVR creator EON (developed by the Technical University of Graz) and Social Internet of Things (SIoT) allow virtual interaction on Facebook. SIoT allows teams to interact in VR and AR with real-world objects through 360° selfie video in order to develop projects together. It also allows realistic 3D student–teacher interaction in the virtual world, making available to the student/teacher VR/AR libraries or the analytical database. In this way, the teacher can grant mixed feedback through the AR application (Fig. 1).

Nowadays, activities in every field become more digital, due to the 5G, cloud computing, and VR revolution. These technologies change the way we work and interact, which is likely to create new opportunities, but also challenges. Some jobs are assimilated and done by computers, but other new digital jobs arise into the new circular and virtual economy. Digital competencies are essential for current and future employees. Lately, high-skilled qualifications and labor have increased. Overall, although VR and AR technologies already exist, they are not adopted in most of the universities or companies because teachers, students or workers aren't qualified to use digital technologies, companies are naturally much more likely to use them (Andrews, Nicoletti, & Timiliotis, 2018). Here comes the university key-role

**Fig. 1** VR and AR immersed in education process (www 1–6)

in adopting new approaches, adding new programs, changing curricula, including data literacy, designing new courses, using the most advanced technologies and creating new learning environments dedicated to research and education as to prepare specialist able to face current and future challenges. The physical learning environment has to be improved with environments for problem-based learning and learning by doing activities for interdisciplinary teams, that use blended learning. Students from different fields such as computer science and social sciences can be brought in the same team to solve problems due to cross-fertilization knowledge (an example on how digital skills promote entrepreneurial and management skills). These specialists will shape the digitally transformed society, that have already started to be present (cyber-marketing is more efficient and more convenient than classic publicity in newspaper and TV, online transactions, distance learning, e-commerce, etc.). For students with digital competences will be fun to use it in their future work and thus the economy will benefit on it, the more so as are in accordance with the green and circular economy principles.

As a practical solution, we recommend a symbioze between universities and VR/AR software development companies. The universities should be able to rethink a new strategic management regarding the request of digital competencies and implement it in the education process. The VR/AR software development companies have the role to develop new VR/AR applications, dedicated to different type of course, to be taught in the university.

The complexity of user data that can be produced or recorded in VR experiences is the main advantage that universities must benefit. It does not really matter whether such information is gathered for user testing/playtesting, for scientific or clinical inquiries or as an outcome measure for training or educational VR/AR applications—fact is that VR/AR applications can produce rich datasets of significant value for the modern student. The immersion of VR/AR technology in education "brings the advantage of covering multiple intelligences, it facilitates the observation and exploration, it drives to consolidation performance and achievement, experimental self-learning, increases motivation and interactivity between students, in contrast to the traditional educational style that present the content in a two-dimensional format. AR learning involves many senses: seeing, touching, hearing, and smelling (Takkaç Tulgar, 2019)." Thus, VR/AR technologies would increase the innovation and would represent a solution to OECD concern regarding the Europe's sovereignty, that will no longer be dependent by American or Chinese technology.

The university has also the role of informing the students regarding new opportunities arising in the labor market based on ICT facilities beyond the walls of higher education institutions. They also have to emphasize the importance of an ethical behavior when using VR/AR technologies.

Furthermore, the gap in generations could also limit the use of this technology, with the millennial students being more used to technology in everyday life, while the more senior academics could be reticent to change of established practices and methods. Nevertheless, all users and developers should be aware and understand the benefits and disadvantages of these virtual technologies and be able to identify when and where these can be effectively used, with the aim of increased efficiency and quality.

## VR/AR Have Positive Factors Leading to Consumer Happiness

Positive psychology has begun to study a new concept: happiness. Happiness can be understood as a function of endogenous factors (biological, cognitive, personality, and ethical subfactors) and exogenous factors (behavioral, sociocultural, economic, geographical, life events, and aesthetic subfactors). According to Farhud, Malmir, and Khanahmadi (2014), biological (endogenous) subfactors were considered significant predictors of happiness. The results showed that genetic factors correspond to efficacy on happiness, mainly associated with emotion and humor (between 35 and 50%). The neuroscience has demonstrated the importance of neurotransmitters (such as dopamine, serotonin, norepinephrine, and endorphine) in the control of happiness, while other studies have defended the role of cortisol and adrenaline (adrenal gland) and oxytocin (pituitary gland) in the control of happiness. These chemicals generate happiness.

In order to briefly explain the role of these chemicals in promoting happiness, a brief concept is presented below: The first one is dopamine, which is the great feeling that favors achieving a goal, and it is turned on when the efforts are rewarded, the result being something like "I have done it!" feeling. On the other hand, it is recommended to be aware that dopamine is addictive; the second is serotonin, which is produced when the individual feels useful, important, belonging to the community; the third is oxytocin, which gives a good feeling, making one feel safe; the final one is endorphin, which is a brief hormone that keeps the goal going, overcoming the physical pain (Breuning, 2012). According to Baixauli Gallego (2017), dopamine was associated with happiness, and serotonin regulates our mood.

Aiming to assess the impact of digital technology on happiness, a survey was developed by Mochón (2018). The scientific approach considered that happiness was measured according to explanatory factors such as well-being, and the benefits were associated with happiness. These benefits were analyzed in terms of: (i) connecting people; (ii) broadening the structure of communities; (iii) crucial intelligence, which allows for evolution in different social dimensions (e.g., health, safety and science); (iv) contentment aiming to empower people to improve; and (v) continuation toward quality. According to the author, technology can have a positive impact on well-being when used well, without excesses.

VR is used with success in tourist consumer behavior. Travel agencies incite human curiosity related to experience new activities, sceneries, and cultures through VR expeditions, offering 360° scenic landscapes, popular museums details, and various attractions. Airline companies invited tourists with nausea and fly sickness to take VR flight experiences. Most of them were likely to try a real flight after the VR experience (Botella, Fernández-Álvarez, Guillén, García-Palacios, & Baños, 2017). A study on students from a technological university in Taiwan reveals that a virtual reality tour-guiding platform is a very useful tool for choosing remarkable tourist experiences. The tourist behavior was analyzed and conceptualized in a technology acceptance model based on the Unified Theory of Acceptance and Use of Technology (UTAUT) model (Chiao, Chen, & Huang, 2018). A study conducted by Marasco, Buonincontri, van Niekerk, Orlowski, and Okumus (2018) investigated whether the perceived visual appeal (PVA) of VR and the emotional involvement (EI) of users had a positive impact on the behavioral intentions to visit a cultural heritage in a destination. The results of the study revealed that this experience had a positive and significant effect on behavioral intentions compared to the site presented in the virtual experience but even more, it had a positive effect on EI (Marasco et al., 2018).

Moreover, an increased degree of satisfaction was noticed for tourists who visit museums that integrate VR experience: Posters were replaced with VR projections and/or videos were inhabited within virtual objects and shaped cross-objects user interfaces (COUIs) (Sun, Zhou, Hansen, Geng, & Li, 2018). Collecting and analyzing data from 36 visitors to a theme park for 12 months suggested that users appreciated their participation through feelings such as control, involvement, vividness, effectiveness, living, temporal association, and the pleasure of the VR experience (Wei, Qi, & Zhang, 2018). An experience that maximizes user satisfaction in the consumption of sports materials is virtual reality spectatorship (VRS). Kim and Ko (2019) show in a study that VRS *"amplified flow experience by vividness, interactivity, and telepresence to the grater extent than the traditional medium (2-D screen)."* Moreover, this experience has strongly influenced those less interested in sports compared to fans of extreme sports (Kin & Ko, 2019).

VR is used for evaluating the negative impact of windfarms on tourist behavior. If the tourists use a VR tool to conceptualize windfarms, their perceptions, attitudes, concerns, and behaviors related to it are altered: Their reactions are rather negative than in case of using text or drawings for conceptualization wind turbines (Teisl, Noblet, Corey, & Giudice, 2018). Technical equipment is so much easier to be sold through VR technology, due to the detailed specifications and distance testing.

Leading companies treat their employees from a consumer perspective, being very sensible to their needs and happiness. These companies create spaces dedicated to relaxation, VR/AR environments for meditation when having a work break, kindergartens for employees' kids, team-building events, special medical insurance, gym membership, yoga and pilates hour during work break, etc. Relaxing spaces enable employees to devote a few minutes to relaxation during the day, the kindergartens offer the mental comfort that their kids are safe and happy, can be visited from time to time, during the day. The VR/AR gaming or relaxing applications used for a short period will stimulate adrenaline or oxytocin that has as consequence the employee happiness. The medical insurance and the gym membership make the employees feel that they are important for the company, that their work counts, and become more responsive to company objectives and more responsible in achieving the performance in the tasks' implementation. These companies have been proven to be the most innovative and productive (Fig. 2).

**Fig. 2** Space for relaxation during working hours (www 7–10)

Furthermore, nowadays there is an expending trend regarding the positive impact of developing the activity/ the job that a person love, alternating working hours with relaxing breaks, eliminating stress, sleeping well, encouraging competition with themselves, and not with other colleagues, called mindfulness (Vonderlin et al., 2020; Wiencke & Cacace, 2016). This trend is expanding in university activities and research: "loving-kindness meditation (LKM) has been shown to improve well-being and positive emotions in clinical and non-clinical populations" (Totzeck et al., 2020; Van Dam et al., 2018).

## Has VR/AR Negative Impacts on Consumer Health?

In the literature, we found concerns regarding the ethics of VR effects on mental health risks. Other concerns refer to the neglect of users' own actual bodies and real physical environments and the use of new technologies affecting personal privacy and manipulation of users' beliefs, emotions, and behaviors.

VR can also blur the distinction between the real and illusory, and it could affect mental health (Klempous et al., 2017; Plechawska-Wojcik & Semeniuk, 2018; Stepanova et al., 2018, Zaidi et al., 2018). Nowadays, the IoT spreads all over the world and brings many advantages, but mobile devices (e.g., cell phones, VR devices) affect children's eyes and their brains absorb substantially higher local radiation doses than adults'. Thus, public education regarding manufacturers' advice to keep phones off the body, and prudent use to limit exposures, particularly to protect the young, are important (Fernandez, de Salles, Sears, Morri, & Davis, 2018; Spiegel, 2018; Teisl et al. 2018). The disadvantages of using VR in therapy are the problems of separating effects of media versus medium, costs, lack of technical standards, and practical in vivo issues (Garrett et al., 2018; Ribeiro et al., 2019). The negative effects associated with VR and AR are cyber-sickness (when eyes and the vestibular system send opposite/disrupted signal—the brain is confused and nausea appears), vertigo effects, derealization/dissociation effect (after removing the VR headset people tend to be detached, using their hand as they would use the controllers and feeling that they are in the virtual world, for another few minutes or hours), strain or fatigue of the eyes (people blink less than normal when using a digital screen device, but this can be compensated by the 20–20–20 rule: every 20 min, shift your eyes to look at an object at least 20 ft away for at least 20 s will be sufficient for long time usage). All these effects have proved to be manifested on short-term (Transon, Verhulst, Normand, Moreau, & Sugimoto, 2017; Vaziri, Liu, Aseeri, & Interrante, 2017; Yu et al., 2018a, 2018b). In any case, the positive impacts of XR overwhelm the negative ones. Even more, if the content of XR experience is positive (e.g., by meditation, holiday experiences, hobby experiences), the impact of XR on human health is positive (Navarro-Haro et al., 2017; Van Kerrebroeck, Brengman, & Willems, 2017).

## Conclusions

Together with AI and automation, VR/AR have been attractive for a few years now as types of technology that can have a profound transformation effect on the way people live and work. Modern technology (e.g., VR, AR, IoT, 5G) has enriched innovation in many research and study areas, having as a result new tools, methods, and implementations in fields such as medical or psychological treatment, product design, database management in industrial companies, marketing campaigns, interactive face-to-face or distance learning, and modern design architecture.

Since each coin has two faces, similarly VR/AR have disadvantages, in addition to having many advantages. It depends on how the technology is applied in each sector or field. Therefore, we should mention some of the negative effects of VR/AR on our health, such as VR disease, eye strain, dizziness, distortion, early myopia for children, security breach, no control over personal details, and addiction. Certainly, in our opinion, the benefits outweigh the weaknesses.

Many examples from different fields of study show that VR is increasingly adopted by enterprises and industries for all the value-chain steps, by research centers for deep insights and simulation, by clinics for complex cutting-edge treatments and innovation and more importantly in universities for creating learning VE that stimulates comprehension and learning through experiment, simulation, and innovation, or by common consumers for fun activities. Thus, VR improves our everyday life though the symbiosis of neuroscience, computer engineering, and social sciences. Tracking the evolution of VR/AR influence in sciences, we assume that future research will bring exponential benefits to humanity.

Universities have a responsibility to ensure that graduates have the skills to make use of new technologies, and they need to prepare them for labor markets where the probability of disruption is high, and the companies have to anticipate their future employees' capacities and abilities. Researchers and students can simulate and experiment under the security of VR environment.

Technology provides the tool, but the impact on happiness depends on how it is used. The sensation of being physically present produces enough adrenaline to influence the perception. The sensation of being transported into the event or place associated with the pleasant and intense feeling of the user's tactile response (haptic) and interactivity, results in the production of some hormones associated with happiness. The ability to focus and concentrate can also be developed. It is suggested to diversify the efforts aiming at achieving the best possible impact on happiness. It is expected that the organism produces enough hormones (dopamine, serotonin, oxytocin, and endorphin) to feel happy during learning.

Nevertheless, the future application of this technology is still in its early stages, and while it is desirable for the VR/AR to be more widely used within the education system, it is necessary for the products to be more cost-efficient and adequate training to be provided. Considering the now established eco-friendly trend, it would be also desirable for the devices to be made from recyclable materials, to be easy to use and carry, to have a powerful processing power, and to cut across several applications and scientific fields, like psychology, medicine, economics, or product design.

If VR is expected to be a future staple, universities either through government grants or partnerships with the tech industry could provide valuable research in the development of better technologies and the overall user experience.

Once improved, and technology, cost and scale is achieved, the VR could become integrated in most aspects of our life. Furthermore, VR/AR have the great advantage that is in accordance with the circular economy principles, protecting the environment, reducing the cost (of planning, designing, scraping).

## References

Akçayır, M., & Akçayır, G. (2017). Advantages and challenges associated with augmented reality for education: A systematic review of the literature. *Educational Research Review, 20*, 1–11. https://doi.org/10.1016/j.edurev.2016.11.002.

Alexander, T., Westhoven, M., & Conradi, J. (2017). Virtual environments for competency-oriented education and training. In J. Kantola, T. Barath, Nazir S., & T. Andre (Eds.), *Advances in human factors, business management, training and education. Advances in intelligent systems and computing* (Vol. 498). Cham: Springer.

Alfalah, S. F. M. (2018). Perceptions toward adopting virtual reality as a teaching aid in information technology. *Education and Information Technologies, 23*(6), 2633–2653. https://doi.org/10.1007/s10639-018-9734-2.

Alonso, R. S., Prieto, J., García, Ó., & Corchado, J. M. (2019). Collaborative learning via social computing. *Frontiers of Information Technology and Electronic Engineering, 20*(2), 265–282. https://doi.org/10.1631/FITEE.1700840.

Ameer, M. A., & Muaidi, Q. I. (2018). Effect of acute static stretching on lower limb movement performance using STABL virtual reality system. *Journal of Sport Rehabilitation, 27*(6), 520–525. https://doi.org/10.1123/jsr.2017-0017.

Andrews, D., Nicoletti, G., & Timiliotis, C. (2018). Digital technology diffusion: A matter of capabilities, incentives or both? In *OECD economics department working papers* (Vol. 1476). Paris: OECD Publishing. https://doi.org/10.1787/7c542c16-en.

Ardiny, H., & Khanmirza, E. (2018). The role of AR and VR technologies in education developments: Opportunities and challenges. In *2018 6th RSI International Conference on Robotics and Mechatronics (ICROM 2018), RSI International Conference on Robotics and Mechatronics ICRoM* (pp. 482–487).

Azevedo, R. T., Bennett, N., Bilicki, A., Hooper, J., Markopoulou, F., & Tsakiris, M. (2017). The calming effect of a new wearable device during the anticipation of public speech. *Scientific Reports, 7*, 2285. https://doi.org/10.1038/s41598-017-02274-2.

Baixauli Gallego, E. (2017). Happiness: Role of dopamine and serotonin on mood and negative emotions. *Emergency Medicine (Los Angeles), 6*(2), 33–51.

Barberia, I., Oliva, R., Bourdin, P., & Slater, M. (2018). Virtual mortality and near-death experience after a prolonged exposure in a shared virtual reality may lead to positive life-attitude changes. *PLoS ONE, 13*(11). https://doi.org/10.1371/journal.pone.0203358.

Bergstrom, I., Papiotis, P., Saldanha, N., Azevedo, A. S., & Slater, M. (2017). The plausibility of a string quartet performance in virtual reality. In: *IEEE Transactions on Visualization and Computer Graphics, 23*(4), 1352–1359. https://doi.org/10.1109/TVCG.2017.2657138.

Botella, C., Fernández-Álvarez, J., Guillén, V., García-Palacios, A., & Baños, R. (2017). Recent progress in virtual reality exposure therapy for phobias: A systematic review. *Current Psychiatry Reports, 19*(42). https://doi.org/10.1007/s11920-017-0788-4.

Boylan, P., Kirwan, G. H., & Rooney, B. (2018). Self-reported discomfort when using commercially targeted virtual reality equipment in discomfort distraction. *Virtual Reality, 22*(4), 309–314. https://doi.org/10.1007/s10055-017-0329-9.

Breuning, L. G. (2012). *Meet your happy chemicals*. System Integrity Press.

Bucea-Manea-Țoniş, R., Pistol, L., & Gurgu, E. (2019). *Plenary speakers on VII traditional science symposium on the topic: "Digital concept in the part of creative economy"*. https://oikosinstitut.org/wp-content/uploads/2019/05/ZBORNIK-APSTRAKTA-BR.-7.pdf.

Chessa, M., Maiello, G., Borsari, A., & Bex, P. J. (2019). The perceptual quality of the oculus rift for immersive virtual reality. *Human-Computer Interaction, 34*(1), 51–82. https://doi.org/10.1080/07370024.2016.1243478.

Chiao, H. M., Chen, Y. L., & Huang, W. H. (2018). Examining the usability of an online virtual tour-guiding platform for cultural tourism education. *Journal of Hospitality, Leisure, Sport & Tourism Education, 23*, 29–38. https://doi.org/10.1016/j.jhlste.2018.05.002.

Chiarovano, E., McGarvie, L. A., Szmulewicz, D., & MacDougall, H. G. (2018). Subjective visual vertical in virtual reality (Curator SVV): Validation and normative data. *Virtual Reality, 22*(4), 315–320. https://doi.org/10.1007/s10055-018-0336-5.

Christopoulos, A., Conrad, M., & Shukla, M. (2018). Increasing student engagement through virtual interactions: How? *Virtual Reality, 22*(4), 353–369. https://doi.org/10.1007/s10055-017-0330-3.

Cipresso, P., Giglioli, I., Raya, M. A., & Riva, G. (2018). The past, present, and future of virtual and augmented reality research: A network and cluster analysis of the literature. *Frontiers in Psychology, 9*, 2086. https://doi.org/10.3389/fpsyg.2018.02086.

Detyna, M., & Kadiri, M. (2019). Virtual reality in the HE classroom: Feasibility, and the potential to embed in the curriculum. *JGHE*. https://doi.org/10.1080/03098265.2019.1700486.

Di Lernia, D., Cipresso, P., Pedroli, E., & Riva, G. (2018a). Toward an embodied medicine: A portable device with programmable interoceptive stimulation for heart rate variability enhancement. *Sensors (Basel), 18*, 2469. https://doi.org/10.3390/s18082469.

Di Lernia, D., Serino, S., Pezzulo, G., Pedroli, E., Cipresso, P., & Riva, G. (2018b). Feel the time. Time perception as a function of interoceptive processing. *Frontiers in Human Neuroscience, 12*, 74. https://doi.org/10.3389/fnhum.2018.00074.

Eckert, M., Volmerg, J., Friedrich, S., & Christoph, M. (2019). Augmented reality in medicine: systematic and bibliographic review. *JMIR mHealth and uHealth, 7*(4), e10967. https://doi.org/10.2196/10967.

Encalada, L. W., & Sequera, J. L. C. (2017). Model to implement virtual computing labs via cloud computing services. *Symmetry (Basel), 9*(7). https://doi.org/10.3390/sym9070117.

Englund, C., Olofsson, A. D., & Price, L. (2017). Teaching with technology in higher education: Understanding conceptual change and development in practice. *Higher Education Research and Development, 36*, 73–87. https://doi.org/10.1080/07294360.2016.1171300.

Erra, U., Malandrino, D., & Pepe, L. (2019). Virtual reality interfaces for interacting with three-dimensional graphs. *International Journal of Human-Computer Interaction, 35*(1), 75–88. https://doi.org/10.1080/10447318.2018.1429061.

Farhud, D., Malmir, M., & Khanahmadi, M. (2014). Happiness & health: The biological factors-systematic review article. *Iranian Journal of Public Health, 43*(11), 1468.

Fernandez, C., de Salles, A. A., Sears, M. E., Morri, R. D., & Davis, D. L. (2018). Absorption of wireless radiation in the child versus adult brain and eye from cell phone conversation or virtual reality. *Environmental Research, 167*, 694–699. https://doi.org/10.1016/j.envres.2018.05.013.

Fernandez, M. (2017). Augmented virtual reality: How to improve education systems. *High Learning Research Communications (HLRC), 7*(1), 1–15.

Fernandez-Caballero, A., Navarro, E., Fernandez-Sotos, P., Gonzalez, P., Ricarte, J. J., Latorre, J. M., & Rodriguez-Jimenez, R. (2017). Human-avatar symbiosis for the treatment of auditory verbal hallucinations in schizophrenia through virtual/augmented reality and brain-computer interfaces. *Frontiers in Neuroinformatic, 11*. https://doi.org/10.3389/fninf.2017.00064.

Fombona, J., Pascual-Sevillano, M. A., & Gonzalez-Videgaray, M. (2017). M-learning and augmented reality: A review of the scientific literature on the WoS repository. *Comunicar, 52*, 63–71. https://doi.org/10.3916/C52-2017-06.

Gârdan, D. A., Andronie, M., Gârdan, I. P., Andronie, I. E., Iatagan, M., & Hurloiu, I. (2018). Bioeconomy development and using of intellectual capital for the creation of competitive advantages by SMEs in the field of biotechnology. *Amfiteatru Economic, 20*(49), 647–666. https://doi.org/10.24818/EA/2018/49/647.

Garrett, B., Taverner, T., Gromala, D., Tao, G., Cordingley, E., & Sun, C. (2018). Virtual reality clinical research: Promises and challenges. *JMIR Serious Games, 6*(4). https://doi.org/10.2196/10839.

Gattullo, M., Scurati, G. W., Fiorentino, M., Uva, A. E., Ferrise, F., & Bordegoni, M. (2019). Towards augmented reality manuals for industry 4.0: A methodology. *Robotics and Computer-Integrated Manufacturing, 56*, 276–286. https://doi.org/10.1016/j.rcim.2018.10.001.

Geng, J., Peng, X., Qiu, B., Wu, Q., Lv, C., Zi, W., & Zhou, D. (2018). Simulation data integration-based approach for motion synthesis in virtual maintenance. *International Journal of Advanced Manufacturing Technology, 99*(5–8), 1481–1501. https://doi.org/10.1007/s00170-018-2560-2.

Gregorio, S., Ortega, L., & Feito, F.R. (2017). Google Tango outdoors. Augmented Reality for underground infrastructure. In F. J. Melero, & N. Pelechano (Eds.), *CEIG—Spanish Computer Graphics Conference.*

Guo, Z., Zhou, D., Chen, J., Geng, J., Lv, C., & Zeng, S. (2018). Using virtual reality to support the product's maintainability design: Immersive maintainability verification and evaluation system. *Computers in Industry, 101*, 41–50 (Special Issue). https://doi.org/10.1016/j.compind.2018.06.007.

Hacmun, I., Regev, D., & Salomon, R. (2018). The principles of art therapy in virtual reality. *Frontiers in Psychology, 9*. https://doi.org/10.3389/fpsyg.2018.02082.

Hasler, B., Spanlang, B., & Slater, M. (2017). Virtual race transformation reverses racial in-group bias. *PLoS ONE, 12*(4). https://doi.org/10.1371/journal.pone.0174965. pmid:28437469.

Horvath, I. (2018). Evolution of teaching roles and tasks in VR/AR-based education. In: *2018 9th IEEE International Conference on Cognitive Infocommunications (COGINFOCOM), International Conference on Cognitive Infocommunications* (pp. 355–360).

Hsu, W. C., Tseng, C. M., & Kang, S. C. (2018). Using exaggerated feedback in a virtual reality environment to enhance behavior intention of water-conservation. *Journal of Educational Technology and Society, 21*(4), 187–203.

Jiang, X. T., Xiao, Z. G., & Menon, C. (2018). Virtual grasps recognition using fusion of leap motion and force myography. *Virtual Reality, 22*(4), 297–308. https://doi.org/10.1007/s10055-018-0339-2.

Karmonik, C., Elias, S. N., Zhang, J. Y., Diaz, O., Klucznik, R. P., Grossman, R. G., & Britz, G. W. (2018). Augmented reality with virtual cerebral aneurysms? A feasibility study. *World Neurosurgery, 119,* E617–E622. https://doi.org/10.1016/j.wneu.2018.07.222.

Kefi, M., Hoang, N., Richard, P., & Verhulst, E. (2018). An evaluation of multimodal interaction techniques for 3D layout constraint solver in a desktop-based virtual environment. *Virtual Reality, 22*(4), 339–351. https://doi.org/10.1007/s10055-018-0337-4.

Kharitonova, N. A., Kharitonova, E. N., & Pulyaeva, V. N. (2019). Prospects for application of new information and communication technologies in contemporary higher economic education. In Popkova, E., & Ostrovskaya V. (Eds.), *Perspectives on the use of new information and communication technology (ICT) in the modern economy. ISC 2017.* Advances in Intelligent Systems and Computing (Vol. 726). Cham: Springer. https://doi.org/10.1007/978-3-319-90835-9_123.

Kim, D., & Ko, Y. J. (2019). The impact of virtual reality (VR) technology on sport spectators' flow experience and satisfaction. *Computers in Human Behavior, 93,* 346–356. https://doi.org/10.1016/j.chb.2018.12.040.

Klempous, R., Kluwak, K., Idzikowski, R., Nowobilski, T., & Zamojski, T. (2017). Possibility analysis of danger factors visualization in the construction environment based on virtual reality model. In *8th IEEE International Conference on Cognitive Infocommunications (CogInfoCom)* (Vol. 201, pp. 00363–000368). Debrecen. https://doi.org/10.1109/CogInfoCom.2017.8268271.

Lin, C. H., & Hsu, P. H. (2017). Integrating procedural modelling process and immersive VR environment for architectural design education. In *MATEC Web of Conferences* (Vol. 104). Les Ulis: EDP Sciences. https://doi.org/10.1051/matecconf/201710403007.

Liu, S. X., Li, Y., Zhou, P. C., Chen, Q. M., Li, S. D., Liu, Y. D., et al. (2018). Full-color multiplane optical see-through head-mounted display for augmented reality applications. *Journal of the Society for Information Display, 6*(12), 687–693. https://doi.org/10.1002/jsid.739.

Marasco, A., Buonincontri, P., van Niekerk, M., Orlowski, M., & Okumus, F. (2018). Exploring the role of next-generation virtual technologies in destination marketing. *JDMM, 9,* 138–148. https://doi.org/10.1016/j.jdmm.2017.12.002.

Martelli, D., Xia, B., Prado, A., & Agrawal, S. K. (2019). Gait adaptations during over-ground walking and multidirectional oscillations of the visual field in a virtual reality headset. *Gait and Posture, 67,* 251–256. https://doi.org/10.1016/j.gaitpost.2018.10.029.

Martinez-Cerda, J. F., Torrent-Sellens, J., & Gonzalez, I. (2018). Promoting collaborative skills in online university: Comparing effects of games, mixed reality, social media, and other tools for ICT-supported pedagogical practices. *Behaviour and Information Technology, 37*(10–11), 1055–1071. https://doi.org/10.1080/0144929X.2018.1476919.

Meerhoff, L. A., Bruneau, J., Vu, A., Olivier, A. H., & Pettre, J. (2018). Guided by gaze: Prioritization strategy when navigating through a virtual crowd can be assessed through gaze activity. *Acta Psychologica (Amsterdam), 190,* 248–257. https://doi.org/10.1016/j.actpsy.2018.07.009.

Mekacher, D. L. (2019). Augmented reality (AR) and virtual reality (VR): The future of interactive vocational education and training for people with handicap. *Pupil: International Journal of Teaching, Education and Learning, 3*(1). Retrieved from https://grdspublishing.org/index.php/PUPIL/article/view/1842.

Mikolajczyk, M. (2019). VR in education—A subjective overview of the possibilities. *E-Mentor, 2,* 33–34.

Milgram, P., Takemura, H., Utsumi, A., & Kishino, F. (1995). Augmented reality: A class of displays on the reality-virtuality continuum. In *Proceedings of the SPIE* (Vol. 2351). Telemanipulator and Telepresence Technologies, December 21, 1995. https://doi.org/10.1117/12.197321.

Mochón, F. (2018). Happiness and technology: Special consideration of digital technology and Internet. *International Journal of Interactive Multimedia and Artificial Intelligence (IJIMAI)*, 5(3), 162–168. ISSN-e 1989-1660.

Mumtaz, K., Iqbal, M. M., Khalid, S., Rafiq, T., Owais, S. M., & Al, A. M. (2017). An e-assessment framework for blended learning with augmented reality to enhance the student learning. *Eurasia Journal of Mathematics, Science and Technology Education, 13*(8), 4419–4436. https://doi.org/10.12973/eurasia.2017.00938a.

Navarro-Haro, M. V., Lopez del Hoyo, Y., Campos, D., Linehan, M. M., Hoffman, H. G., Garcia-Palacios, A., et al. (2017). Meditation experts try virtual reality mindfulness: A pilot study evaluation of the feasibility and acceptability of virtual reality to facilitate mindfulness practice in people attending a mindfulness conference. *PLoS ONE, 12*(11). https://doi.org/10.1371/journal.pone.0187777.

Oyekan, J. O., Hutabarat, W., Tiwari, A., Grech, R., Aung, M. H., Mariani, M. P., et al. (2019). The effectiveness of virtual environments in developing collaborative strategies between industrial robots and humans. *Robotics and Computer-Integrated Manufacturing, 55,* 41–54. https://doi.org/10.1016/j.rcim.2018.07.006.

Plechawska-Wojcik, M., & Semeniuk, A. (2018). Usage analysis of VR headset in simulated stress situations. In *12th International Technology, Education and Development Conference (INTED Proceedings)* (pp. 7217–7224). https://doi.org/10.21125/inted.2018.1690.

Postolache, O. (2017). Remote sensing technologies for physiotherapy assessment. In *International Symposium on Advanced Topics in Electrical Engineering* (pp. 305–312). Bucharest: IEEE.

Ribeiro, J. M. T., Martins, J., & Garcia, R. (2019). Augmented reality technology as a tool for better usability of medical equipment. In *World Congress on Medical Physics and Biomedical Engineering. IFMBE Proceedings* (Vol. 68, No. 3, pp. 341–345). https://doi.org/10.1007/978-981-10-9023-3_61.

Riches, S., Garety, P., Rus-Calafell, M., Stahl, D., Evans, C., Sarras, N., et al. (2018). Using virtual reality to assess associations between paranoid ideation and components of social performance: A pilot validation study. *Cyberpsychology, Behavior, and Social Networking.* https://doi.org/10.1089/cyber.2017.0656.

Rigutti, S., Straga, M., Jez, M., Baldassi, G., Carnaghi, A., Miceu, P., & Fantoni, C. (2018). Don't worry, be active: how to facilitate the detection of errors in immersive virtual environments. *PEERJ, 6.* https://doi.org/10.7717/peerj.5844.

Riva, G., Wiederhold, B. K., & Mantovani, F. (2018). Neuroscience of virtual reality: From virtual exposure to embodied medicine. *Cyberpsychology, Behavior, and Social Networking.* https://doi.org/10.1089/cyber.2017.29099.gri.

Saadatzi, M. N., Pennington, R. C., Welch, K. C., & Graham, J. H. (2018). Small-group technology-assisted instruction: Virtual teacher and robot peer for individuals with autism spectrum disorder. *Journal of Autism and Developmental Disorders, 48*(11), 3816–3830. https://doi.org/10.1007/s10803-018-3654-2.

Saeedi, S., Bodin, B., Wagstaff, H., Nisbet, A., Nardi, L., Mawer, J., et al. (2018). Navigating the landscape for real-time localization and mapping for robotics and virtual and augmented reality. In *Proceedings of the IEEE, 106*(11), 2020–2039 (Special Issue). https://doi.org/10.1109/JPROC.2018.2856739.

Schmidt, M., Beck, D., Glaser, N., & Schmidt, C. (2017). A prototype immersive, multi-user 3D virtual learning environment for individuals with autism to learn social and life skills: A virtuoso DBR update. In *International Conference on Immersive Learning.* Cham: Springer. (pp. 185–188). https://doi.org/10.1007/978-3-319-60633-0_15.

Schuster-Amft, C., Eng, K., Suica, Z., Thaler, I., Signer, S., Lehmann, I., et al. (2018). Effect of a four-week virtual reality-based training versus conventional therapy on upper limb motor function

after stroke: A multicenter parallel group randomized trial. *PLoS ONE, 13*(10). https://doi.org/10.1371/journal.pone.0205145.
Seinfeld, S., Arroyo-Palacios, J., Iruretagoyena, G., Hortensius, R., Zapata, L. E., Borland, D., et al. (2018). Offenders become the victim in virtual reality: Impact of changing perspective in domestic violence. *Scientific Reports, 8*(1). https://doi.org/10.1038/s41598-018-19987-7.
Seo, J. H., Smith, B. M., Cook, M., Malone, E., Pine, M., Leal, S., et al. (2018). Anatomy builder VR: Applying a constructive learning method in the virtual reality canine skeletal system, advances in human factors in training, education, and learning sciences, AHFE 2017. *Advances in Intelligent Systems and Computing, 596,* 245–252. https://doi.org/10.1007/978-3-319-60018-5_24.
Skarbez, R., Neyret, S., Brooks, F. P., Slater, M., & Whitton, M. C. (2017). A psychophysical experiment regarding components of the plausibility illusion. *IEEE Transactions on Visualization and Computer Graphics, 23*(4), 1369–1378. https://doi.org/10.1109/TVCG.2017.2657158.
Son, B., & Park, J. (2018). Haptic feedback to the palm and fingers for improved tactile perception of large objects. In *Conference: ACM User Interface Software and Technology Symposium (UIST) 2018.* At: Berlin, Germany. https://doi.org/10.1145/3242587.3242656.
Song, H., Chen, F., Peng, Q., Zhang, J., & Gu, P. (2018). Improvement of user experience using virtual reality in open-architecture product design. *Journal of Engineering Manufacture, 232*(13), 2264–2275 (Special Issue). https://doi.org/10.1177/0954405417711736.
Sousa, M. J., & Rocha, A. (2019). Digital learning: Developing skills for digital transformation of organizations. *Future Generation Computer Systems, 91,* 327–334. https://doi.org/10.1016/j.future.2018.08.048.
Spiegel, J. S. (2018). The ethics of virtual reality technology: Social hazards and public policy recommendations. *Science and Engineering Ethics, 24*(5), 1537–1550. https://doi.org/10.1007/s11948-017-9979-y.
Steiniger, B. S., Wilhelmi, V., Berthold, M., Guthe, M., & Lobachev, O. (2018). Locating human splenic capillary sheaths in virtual reality. *Scientific Reports, 8.* https://doi.org/10.1038/s41598-018-34105-3.
Stepanova, E. R., Quesnel, D., & Riecke, B. (2018). Transformative experiences become more accessible through virtual reality. In *IEEE Workshop on Augmented and Virtual Realities for Good (VAR4Good).* Reutlingen, Germany: IEEE. https://doi.org/10.1109/VAR4GOOD.2018.8576881.
Suarez, R., Alonso, A., & Sendra, J. J. (2018). Virtual acoustic environment reconstruction of the hypostyle mosque of Cordoba. *Applied Acoustics, 140,* 214–224. https://doi.org/10.1016/j.apacoust.2018.06.006.
Sun, L. Y., Zhou, Y. Z., Hansen, P., Geng, W. D., & Li, X. D. (2018). Cross-objects user interfaces for video interaction in virtual reality museum context. *Multimedia Tools and Applications, 77*(21), 29013–29041. https://doi.org/10.1007/s11042-018-6091-5.
Takkaç Tulgar, T. A. (2019). In between reality and virtuality: Augmented reality in teaching english to young learners. *Selçuk Üniversitesi Sosyal Bilimler Enstitüsü Dergisi, 41,* 356–364.
Tang, G., Webb, P., & Thrower, J. (2019). The development and evaluation of robot light skin: A novel robot signalling system to improve communication in industrial human-robot collaboration. *Robotics and Computer-Integrated Manufacturing, 56,* 85–94. https://doi.org/10.1016/j.rcim.2018.08.005.
Tatic, D. (2018). An augmented reality system for improving health and safety in the electro-energetics industry. *Facta Universitatis Series Electronics and Energetics, 31*(4), 585–598. https://doi.org/10.2298/FUEE1804585T.
Teisl, M. F., Noblet, C. L., Corey, R., & Giudice, N. A. (2018). Seeing clearly in a virtual reality: Tourist reactions to an offshore wind project. *Energy Policy, 21*(122), 601–611. https://doi.org/10.1016/j.enpol.2018.08.018.
Totzeck, C., Teismann, T., Hofmann, S. G., et al. (2020). Loving-kindness meditation promotes mental health in university students. *Mindfulness, 11,* 1623–1631. https://doi.org/10.1007/s12671-020-01375-w.
Transon, A., Verhulst, A., Normand, J. M., Moreau, G., & Sugimoto, M. (2017). Evaluation of facial expressions as an interaction mechanism and their impact on affect, workload and usability in

an AR game. In *Proceedings of the International Conference on Virtual Systems & Multimedia (VSMM)* (Vol. 217, pp. 116–123).

Tsai, C. F., Yeh, S. C., Huang, Y. Y., Wu, Z. Y., Cui, J. J., & Zheng, L. R. (2018). The effect of augmented reality and virtual reality on inducing anxiety for exposure therapy: A comparison using heart rate variability. *Journal of Healthcare Engineering, 18.* https://doi.org/10.1155/2018/6357351.

Tsatsis, C. G., Rice, K. E., Protopopova, V., Ramos, D., Jadav, J., Coppola, J. F., et al. (2017). Lateropulsion rehabilitation using virtual reality for stroke patients. In *IEEE Long Island Systems, Applications and Technology Conference (LISAT)*. Farmingdale, NY. Ed. IEEE.

Van Dam, L. C. J., & Stephens, J. R. (2018). Effects of prolonged exposure to feedback delay on the qualitative subjective experience of virtual reality. *PLoS ONE, 13*(10). https://doi.org/10.1371/journal.pone.0205145.

Van Dam, N. T., van Vugt, M. K., Vago, D. R., Schmalz, L., Saron, C. D., Olendzki, A., et al. (2018) Mind the hype: A critical evaluation and prescriptive Agenda for research on mindfulness and meditation. *Perspectives on Psychological Science, 13*(1) 36–61. https://doi.org/10.1177/1745691617709589. https://journals.sagepub.com/doi/pdf/10.1177/1745691617709589.

Van der Veer, A. H., Alsmith, A. J. T., Longo, M. R., Wong, H. Y., & Mohler, B. J. (2018). Where am I in virtual reality? *PLoS ONE, 13*(10). https://doi.org/10.1371/journal.pone.0204358.

Van Kerrebroeck, H., Brengman, M., & Willems, K. (2017). Escaping the crowd: An experimental study on the impact of a virtual reality experience in a shopping mall. *Computers in Human Behavior, 77,* 437–450. https://doi.org/10.1016/j.chb.2017.07.019.

Vaziri, K., Liu, P., Aseeri, S., & Interrante, V. (2017). Impact of visual and experiential realism on distance perception in VR using a custom video see-through system. In *SAP'17. Proceedings of the ACM Symposium on Applied Perception*. New York: ACM https://doi.org/10.1145/3119881.3119892.

Verhulst, A., Normand, J. M., Lombart, C., Sugimoto, M., & Moreau, G. (2018). Influence of being embodied in an obese virtual body on shopping behavior and products perception in VR, front. *Robotics and Artificial Intelligence, 5.* https://doi.org/10.3389/frobt.2018.00113.

Voinescu, A., & David, D. (2019). The effect of learning in a virtual environment on explicit and implicit memory by applying a process dissociation procedure. *International Journal of Human-Computer Interaction, 35*(1), 27–37. https://doi.org/10.1080/10447318.2018.1424102.

Vonderlin, R., Biermann, M., Bohus, M., et al. (2020). Mindfulness-based programs in the workplace: A meta-analysis of randomized controlled trials. *Mindfulness, 11,* 1579–1598. https://doi.org/10.1007/s12671-020-01328-3.

Wefers, F., & Vorlaender, M. (2018). Flexible data structures for dynamic virtual auditory scenes. *Virtual Reality, 22*(4), 281–295. https://doi.org/10.1007/s10055-018-0332-9.

Wei, W., Qi, R., & Zhang, L. (2018). Effects of virtual reality on theme park visitors' experience and behaviors: A presence perspective. *Tourism Management, 71,* 282–293. https://doi.org/10.1016/j.tourman.2018.10.024.

Wiederhold, B. K., Miller, I., & Wiederhold, M. D. (2018). Augmenting behavioral healthcare: Mobilizing services with virtual reality and augmented reality. In H. Rivas & K. Wac (Eds.), *Digital health: Scaling healthcare to the world.* Health informatics series (pp. 123–137). New York: Springer. https://doi.org/10.1007/978-3-319-614465_9.

Wiencke, M., & Cacace, M. (2016). *Sebastian fischer, healthy at work: Interdisciplinary perspectives.* Springer. https://doi.org/10.1007/978-3-319-32331-2.

Xu, X. H., Mangina, E., Kiroy, D., Kumar, A., & Campbell, A. G. (2018). Delaying when all dogs to go heaven: Virtual reality canine anatomy education pilot study. In *IEEE Games, Entertainment, Media Conference (GEM)*. Galway, Ireland, ED. New York, USA: IEEE.

Yeh, Y. L., Lan, Y. J., & Lin, Y. T. R. (2018). Gender-related differences in collaborative learning in a 3D virtual reality environment by elementary school students. *Educational Technology and Society, 4,* 204–216.

Yip, J., Wong, S. H., Yick, K. L., Chan, K., & Wong, K. H. (2019). Improving quality of teaching and learning in classes by using augmented reality video. *Computers and Education, 128,* 88–101. https://doi.org/10.1016/j.compedu.2018.09.014.

Yu, C. P., Lee, H. Y., & Luo, X. Y. (2018a). The effect of virtual reality forest and urban environments on physiological and psychological responses. *Urban Forestry and Urban Greening, 35,* 106–114. https://doi.org/10.1016/j.ufug.2018.08.013.

Yu, W. J., Wen, L., Zhao, L. A., Liu, X., Wang, B., & Yang, H. Z. (2018b). The applications of virtual reality technology in medical education: a review and mini-research. In *2018 International Seminar on Computer Science and Engineering Technology (SCSET 2018). IOP, Journal of Physics Conference Series* (Vol. 1176). https://doi.org/10.1088/1742-6596/1176/2/022055.

Zaidi, S. F. M., Duthie, C., Carr, E., & Abd El Maksoud, S. H. (2018). Conceptual framework for the usability evaluation of Gamified virtual reality environment for non-gamers. In *Proceedings of the 16th ACM SIGGRAPH International Conference on Virtual-Reality Continuum and Its Applications in Industry (VRCAI 2018).* Spencer. https://doi.org/10.1145/328.

Zhihan, L., Xiaoming, L., & Wenbin, L. (2017). Virtual reality geographical interactive scene semantics research for immersive geography learning. *Neurocomputing, 254,* 71–78. https://doi.org/10.1016/j.neucom.2016.07.078.

1. VR/AR in education. https://www.indiatoday.in/education-today/featurephilia/story/role-of-augmented-virtual-reality-in-education-1417739-2018-12-26.
2. VR/AR in education. https://education.viewsonic.com/ar-vr-and-mixed-reality/.
3. VR/AR in education. https://www.simlabit.com/virtual-reality-technology-in-search-of-workers/.
4. VR/AR in education. https://appreal-vr.com/blog/augmented-reality-in-education/.
5. VR/AR in education. https://www.medgadget.com/2016/06/holoanatomy-app-previews-use-of-augmented-reality-in-medical-schools.html.
6. VR/AR in education. https://edtechmagazine.com/k12/article/2019/03/k-12-teachers-use-augmented-and-virtual-reality-platforms-teach-biology-perfcon.
7. VR's answer to relaxation? Meditation experiences. https://www.theverge.com/2017/8/6/16094490/vr-meditation-experiences-mindfulness.
8. 10 Relaxing VR apps to calm your stress and improve sleep. https://www.thedailymeditation.com/10-top-relaxing-vr-apps-create-meditative-calm.
9. *Guided meditation—Relaxing VR experience for oculus rift.* https://i.ytimg.com/vi/KgloGxiTtZA/maxresdefault.jpg.
10. Relaxation space room/acerting art VR app for oculus & steamVR. https://i.ytimg.com/vi/Ovpb34gizo/maxresdefault.jpg.

# The Path Less Traversed: Neuroscience and Robots in Nudging Consumer Happiness

### Elena Gurgu and Rocsana Bucea-Manea-Țoniș

**Abstract** Brain scanning of clients will help strategy marketing specialists understand the human brain and its consequent behaviour. Thus, the marketing specialists will know how to change the customers behaviour. The brain behaviour is measured through neuro-marketing techniques: from physiological aspects such as perspiration, the electrical conductivity of the skin, hormonal and neurotransmitter changes, movement and dilation of the pupil, movements of muscles (body and face), to even the understanding of complex cognitive aspects, such as the functional activity of specific regions of the brain through the analysis of different markers such as electrical waves, cerebral metabolism and its blood flow. We believe that smart robots will play a significant role in physical retail in the future. In the last decade, companies have developed a large number of intelligent products. Due to the use of information technology, these products operate somewhat autonomously, cooperate with other products or adapt to changing circumstances. Robotics is a growing industry with applications in numerous markets, including retail, transportation, manufacturing and even as personal assistants. Consumers have evolved to expect more from the buying experience, and retailers are looking at technology to keep consumers engaged. In today's highly competitive business climate, being able to attract, serve and satisfy more customers is a key to success. Consumer behaviour control and paternalism represent a central role in both behaviour analysis and nudging, and they are needed to elaborate on ethical considerations in this regard. Nudging can profit from behaviour analysis by getting a better understanding of the underlying mechanisms of behaviour change. Countless specialized studies show us that the new techniques used by companies to keep the consumer happy are sooner necessary and effective than expensive. A satisfied and happy customer will always return to purchase products and services of the company he already knows and trusts, which

E. Gurgu (✉)
Springer Nature Singapore Pte Ltd., 152 Beach Road, #21-01/04 Gateway East, Singapore 189721, Singapore
e-mail: elenagurgu@yahoo.com

R. Bucea-Manea-Țoniș
Doctoral School, University of Physical Education and Sports, Bucharest, Romania

© The Author(s), under exclusive license to Springer Nature Singapore Pte Ltd. 2021
T. Dutta and M. K. Mandal (eds.), *Consumer Happiness: Multiple Perspectives*,
Studies in Rhythm Engineering,
https://doi.org/10.1007/978-981-33-6374-8_13

continuously supports and motivates him. A company should never ignore the importance of customer satisfaction. There are dozens of factors contributing to the success (or failure) of a business, and customer satisfaction is one of them. Companies need to track this factor and work on improving it to make their customers more loyal and eventually turn them into brand ambassadors. And, the modern techniques used are quite effective and not as expensive as one would think.

## Introduction

Any company should identify consumers' preferences and march on them in their management and marketing strategy. They have to understand that the more happiness offer to clients through their products, the more fidel will be the client (Hao et al. 2019). Happiness is like a drog. The clients will look for it over and over, but the companies have to take into account that consumer preferences will change continuously (Béjar 2018).

The second issue that a company have to take into account for customers satisfaction is the close relationships that are indeed related to happiness, 'meaning that a happier experience for the consumer will be obtained if the product or service brings benefits to the whole family or close relationships (Saphire-Bernstein and Taylor 2013).

The third issue that a company have to take into account for consumer happiness is to help consumer cultivate emotional balance that will make consumers happy and stuck or stick with/to the brand (Mohanty 2014). These balance combines methods extracted from Buddhism, like mindfulness, with synergistic training from modern psychology, like reading microexpressions, and seeks to help people better manage their emotions and relationship (Goleman 2003).

The pursuit of consumer happiness is a preoccupation for many companies (Petersen et al. 2018). Yet only the pursuit can be promised, not consumer happiness itself (Pham 2015). Can neuroscience help? We try to focus on the most tractable ingredient, hedonic or positive affect. A step towards consumer happiness might be gained by improving the pleasures and positive moods in consumer daily life (Pozharliev et al. 2017). The neuroscience of pleasure and reward provides relevant insights, and we discuss how specific hedonic mechanisms might relate to happiness (Balconi and Molteni 2016). Although the neuroscience of happiness is still at the beginning, further advances might be made through mapping overlap between brain networks of hedonic pleasure with others, such as the brain's default network, potentially involved in the other happiness ingredient, eudaimonia or life meaning and engagement (Kringelbach and Berridge 2009, 2010a, 2010b).

Interactive communication and collaboration in human/robots will be a crucial issue to enrich human welfare and happiness in the society (Schellong et al. 2019). Robotics can be used for welfare/rehabilitation, entertainment, education, service, medical and defence (Vaufreydaz et al. 2015; Gupta and Bala 2013; Jones and

Deeming 2008). Robotics is used with success in tourism and hospitality operations, where it is important to understand consumer responses to hotel service robots.

The key factors that affect consumers propensity to interact with robots are functionality, appearance and construction, terminology, status and pride, real and perceived value, personalization and personal expression, as well as support material accompanying the product (Wyatt et al. 2008).

This chapter explores the link of product intelligence to consumer satisfaction through the innovation attributes of relative advantage, compatibility and complexity (Saleem et al. 2015). Also, this chapter considers practical and theoretical implications and identifies future research directions.

Nudging focuses primarily on changing consequences. Studying the effect of behaviour on the environment (changes in social contact or the access to resources are important events in the sense that they have the power to influence behaviour) is another way to influence behaviour and how can be nudged. Results from nudging experiments support the refinement of behaviour analytic thinking (Simon and Tagliabue 2018; Weiss et al. 2013). Nudge is anti-regulation, but behavioural economics is not. Thus, it is proposed that a new approach be developed–budge–or in longhand, behavioural economic-informed regulation designed to budge the private sector away from socially harmful acts. People have systematic cognitive biases that are not only as persistent as visual illusions, but also costly in real life—meaning that governmental paternalism is called upon to steer people with the help of "nudges." (Oliver 2013a, b; Gerdes et al. 2011). In the era of IoT, digital nudging will help the consumer to make decisions in more situations and sectors, due to the new interaction and interface design elements, such as kinetics, virtual reality and hologram. The company has to understand the potential behavioural effects of these new technologies on consumer's judgment and decision-making (Weinmann et al. 2016a, b).

Brands can and must embrace behavioural science to connect with customers—rather than deceive them. With over nearly two decades of helping brands understand the human truth of their customers, we know that any sustainable strategy will be rooted in shared values. Modern consumers are increasingly interested in the principles that brands embrace. Beyond transactional concerns like price, customers invest in companies that align with their convictions and aspirations. This alignment is the basis of brand promise: The most powerful brands in the world help customers realize some aspect of their potential. But if a brand aspires to help its customers progress towards common goals, it must respect their reality. That means abandoning the purely rational view of human behaviour and embracing nudges as a means to encourage and support consumers' positive choices.

The future of keeping a customer satisfied and happy belongs to the experience around company's product or service. Experience will be *the key* differentiator by now on. Not the price nor the product itself. It will first and foremost be about the experience—and customer satisfaction—a company can provide. Customer experience become more than content marketing, more than mobile, more than personalization and more than social. Experience—and by extension, satisfaction—challenged some very heavy hitters. Businesses are going to focus on customer satisfaction and experience, and consumers are actively looking for those brands that deliver on the promise. Companies that prioritize customer satisfaction grow and increase revenue. Those that do not, don't (Pappas et al. 2014).

## *Brain–Behaviour Relationship: Use of Neuroscience Techniques to Measure Happiness and Translating Them to Experience for the Consumer*

**The main trends in behavioural research for online consumer**

The concern of specialists for consumer understanding has adapted to the trend given by the evolution of the Internet and technology. The online consumer is much more demanding, more informed, more numerous and more diverse. It continuously adapts to the flow of information with which it is bombarded. At the same time, there is also an increase in the number of brands and the diversification of the range of products and services offered. Increasing the complexity of the online business environment, online marketing research specialists have adopted new tools and methods of analysing the online consumer's behaviour (Shaffer 2016).

New research methods in online marketing combine elements of classical quantitative and qualitative research with discoveries in the field of neuroscience and social media (Ghosh 2018). We identify as the main trends in the research of the behaviour of the online consumer:

- Visual Marketing;
- Neuro-Marketing;
- Analytics Services;
- Mobile Marketing;
- Search Marketing;
- Multimedia Marketing;
- Online Social Networks.

1. **Visual Marketing**

Visual marketing stats from the byword "An image counts as a thousand words", because as psychologists say "a person can remember 10% of what they read, 30% of what they see; 80% of what they say and 90% of what they say and do at the same time". The online marketers understood very well this principle and integrated in their online campaigns very short videos (no more than 120 s), with impactful images.

Having in mind that the experience and interactivity is very important, they decided to add interactive virtual reality (VR)/augmented reality (AR) applications and online contests.

An important nudge for online consumer behaviour is the testimonials that emphasize what their friends prefer and what is the trend, what the majority like. Testimonials is a very important component of visual marketing, and it has the great advantage that is free. It is the more valuable form of publicity, and thus, the companies has to maintain, open and continuous communication with the consumer, even in real-time, through chatbots.

To measure the impact of the visual marketing campaign, scientist developed two new methods of analysing the behaviour of the online consumer: eye tracking and mouse tracking. They track the consumers' eyes and mouse movements on the web page and observe what is important for the consumer, what is attracting his eyes and what the visual impact signals are.

**Eye tracking** is the process that measures the movement of an eye relative to the head or their position when it snaps a point with the gaze through a device.

Eye-tracking device emphasizes the visual items from the web page content capture the consumers' attention. This device follows the consumer experience on the web pages and collects data regarding this process. The collected data is used to design eyes tracking graph, able to extract the customers reading patterns. The results can be used to design different dynamic visual exploration interfaces for each person. Thus, companies can use specific design elements on the web pages to attract direct attention and increase the ease of processing (Khan 2017)

**Mouse tracking** is the process of tracking the cursor's position on the consumer's computer screen. Mouse tracking records and analyses mouse movements on the screen in search of potential responses (by continuously recording the **x-** and **y-** coordinates of the mouse). By analysing the dynamics of consumer mouse movements, the companies can offer different to change the points of interest. It is a technique to sustain and evaluate real-time cognitive processing and offer valuable information on these facts.

Mouse tracking is more powerful than eye tracking due to the very high temporal resolution. It can be applied in the process of analysing images, strings of letters, sounds and videos. Mouse trails are analysed to evaluate the magnitude of attraction/trails, intricacy, velocity and speeding up of trails. Then psychology, cognitive sciences and marketing specialist can offer research theme regarding consumer behaviour (Gigerenzer 2018). Their conclusions have to be used in the process of testing the site performance, accessibility, handiness and to establish a strategy of enhancing the consumer–computer interface.

2. **Neuro-marketing**

Neuro-marketing is a new discipline that integrates the discoveries of neuroscience to understand the consumer brain functionality (Davidson and Schuyler 2015; Infantolino and Miller 2017). The brain neurological studies offer valuable insight regarding consumer preferences and an authentic image of the brain reactions to various marketing impulse and stimuli (Fortunato et al. 2014).

In neuro-marketing are used two types of examination approaches and techniques: neurological and non-neurological (Sporns 2014).

The **neurological techniques** are:

– *fMRI* (magnetic resonance)—the consumer attraction and implication is measured by the blood–oxygen-level-dependent (BOLD) signal, meaning: the higher concentration of oxygenated haemoglobin in brain blood, the highest interest, emotion for the consumer. The neural activity shows more streaming zones of the brain, on behalf of oxygen consumption, and then, the psychologist can interpret the human response to the marketing stimuli (Zachary).

- **EEG** (electroencephalogram)—electrodes (from 2 to 256) are applied on consumers' head to measure the electrical activity (the difference in electrical charge between pairs of points on the head) of the brain. It is the most competitive economical method used in neuro-marketing research for measuring neural activity directly.
- **MEG** (magnetoencephalogram)—sensors placed on consumers scalp records neuronal activity through magnetic potential. Magnetic fields go through tissues without consistent changes, being more accurate regarding spatial resolution than EEG. The MEG technology is more costly than EEG, so in neuro-marketing do not have a large spread.

EEG and MEG are both excellent for elucidating the temporal dynamics of neural processes.

- *Single Cell Recording*—specialized neurons activity can be also recorded and evaluated to extract important information on brain activity nudged by different stimuli and consumer preferences
- **PET** (positron emission tomography)—is a medical approach based on positron-emitting tracer atom tracking. The molecule is immersed into the bloodstream, such as glucose, water or ammonia. If fludeoxyglucose is immersed in the blood, it will be concentrated into the zone with higher metabolic needs. In time, the fludeoxyglucose will emit positrons (particles with a positive charge) that can be recorded by a sensor. Thus, it can be found the spatial location (3D images for neural areas with the highest metabolic needs) of the fludeoxyglucose and make specific interpretation. For an accurate result, PET has to be combined with computed tomography (CT) images.
- **Study of lesions**—it is rather a medical approach used to analyse the pathology of the brain and their consequences each individual's behaviour. It can be used in marketing campaigns for consumers with different pathologies (Lamichhanea et al. 2020)
- **TMS** (magneto-transcranial simulation)—this approach is based on depolarization or hyperpolarization in neurons near the scalp; electromagnetic induction records the activity of certain areas of the brain that are interested in research.
- **SST** (Steady-State Topography)—is a widely spread approach in brand communication, media research and entertainment. The brain activity of consumer is examined through EEG during watching a spot, video or during a psychological task, while a dim sinusoidal visual flicker is activated in the visual circumference.

The **non-neurological investigation techniques** contain biometrics that evaluate the peripheral nervous system concerning marketing stimuli. Some examples are (Zachary):

- *skin conductance*—measuring the galvanic level of the skin—measures the changes of the electrical properties of the skin, which are established according to the level of its perspiration
- *heart rate*—measures heart rate per minute, the cardiovascular responses to marketing stimuli
- *respiration rate*—measures the frequency of respiration per minute as a response to external stimuli
- *pupil diameter*, *eye blinks* and *eye movements* as a response to marketing stimulation
- *muscle activity* and *voice analyzer*—records the psychophysiological changes of the consumer as a response to certain marketing stimuli.

All these indicators change in association with certain images, events that trigger subjective experiences.

3. **Analytics Services**

Data analytics services start with mouse-tracking technology. Various services allow to analyse the visibility of web pages and traffic on the site of which Google analytics is the most complex, efficient and with international recognition.

These services provide statistics and information on the number of hits of a web page, the geographic area of the customer, the time spent on each web page, the website from which it was redirected, the browser used, the Windows or android operating system, etc. One can analyse which are the most effective advertising spots or banners to optimize sales and increase the conversion rate of visitors into consumers. It also calculates the number of unique visits to the site, can analyse whether a potential customer enters a page of the site and exits without initiating an order, etc. In these cases, it is recommended to implement a site search module for a more enjoyable experience. It is also recommended to use messages that prompt action and to implement *shopping cart*-mode. You can discover which keywords are most used by buyers, conversion rate, order size in value and quantity, loyalty and other items important to the business.

Based on these statistics, companies can develop their marketing strategies and online marketing campaigns. Finally, one can analyse which marketing campaign provides the highest ROI (profit) and update the online marketing strategy according to new data.

4. **Mobile Marketing**

Heart rate increases when people relive experiences related to anger or happiness (Suardi et al. 2016). These results demonstrate the usefulness of smartphones for scientific research: smartphones offer an easy-to-use and cost-effective way to measure heart rate, even by relatively inexperienced researchers. The stronger increase of the heart rate in anger compared to the state of happiness reproduces the previous findings (Warriner et al. 2017). Because the use of smartphones allows researchers to collect physiological measures from a larger group of participants with relative ease, mobile devices could be a way to more easily and effectively examine human resource differences between smaller emotions. of the effect, while gathering enough participants to have sufficient statistical power (Lakens 2013).

Mobile devices bring even more benefits to online consumers, giving them even greater mobility and gains below the time ratio. Their behaviour differs from that specific to navigation on fixed computers or laptops. Companies need to adapt their online marketing strategy by listing the company in the free Google My Business service. After receiving the company verification code, potential customers who pass through the company's headquarters will be notified of the existence and services offered by the company in question. The company may also use "Advertise with Adwords" advertising services to send offers, discounts to certain categories of customers, considered the target audience.

A form of free advertising for the company is customers' decision to provide relationships about the company's services and to post images from inside the company when asked about the Googe Maps service. The advantage is mutual and based on feedback: on the one hand, customers are notified about the existence of a company that could offer them services that interest them, and on the other hand, they provide feedback about the services of the companies visited, informing other potential customers.

Market surveys can also be done through mobile devices. It is recommended to use a set of questions as small and clear as possible so that the consumer answers quickly, when he has a respite, during lunch break or between two more complex activities, etc.

It is observed how active involvement and direct communication with the client gains ground. There are no secrets in the virtual environment. In this respect, marketing campaigns and online market studies can be done more efficiently and quickly, being adapted to the consumer profile.

5. **Search Marketing**

Search marketing methods refer in particular to Search Engine Optimization (SEO)—site optimization for search engines and Search Engine Marketing (SEM)—dedicated search engine marketing. In particular, these methods aim to increase the visibility of web pages for search engines and site traffic, which is associated with page rank growth and site return among the top in the search engine list. The quality of the web content and its continuous updating is quoted.

SEO refers in particular to two types of free online marketing techniques: on-page SEO (elements related to metadata on the page, title, URL, HTML elements, quality content, great design, etc.) and off-page SEO (links between sites, social media—representing customers' opinion about the site). Hackers have discovered illicit ways to increase the number of clicks per page through snippets of code that activate banners, pop-ups or other very aggressive forms of advertising. Those who resort to unfair competition methods or violate certain search engine rules (White Hat SEO) are penalized.

SEM refers in particular to paid advertising methods to increase traffic on a site such as cost-per-click (CPC) or pay-per-click (PPC), cost-per-thousand-views (CPM) and cost-per-acquisition (CPA).

Firms must invest in all forms of online advertising and perhaps offline ones. Of all forms of online advertising, social media advertising seems to be the best on efficiency–price ratio. Added to this is the promotion of online products and services by including in searches, email marketing and advertising on display.

6. **Multimedia Marketing**

This technique refers to the choice of the most appropriate combination of the quality of web content, video, audio and active consumer involvement, through active exhortations to action (attracting the consumer through games, challenging questions, motivation with discount coupons or the accumulation of points that can be converted into money/products/services/advantages). The effectiveness of this technique was demonstrated in the visual marketing section, when we discussed different percentages of information retention, depending on how the advertising message is transmitted and received.

Watch high-impact videos that do not exceed 120 s which are to easily download to mobile devices and respond to limited customer time.

A recent study shows that 74% of marketers said that generating original multimedia content is the most effective way to get the most benefits (http://corporatesolutions.thomsonreuters.com/marketing/).

7. **Online social networks**

Social networks have become the cheapest and most active form of customer communication and a tool for developing and conducting online businesses. Once again, it turns out that technology does not help us if we forget about the main asset of companies, the human resource. The success of social media is based on the social character of the human/brain. Through communication, each of us look in the mirror offered by the society in which we live and expect continuous confirmations of our way of thinking, behaving, speaking, etc.

Social media allows for global communication, not just with friends or followers. Social media has become a recruitment channel, but it also brings the disadvantage of rejecting the candidate, depending on personal posts on social media accounts.

A study by Eurocom Worldwide, involving 318 multinational companies in Europe and the USA, revealed that a fifth of company executives admit to rejecting applications because of social media postings (Reynolds, EWAS, 2012).

Social media advertising has proven to be very effective and less expensive than radio, TV or print media advertising and has been adopted by companies, government and non-profit organisations. It should be noted that, to be effective, the message must be adapted to the market segment to which the company is addressed.

## *Present Versus Future: Next Is the Use of Robots. Will the Use of Robots Have an Impact on Consumer Happiness?*

Interactive communication and collaboration in human/robots will be a crucial issue to enrich human welfare and happiness in the society (Bell 2018). Robotics can be used for welfare/rehabilitation, entertainment, education, service, medical and defence robots (Vaufreydaz et al. 2015; Gupta and Bala 2013; Jones and Deeming 2008). Robotics is used with success in tourism and hospitality operations, and we believe that it is important to understand consumer responses to hotel service robots.

The key factors that affect consumers' propensity to interact with the robot are functionality, appearance and construction, terminology, status and pride, real and perceived value, personalization and personal expression, and support material accompanying the product (Wyatt et al. 2008).

As it is expected, nowadays robots are used especially in industry and defence activities (Bertacchini et al. 2017). It is followed by health care and a very few percentages for domestic activities, entertainment and education (Strulika and Trimbornb 2018).

Industrial robots are appliances, preprogrammed to execute complex specific tasks, without human intervention and with high autonomy in different industrial fields (Castelo et al. 2019). They can perform repetitive tasks such as assembling (electronics, machine, food packaging, small components of toys, etc.), painting by spray methodology, embellishment of 3D plastic objects, casting, forging, polishing or finishing objects, spot fuse, arc fuse, process investigation, surveillance and supervision, loading and unloading tasks, etc. The industrial robots are used in agriculture for processing cereals, collecting fruits and vegetables, milking tasks in farms, cleaning and logistics tasks in hospitality (Chung 2018) (Fig. 1).

Talking about industrial robots, they have a great impact on consumer happiness, if we are talking about consumers both as a company and a physical person (Cooney et al. 2014). If it is a company, the consumer is very content to use robots because they execute very accurate tasks, quicker and without rebuts (Gonzalez-Jimenez 2018). But they are very expensive, and the cost of capital is not usually fair (Pakrasi 2018). If we are talking about the final consumer, they are happy to use/consume a perfect product, but sometimes, it is more expensive (McColl and Nejat 2014). Some products or services can be done by a robot, but the clients would prefer to interact with a human, because of their need for social interaction (Kiesler and Goodrich 2018).

Another category of robots is the one for personal use, such as domestic robots used for cooking, cleaning, lawn mowing, or in entertainment, hobby, pet activities (Fig. 2).

There are also service robots, used to execute commended tasks or repetitive task for human well-being (Piçarra and Giger 2018). These robots are controlled by humans: it can be programmed for different tasks that are unpredictable (Lum 2020). They are multifunction robots, adaptable for domestic activities, delivering

**Fig. 1** Industrial robots in different fields

**Fig. 2** Domestic robots

mail, offering common information in a museum, hotel, school, arranging packages in warehouses, making coffee, entertain, etc (Piçarra et al. 2016).

Some of the service robots can be manipulated from distance, having a remote control to protect human life: some of them are used to explore and expose underground tunnels, caves, deep ocean areas, spatial traffic, events, for example, planetary rovers, drones, remote vehicles, etc. Other service robots read and walk on trails on the floor or use vision landmarks and laser restriction to change position, to turn around. They are used in domestic tasks (vacuum cleaner) or military applications.

A very interesting kind of robot is the modular one. It has the great advantage that can take different shapes and reconfigure itself by the task that has to be done. It is frequently encountered in hybrid transportation, tube cleaning, industrial automation and open manipulation. "Modular robots architectural structure allows hyper-redundancy for modular robots, as they can be designed with more than 8 degrees of freedom (Stahl et al. 2014). Creating the programming, inverse kinematics and dynamics for modular robots is more complex than with traditional robots" (Gupta and Bala 2013).

The robots used in health care, for persons with disabilities or in hospitals for surgeries, seem to bring the grates value-added regarding human happiness and social interaction (Van Pinxteren et al. 2019). It is an ethical concern to employ a person to take care of a patient with disabilities for his treatment, rehabilitation and personal care, because this medical sister can be mentally and emotionally affected by the patient's condition. This is a great reason for which patients should be helped by robots (Wu et al. 2016).

Robots are very useful in complex surgeries, performing very accurate tasks, with no tremor, sometimes more rapidly and performant than a doctor. Even more, a doctor can use the robot to operate himself. In medicine, robots are used for investigation (different blood testing, vital signals scanning, health monitoring, etc.), for diagnostic, for matching and sharing medicine, for caring and arranging the patient in the right position. A miracle seems to be the manipulation of a robot from distance to make surgeries. They are manipulated by a surgeon in real time through an application that offers remote control. They are doing these tasks over and over, without getting tired, but with energy consumption and high costs.

In education robots are used even in primary school for kids to learn how to assemble and program them, using different languages and interfaces: Logo language, Scratch or OpenRoberta interface. They are used for simulations in laboratory classes or research experiments. Robotic kits (Lego Mindstorms, BIOLOID, OLLO, BotBrain Educational Robots) have the scope to let the pupil experiment the real phenomena in a safe environment. Virtual tutor robots are helping teachers and pupils with homework or other comprehending tasks, for increasing motivation and emphatic activities.

**The affective relation between human and robots**
Empathic human–robotic interaction seems to be very important for the consumer, when talking about entertainment robots or robots designed to solve different tasks

at home, in the hospitality industry and other lucrative tasks that imply a human-robot interaction, especially by command. The empathy is a subconscious cognitive process very important human interaction and communication, due to the feeling of security as a result of anticipation (Vaufreydaz et al. 2015). The entertainment robots are sometimes materialized in playful dog companions. These robots are capable to recognize owner emotion. For example, if the owner is angry, the robot dog (AIBO) playback a moaning noise and anticipate cringe by moving its front legs forward to its face, while keeping its rear legs in the same position as well as moving its head and neck downwards. If the owner is sad the dog robots groan by arching up and raising its head, while opening and closing mouth. If the owner is happy, the dog shake, waggle tail, then bark again. If the owner is bored, the dog stretch and open its mouth as if it were yawning. If the owner is surprised, the dog move its head up and then move it from side to side while making a sniffing noise as if to see what caused a "surprise" (Yamazaki et al. 2010; Jones and Deeming 2008). This dog make real the non-verbal signals of the human intention to start an interaction.

This AIBO companion dog is paying attention to the owner's emotional state and voice and adapt his behaviour on the context like acoustic emotion recognition and assertion speech. Due to this emotional component companion dog offer emotional experiences, empathic pleasant interaction and compassion if appropriate (Jones and Deeming 2008).

Another type of companion robot was designed for an apprentice for everyday activities such as teacher assistant, personal trainer, desktop manager, activity monitoring, deficiency assessment, therapy and rehabilitation assistance, such as Kompai Robot, etc. (Vaufreydaz et al. 2015) Their interaction with the human being is based on observing, reasoning and communicating affective interactivity, through face-to-face and non-verbal communication.

"The set of relevant features for starting of interaction detection can be reduced without loss of performance using a feature space reduction process using the Minimum Redundancy Maximum Relevance (MRMR) method" (Hanchuan et al. 2005). This approach provides essential and ergonomic assistance for persons with disabilities due to the facility of detecting engagement and making it socially acceptable.

This empathy and detecting the interaction intentions are also important in tourism service. The NAO and Relay robots were designed for hotel room delivery. The higher the anthropomorphism, anticipated intelligence and security characteristics of the robots, the higher probability the clients to be happy to use robot services. These robots were employed for the first time in Hennna Hotel for different services, such as check-in at the front desk, room service to luggage delivery, press delivery or in-room companion (Tussyadiah and Park 2018a, b). This hotel also integrated a lot of supportive technologies, such as facial recognition, automatic payment, drone delivery and self-driving cars, meaning that for their clients, companion robots are not odd. The hotel adopted that solution as being more cost-effective (Osawa et al. 2017), but they border-line between satisfied clients by the robot services and human services is very thin. Thus, the management has to be aware of not losing clients. Some clients do not feel comfortable and safe when receiving services or advice from a robot. This problem might be surpassed by anthropoid characteristics of robots and adding more functionality as to surprise the clients with reasoning and intelligent answers.

Overall men are more likely to adopt robot in their everyday life assistance activities. On the other hand, aged people with different conditions and disabilities understand the potential of robots' contribution to their contentment, independence in fulfil ergonomic the daily tasks. They reject the idea of neuro-prosthesis but agree with the immersion of the artificial organ to improve their quality of life. They do not like the idea of feeling lonely or isolated with a robot. This is why an emphatic robot, one that provides social facilities and interaction, is more likely to be accepted by the elderly. The robots should not be just reliable useful tools, but they have to become companion, provide breakthrough interactive and communication skills, reasoning and intelligent reactions (Block and Kuchenbecker 2018a, b; Arras and Cerqui 2005).

The robots and human can identify eight social emotions: sadness, fear, related joy, surprise, anger, boredom, interest and happiness. These emotions were identified due to specific human-body actions and postures and can be spread out by a life-sized human-like robot. The future robot should be able to recognize also facial expressions and vocal intonation (Block and Kuchenbecker 2018a, b; McColl and Nejat 2014; Young et al. 2011).

Having in mind that the anthropomorphic characteristics and the human–robot interaction are so important for greater adoption of robots in different industries, a humanoid robot was designed. It looks like a human woman; it has artificial skin made of soft silicon texture and is called EveR-1; their eyes are cameras able to distinguish, recognize and record human faces; it can assert its emotion through eyes and lip synchrony (Fig. 3). This cute robot can change the facial expression and move the upper body and hand during communicating information. It is appropriate for advice and instruction information within exhibitions, conferences, tourism customer service, theatre or movie representations, narration of fairy tales, playing cultural and artistic content, common conversation and teacher assistant in education (Klatt et al. 2020).

In the future, this robot will be equipped with different types of personality, such as aggressive, passive, cheerful personality, and so on (Ahn et al. 2013).

These technologies form the basis for building systems that will interact with students in more natural ways, even bootstrapping the machine's own ability to learn from humans (Parasuraman and Rizzo 2007, 2008). Broad neuroscience topics and new technologies may enhance the understanding of how to create useful vision for blind persons and related neuro-ergonomic applications (Parasuraman and Rizzo 2007, 2008).

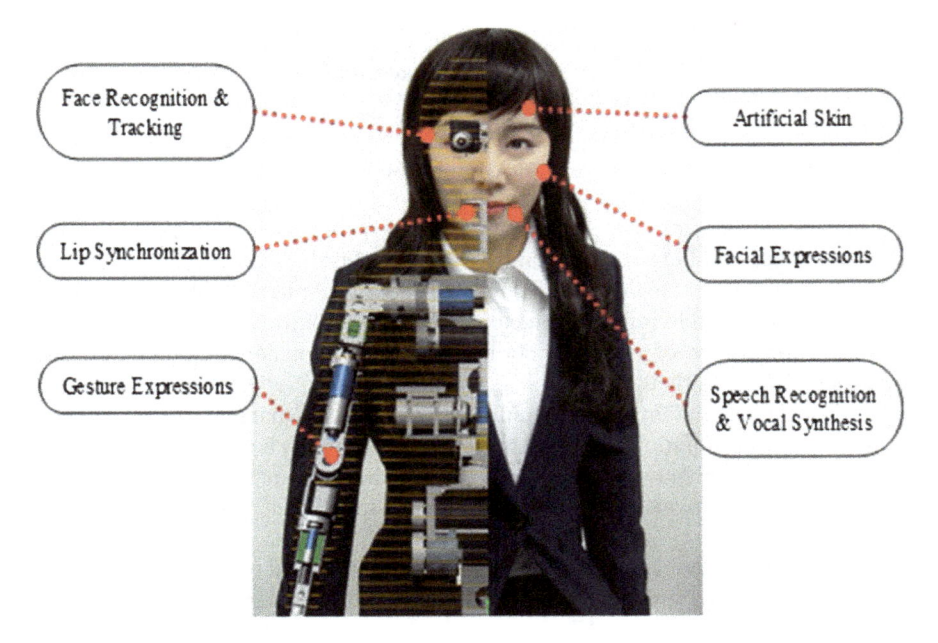

**Fig. 3** Features of EveR-1

## Nudging Happiness: Effective Use of Nudges to Promote Consumer Happiness

Nudges are intended transformation in the choice architecture, to modify peoples' behaviour with by changing the environment, by offering scientific useful information to prevent mistakes, by simplifying and offering information when is difficult to extract or understand, by encouraging positive behaviours through opt-in solutions. Anticipating human mistakes, a constant feedback on consumer behaviour impact on environment, on society, on his own health, etc., is proved to change consumer behaviour in some contexts. Nudges can be done with more salient incentives as to influence consumer choices that are almost automatically. It is important to specify that nudges have the aim to enforce positive behaviour by personal value systems, helping persons to make better decisions for themselves, through a better understanding, rationalization or biases, in the beneficial of society and individual's long-term interests (Thaler and Sunstein 2008).

According to Hansen (2014: 2): "A nudge is … any attempt at influencing people's judgment, choice or behavior in a predictable way (1) made possible because of cognitive biases in individual and social decision-making posing barriers for people to perform rationally in their own interest and (2) working by making use of those biases as an integral part of such attempts".

A nudge is any aspect of the choice architecture that predictably alters people's behaviour without forbidding any options or significantly changing their economic consequences. To count as a mere nudge, the intervention must be easy and cheap to avoid. Nudges are not mandates (Kim et al. 2013).

According to *Merriam-webster.com* nudge aims "to seek the attention by a push of the elbow, to prod lightly, urge into action".

Although technocrats may offer viable solutions, people have the capacity and the freedom to reject the nudges, they are not agreed with. Furthermore, the autonomy and decoupling are often associated with some psychological phenomena, such as incubation and mind wandering. This phenomena/state is confirmed by neural network activity in the course of mind wandering that seems to become autonomous and connected with hypothetical reasoning and decoupling (Viale 2019). The nudges process capacitates and favorizes the rationalization process that allows the consumer to choose the right solution for him. The nudges process can be used as a libertarian paternalism that ethically help consumer to evaluate information and offer effective cognitive instruments to simplify the problem-solving activity in the process of making decisions.

Consumer behaviour control and paternalism have a central role in both behaviour analysis and nudging, and they are needed to elaborate on ethical considerations in this regard (Benartzi et al. 2017). Nudging can profit from behaviour analysis by getting a better understanding of the underlying mechanisms of behaviour change. Nudging focuses primarily on changing consequences (Bucher et al. 2016). Studying the effect of behaviour on the environment (changes in social contact or in the access to resources are important events in the sense that they have the power to influence behaviour) is another way to influence behaviour and how it can be nudged. Results from nudging experiments support the refinement of behaviour analytic thinking (Simon and Tagliabue 2018; Weiss et al. 2013).

So that the nudge to have impact on consumer behaviour, some elements has to be accomplished: the public information should be very clear, coherent and apprehensible; constant evaluation and management of the consumer feedback must be done; making a bridge between actual choices and expected consequences; counterbalance consumer mistakes regarding personal choices; simplifying the problem-solving activity by offering guidance in probabilistic and deductive thinking and interpretation; consumer can also be guided in deciding uncertain conditions with heuristics and debiasing techniques. These techniques of systematic and rational thinking will improve the process of autonomous decision. This kind of nudge will have a great impact on consumer because it drives to neuroplasticity changes in consumer brain.

In the era of IoT digital nudging will help consumers to make decisions in more situations and sectors, due to the new interaction and interface design elements, such as kinetics, virtual reality and hologram (Chriss 2016). The company has to understand the potential behavioural effects of these new technologies on consumer's judgment and decision-making (Weinmann et al. 2016a, b).

Providing simple information, changing accessibility and visibility, infusing size, presenting social norms and ideal-type behaviours are methods to influence consumer behaviour. This happens because people often based their decision on behavioural biases, they do not make rational decision, but behave on behalf of mental shortcuts and habits.

As people are often unaware of the effects of the decision environment on their actions, nudges mostly work on changing non-deliberative aspects of individuals' actions (House of Lords 2011). Nudge tools include defaults, working with warnings of various kinds, changing layouts and features of different environments, reminding people about their choices, drawing attention to social norms and using framing to change behaviour. Coercive policy instruments such as laws, bans, jail sentences or economic and fiscal measures, e.g. taxes or subsidies, are not nudges according to Thaler and Sunstein (2008).

The scientific literature presents the nudge principles and methods as: (based on Thaler and Sustein 2008)

1. **Managing the marketing stimulus**, the incentive, to be more pertinent, convincing to increase the impact on the emotional and rational consumer behaviour.

A VR/AR demo of a game, a VR/AR tour in a museum or in a library will incline the consumer decision easier.

2. **Making the information more comprehensible** if the evaluation is difficult. The information must be simplified, quantified and framed.

For example, explaining technical information in a graphical form, using images, schemes, scratches, graphs, dashboards, etc.

3. **Influencing consumer behaviour by default options**.

More than 95% of customers accepted the default opt-in solution and agreed with "green electricity" consumption although is more expensive. Some of them even accepted the opt-in default solution of smart grid trial (accepted the installation of energy control consumption technology). They were free to check the opt-out solution.

4. **Constant feedback to consumer actions**, to understand the consequences of their actions.

For example: The consumer can receive a smiley face if respects the rules in a natural reservation/exposition/botanical garden/zoological garden or an attention sign if it is doing a mistake (waking in restricted area, feeding animals, etc.). A constant feedback on energy consumption, through informative energy bills, metering and energy labelling of appliances and buildings may conduct to 20% savings.

5. **Anticipating user mistakes,** explaining the consequences and making recommendation to prevent and avoid it. For example: displaying an image with an example of buying card and the numbers to be entered in the form.

6. **Presenting all the characteristics of all the alternatives** and letting people make trade-offs when necessary.

In online selling, the complementary products are associated (the computer information is associated with software recommendations).

7. **Boost costumer self-control and not by activating a desired behaviour** (Kim et al. 2013):

A nudge was concepted to stimulate self-control and help consumers to implement it. For example, a fitness pass for one year will nudge the consumer to exercise more and strengthen the health condition.

The shipment of products raises the cost of online products/services and may cause delays or deliveries cannot be made on time. A solution to nudge the consumer behaviour and determine him to buy products is the development of pick-up and drop-off outlets (P&D), designed to overcome the barriers of online shopping, i.e. the payment of transportation costs, as well as the inconvenience of fixed delivery windows (between 8 a.m. and 5 p.m.) These solutions will also reduce traffic pollution, urban congestion, shop crowding, traffic bottleneck, etc. P&D are small local shops in high-traffic areas or locations on highways (railways or gas stations), warehouses, neighbourhood supermarkets or even postboxes and street mailboxes (DHL) warehouses, in malls (e.g. De Buren) or in urban centres. The format can be a service desk for in-store customers, a drive-through concept or a wall with cabinets and codes.

8. **Self-imposed and not by external imposed**

The consumer willpower is important in the decision of adopting one behaviour, buying a product, a service, or make a specific activity. For example, the consumers are nudged to protect the environment through battery collection. For each five used batteries, the consumer receives one new battery.

Companies can opt only for online stores to reduce the biological footprint. They can create medium-sized virtual showrooms: some stores have digital interaction through virtual and augmented reality applications, touchscreen walls and other devices to present the rational reasons of buying a product and the positive consequences. This format allows merchants to display their complete collection on a limited space that is particularly attractive to big box retailers and car brands because it gives them a way to enter urban areas. For customers, virtual showrooms provide the benefit of facing a custom product, to their requests and needs (such as a car, kitchen or shoes, all customized). An additional feature of virtual showrooms is the high level of connectivity with customers' social network, providing a perfect online and offline shopping experience. The initiators of this trend are car manufacturers and clothing retailers, for example American Apparel, Adidas, John Lewis and Marks & Spencer.

9. **Mindful not Mindless**

As we explained before, the rational arguments and specific decision-making techniques are important in consumer choices. The images with the consequences of long run smoking that are presented on the packages of cigarettes may reduce the smoking, at least for the persons that are not dependent.

The "halo effects" of interactive technology change the role and experience of the physical environment and has an important role in making the incentive more salient. Some museums are not very populated because of high degree of technical information. Virtual and augmented reality will partially solve this impediment by creating VR/AR apps to make the information clearer, to increase the interactive customer experience and nudge the client to visit or come back again.

Nudges are very important if consumers have a positive attitude or desire for a particular behaviour but is not able to achieve it in daily activities. Nudges are not very effective, if a person is not keen on it or have opposite ideas. People may choose which nudges are good for them, are under their rational or are preferred. Often nudges work for persons ready or open to adopt them. Nudges should follow information and education campaigns to support the underlying policy (Hansen 2014) and to enhance desired positive social norms.

10. **Encourage not discourage**: the fine, the taxes are tools able to discourage a negative behaviour, while the reword will encourage a pozitive behaviour.

Sometimes physical environment adaption encourages a good behaviour. Placing the good food in the reach of the consumer, on the eyes level and the unhealthy food on an upper or lower shelve will encourage the consumer to choose the healthy food. Another example is the decision to use standards in some environments: adding bumpers on streets to reduce speed, using key cards in institutions to turns of lights when removed or nudging clients to shop online, for the sake of interactive virtual reality application (Oliver 2015). Advanced mobile technologies (online payments, virtual reality) will allow consumers to scan and buy everything they see in real life through the online applications, reducing the ecological footprint. Mobile devices are more than just communication tools these days, but they become nudging tools (Viale 2019). They allow the share of safety recommendations, spread of cultural and economic principles, the share of good practices or good example.

The restaurants that reduced the portion size and the plate size, spoon size and package size have strong positive impact on the environment protection, recording savings up to 20% (Bucher et al. 2016).

A good behaviour can be encouraged by de social norms' presentation: a billing feedback on social comparison has large effects in small-scale trials, recoding 11% savings (Bucher et al. 2016).

The consumer can be encouraged to choose products through heart-to-heart online communication using social media and new artificial intelligence technology, such as Internet of Things (IoT—multitude of different types of devices interconnected in real time) and blockchain technology (Wang et al. 2018). These technologies are useful tools for encouraging the clients with fun solutions, such as gamification.

Blockchain became a valuable marketing appliance that facilitates the conversion of fidelity cards, points in tokens. Then, the tokens may be used perse in nudging the clients or can be transformed in cryptocurrencies. Cryptocurrencies also can be exchanged in real currency (Choi 2018; Wang et al. 2018). The tokens often nudge the consumer behaviour, for double benefits: consumer and organizational benefits.

The company loyalty schemes, based on association between tokenization and gamification, are useful tools in nudging the consumer (Whitehead et al. 2014). Through these games, the company can fulfil the first two nudging principles, emphasizing the advantages, the uniqueness and differentiation quality of the products and brands. Some example of nudging tourist consumers with tokens or cryptocurrencies are Ethereum or GOZO token (Antoniadis et al. 2019). Other examples of nudging clients in financial world are the use of blockchain to convert the token in digital wallets.

The best results of applying nudging appears in public environments without counteracting effect of marketing, e.g. in school canteens, where the consumer acts in a very controlled environment. The effect is not similar in people's homes. Furthermore, changes in physical environment are also very effective, because reduce the decision options to social norms but are dependent on the context, and the policy makers' control of the environment.

Nudge is anti-regulation, but behavioural economics is not. Thus, it is proposed that a new approach be developed—budge—in longhand, behavioural economic-informed regulation designed to budge the private sector away from socially harmful act; people have systematic cognitive biases that are not only as persistent as visual illusions but also costly in real life—meaning that governmental paternalism is called upon to steer people with the help of "nudges" (Oliver 2013a, b; Gerdes et al. 2011).

All experts are in consens that the nudges are suplementary tools of traditional politics and not an alternative of coercitive measures (laws and reglemetations) and economic tools (fiscal nudges, subventions, taxes, etc.). They are also used to change automate, intuitive and non-deliberative bahaviours. The consumer acceptance empower the nudges, and in this case, the consumer targeted behaviours are non-controversial and is following social norms and values. They help customers prevent mistakes, due to lack of information and understanding. But, there are researchers that say that nudges could be manipulative, and only an open debate can establish which actions that change behaviour are in the common advantage.

It is important to have confidence in the person/institution authorized to nudge people. It is important to understand the criteria used in the decision process and how each person can improve his own decisions and the types of incentives used to change behaviour (Sugden 2009a, b).

## Effectiveness Versus Expensiveness: Are the Techniques that Could Be Used to Promote Consumer Happiness Effective or Expensive?

Countless specialized studies show us that the new techniques used by companies to keep the consumer happy are sooner necessary and effective than expensive. A satisfied and happy customer will always return to purchase products and services of the company he already knows and trusts, which continuously supports and motivates him. A company should never ignore the importance of customer satisfaction. There are dozens of factors contributing to the success (or failure) of a business, and customer satisfaction is one of them (Oliver 2014). Companies need to track this factor and work on improving it to make their customers more loyal and eventually turn them into brand ambassadors. And the modern techniques used are quite effective and not expensive.

The future of keeping a customer satisfied and happy belongs to the experience around the company's product or service. Experience will be *the key* differentiator by now on, not the price nor the product itself. It will first and foremost be about the experience—and customer satisfaction—a company can provide. Customer experience become more than content marketing, more than mobile, more than personalization and more than social (Rouhiainen 2016). Experience—and by extension, satisfaction—challenged some very heavy hitters. Businesses are going to focus on customer satisfaction and experience, and consumers are actively looking for those brands that deliver on the promise. Companies that prioritize customer satisfaction grow and increase revenue. Those that do not, don't (Pappas et al. 2014).

Growth and revenue are key elements of a successful business (Fig. 4). Beyond the growth correlation, if a company actively works to increase customer satisfaction, it is more likely to see an increase in revenue, there are plenty of other reasons to make

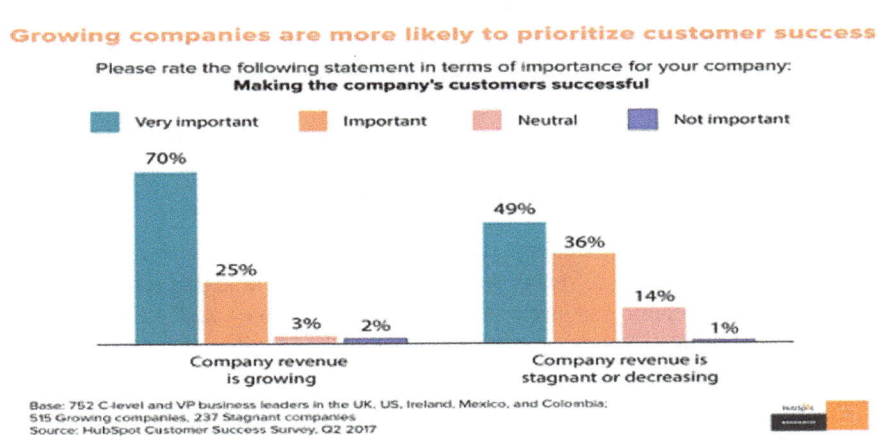

**Fig. 4** Prioritizing customer success

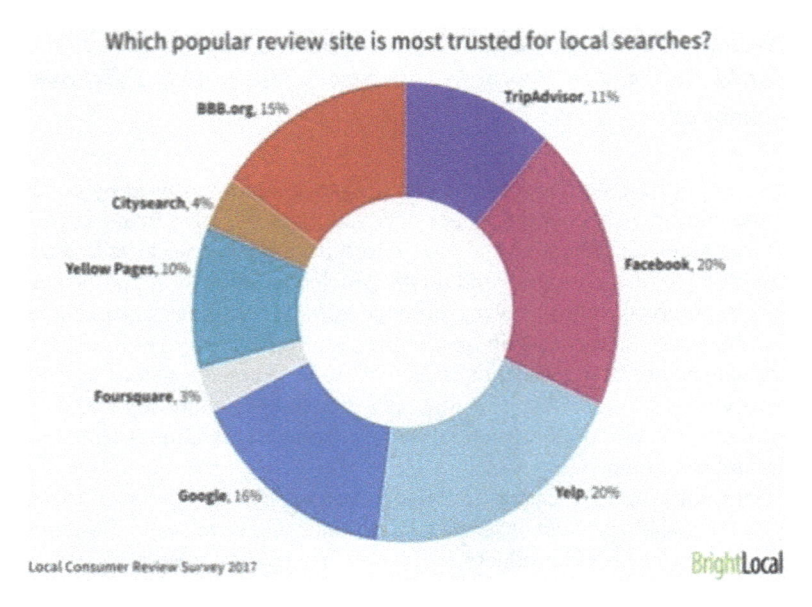

**Fig. 5** Popular review sites

it a top priority. Let us take *word-of-mouth*, for example. It matters, especially in the ultra-connected and always-on digital world we call home. We can instantly share our experience with a brand with thousands of others on social media and review sites like Yelp (Fig. 5), (Chow and Shi 2015).

That is a lot of potential goodwill and positive publicity. But it works both ways. Sixty percentage% of consumers share a bad experience with others, and they tell 3 times as many people, compared to only 40% who share the good ones. A company would better do its best to ensure each customer interaction is a positive one. If the company do not place a premium on relationship marketing and customer satisfaction, then it would not be aware of problems or complaints until it is too late. If a company prioritize keeping its customers happy, it will reduce the number of unhappy ones and will know about and work to resolve dissatisfaction that much faster, being a win-win situation (Cazier et al. 2007).

But, the benefits of a customer-first approach do not stop there:

**Brand loyalty** Why would a happy, satisfied customer ever look elsewhere or want to leave its company? The company will see lower churn and higher retention. And a 5% increase in retention can increase profitability by as much as 25–95%. "*Return customers tend to buy more from a company over time. As they do, your operating costs to serve them decline. What's more, return customers refer others to your company. And they'll often pay a premium to continue to do business with you rather than switch to a competitor with whom they're neither familiar nor comfortable*". *~Fred Reichheld, Creator of Net Promoter* (Groysberg 2019).

**Fig. 6** SMART goal system

> Specfic ✓
> Measurable ✓
> Achievable ✓
> Relevant ✓
> Time-bound ✓

**Brand buzz** A platoon of happy advocates and cheerleaders singing company's praises on social media and review sites is the absolute best publicity that money can or cannot buy. Companies must work to make it happen (Hewett et al. 2016).

**Brand trust** Consumers trust people, even strangers, more than they do advertising and marketing. *"Measurement is the first step that leads to control and eventually to improvement. If you can't measure something, you can't understand it. If you can't understand it, you can't control it. If you can't control it, you can't improve it". ~H. James Harrington, CEO of Harrington Management Systems.* So, for companies, it is very important to measure, understand, control and improve (Charron et al. 2014).

**Customer Satisfaction Goals** No one knows perfectly what should a customer satisfaction goals include and because the owner knows his/her business better than anyone else and the company's goals may not be the client's goals and vice versa. So, generally speaking, the company wants to keep things simple. Using the SMART goal system (Specific, Measurable, Achievable, Relevant, Time-limited) (Fig. 6) and set only 1–2 at a time (otherwise, they start competing with each other), companies can be realistic, as studies shown that they are 2–3 times more likely to follow-through (Bennett and Rundle-Thiele 2004).

Customer satisfaction is any company's business, regardless of its product, industry or niche. Companies must make it a priority. That is true today and will only increase in importance in the years to come.

Companies must collect, analyse and use data on customer satisfaction for every stage of their funnel, every interaction and touch-point, every product launch, and more. Every company has to pick and choose its moment, of course, as customers do not want to be inundated with surveys all the time. But no area is off-limits for selectively surveying and asking for feedback. By keeping the customers satisfied, with limited costs, that is how companies improve. That is how companies grow. And that is how a company turn customers into repeat customers and repeat customers into cheerleaders (Jyoti and Sharma 2012).

Companies are obsessed with customer satisfaction, and high-standard customer service can win companies clients' hearts and make them recognizable within their target group. Nowadays, when social media play such an important role in making decisions it is crucial to keep an eye on the quality of customer service a company can provide.

There are solid reasons why customer satisfaction is not only important but also beneficial for any company's brand.

1. A loyal customer is a treasure a company should keep and hide from the world. According to the White House Office of Consumer Affairs, on average, loyal customers are worth up to 10 times as much as their first purchase. Some research says that it is 6–7 times more expensive to acquire a new customer than it is to keep a current one. Banks or mobile providers know it best, so they do not have any problem with going the extra mile for a customer who is not quite satisfied and often offer him something special. Not only it is more expensive but also much more difficult to keep existing and loyal clients—let alone keeping them fully satisfied and happy—than to gain some new ones (Mayer 2012).
2. Customers can stop being a company's clients in a second. Clients easily switch their love brands. It is often caused by terrible customer service. Clients wait for ages to get feedback or comment from a brand, which is unacceptable. But it still happens. And gaining clients' trust takes up to 12 positive experiences to make up for one unresolved negative experience.*"When customers share their story, they're not just sharing pain points. They're actually teaching you how to make your product, service and business better. Your customer service organization should be designed to effectively communicate those issues"* (Smaby 2011).

A company cannot gain customers' satisfaction forever, it needs to look after them all the time. Any company has to teach how to talk to them, ask questions, offer constant support, send personalized messages or offers, use dedicated customer satisfaction survey tool or any other technique that will help the company to communicate with its customers and collect insights. Taking care of each of its clients' needs, any company will be rewarded with their gratitude and loyalty. Brands often take their audience for granted, and they have never been so wrong—one decision, or lack of it, can result in losing a lot of clients and their respect (Dewani et al. 2016).

3. The money is very important, too. It should not be surprising, but customer satisfaction is also reflected in a company's revenue. Customers' opinion and feelings about the brand can affect, in both positive and negative way, the essential metrics – such as the number mentions and repeated transactions and also customer lifetime value or customer churn. Happy customers would not look at the competitors offers—they will happily interact with the same brand again, make a purchase and recommend the product further. If a company can meet all of clients requirements and answers, their needs while delivering the best quality of services, they will be fully satisfied. And the brand will increase sales revenue. *"Loyalty is when people are willing to turn down a better product or price to continue doing business with you"* (Sinek 2009).
4. Customer satisfaction is a factor that helps any company stand out of the competition. Kate Zabriskie once said that *"Although your customers won't love you if you give bad service, your competitors will."*, and we could not agree more. (Hasan 2019). Competitive rivals are just waiting for any company to make a wrong move. What is more, they can often play the role of an instigator. However, if companies know how to provide their customers with amazing customer service, they will gain arguments to convince those uncertain of their services.

5. Great customer experience can take any brand places. The importance of customer satisfaction should never be neglected. Satisfied customers are more likely to share a company's content across social media. They will also more keenly interact with the company's posts, leaving some delightful and admirable comments. Later, companies can use it as the source for case studies and success stories.

Providing great customer service will satisfy both the company and its customers. The clients get proper service, the company gets a proper revenue, and everyone is happy.

Instead of running expensive and time-consuming market research, simply asking questions by email, on social media or by monitoring what people have to say can offer valuable insight into what is working and what is not for a brand. A company can even take this a step further and share the survey results with its customers. Here is what Rocco Baldassarre said in an Entrepreneur article about using transparency to keep clients happier: *"Customer loyalty increases also based on how mistakes are being handled. Studies show that up to 70 per cent of unhappy customers transform into loyal customers if the mistake has been fixed exceeding their expectations"* (Rouhiainen 2016). People appreciate this level of transparency, and when they know their voice is being heard, it gives them more incentive to share their opinion. Not only that, but when a company makes changes based on their feedback it shows that a company genuinely care to deliver the best product and customer experience possible.

## Conclusion

To conclude, customer satisfaction has long been considered a milestone in the path towards companies profitability. Although it is widely acknowledged that customer satisfaction leads to higher and more stable revenues, the relationship between customer satisfaction levels and the costs that the company incurs in producing and delivering customer services has received far less attention, and the research results vary significantly across sectors. There seems to be little guidance for linking company costs to the key elements involved in providing customer satisfaction in services, thereby diminishing the ability of a company to manage its activities accordingly.

## References

Ahn, H. S., Lee, D.-W., Choi, D., Lee, D.-Y., Lee, H.-G., & Baeg, M.-H. (2013). Development of an incarnate announcing robot system using emotional interaction with humans. *International Journal of Humanoid Robotics, 10*(2), 1350017 (24 pages). World Scientific Publishing Company. https://doi.org/10.1142/s0219843613500175.

Antoniadis, I., Kontsas, S., & Spinthiropoulos, K. (2019). *Blockchain and brand loyalty programs: A short review of applications and challenges*. In International Conference on Economic Sciences and Business Administration (Vol. 5, No. 1, pp. 8–16). Spiru Haret University.

Arras, K. O., & Cerqui, D. (2005) *Do we want to share our lives and bodies with robots?* A 2000 people survey A 2000-people survey. https://doi.org/103929/ethz-a-010113633.

Balconi, M., & Molteni, E. (2016). Past and future of near-infrared spectroscopy in studies of emotion and social neuroscience. *Journal of Cognitive Psychology, 28*(2), 129–146.

Béjar, H. (2018). El Código Espiritualista De La Autoayuda: La Felicidad Negativa the spiritual code of self-help. *The Negative Happiness, Athenea Digital, 18*(3), e2339. ENSAYOS. ISSN: 1578-8946.

Bell, G. (2018). Making life: A brief history of human-robot interaction. *Consumption Markets & Culture, 21*(1), 22–41.

Benartzi, S., Beshears, J., Milkman, K. L., Sunstein, C. R., Thaler, R. H., Shankar, M., ... Galing, S. (2017). Should governments invest more in nudging?. *Psychological Science, 28*(8), 1041–1055.

Bennett, R., & Rundle-Thiele, S. (2004). Customer satisfaction should not be the only goal. *Journal of Services Marketing*.

Bertacchini, F., Bilotta, E., & Pantano, P. (2017). Shopping with a robotic companion. *Computers in Human Behavior, 77*, 382–395.

Block, A. E., & Kuchenbecker, K. J. (2018). *Emotionally supporting humans through robot hugs*. HRI'18 Companion, Chicago, IL, USA, March 5–8, 2018 © 2018 Copyrightheldbytheowner/author(s).

Block, A. E., & Kuchenbecker, K. J. (2018, March). Emotionally supporting humans through robot hugs. In *Companion of the 2018 ACM/IEEE International Conference on Human-Robot Interaction* (pp. 293–294).

Bucher, T., Collins, C., Rollo, M. E., McCaffrey, T. A., De Vlieger, N., Van der Bend, D., et al. (2016). Nudging consumers towards healthier choices: A systematic review of positional influences on food choice. *British Journal of Nutrition, 115*(12), 2252–2263.

Castelo, N., Schmitt, B., & Sarvary, M. (2019). Human or robot? Consumer responses to radical cognitive enhancement products. *Journal of the Association for Consumer Research, 4*(3), 217–230.

Cazier, J. A., Shao, B. B., & Louis, R. D. S. (2007). Sharing information and building trust through value congruence. *Information Systems Frontiers, 9*(5), 515–529.

Charron, R., Harrington, H. J., Voehl, F., & Wiggin, H. (2014). *The lean management systems handbook* (Vol. 4). Boca Raton: CRC Press.

Choi, J. (2018). Modeling the integrated customer loyalty program on blockchain technology by using credit card. *International Journal on Future Revolution in Computer Science & Communication Engineering, 4*(2), 388–391.

Chow, W. S., & Shi, S. (2015). Investigating customers' satisfaction with brand pages in social networking sites. *Journal of Computer Information Systems, 55*(2), 48–58.

Chriss, J. J. (2016). Influence, nudging, and beyond. *Society, 53*(1), 89–96.

Chung-En, Y. (2018, March). Humanlike robot and human staff in service: Age and gender differences in perceiving smiling behaviors. In *2018 7th International Conference on Industrial Technology and Management (ICITM)* (pp. 99–103). IEEE.

Cooney, M., Nishio, S., & Ishiguro, H. (2014). Affectionate interaction with a small humanoid robot capable of recognizing social touch behavior. *ACM Transactions on Interactive Intelligent Systems (TiiS), 4*(4), 1–32.

Davidson, R. J., & Schuyler, B. S. (2015). Neuroscience of happiness. *World Happiness Report* (pp. 88–105).

Dewani, P. P., Sinha, P. K., & Mathur, S. (2016). Role of gratitude and obligation in long term customer relationships. *Journal of Retailing and Consumer Services, 31*, 143–156.

Esmark, A. (2019). Communicative governance at work: How choice architects nudge citizens towards health, wealth and happiness in the information age. *Public Management Review, 21*(1), 138–158.

Fortunato, V. C. R., Giraldi, J. D. M. E., & de Oliveira, J. H. C. (2014). A review of studies on neuromarketing: Practical results, techniques, contributions and limitations. *Journal of Management Research, 6*(2), 201.

Fortunato, V. C. R., de Moura Engracia Giraldi, J., & de Oliveira, J. H. C. (2014). *A review of studies on neuromarketing: Practical results, techniques, contributions and limitations.* doi:105296/jmrv6i25446.

Gerdes, K. E., Lietz, C. A., & Segal, E. A. (2011). Measuring empathy in the 21st century: Development of an empathy index rooted in social cognitive neuroscience and social justice. *Social Work Research, 35*(2), 83–93.

Ghosh, S. K. (2018). Happy hormones at work: Applying the learnings from neuroscience to improve and sustain workplace happiness. *NHRD Network Journal, 11*(4), 83–92.

Gigerenzer, G. (2018). The Bias Bias in behavioral economics. *Review of Behavioral Economics, 5*(3–4), 303–336.

Goleman, D. (2003). Finding happiness: Cajole your brain to lean to the left. *New York Times*, Vol. 4, No. 3.

Gonzalez-Jimenez, H. (2018). Taking the fiction out of science fiction: (Self-aware) robots and what they mean for society, retailers and marketers. *Futures, 98,* 49–56.

Groysberg, B. (2019). Fred Reichheld-creator of net promoter score.

Gupta, V., & Bala, R. (2013). Advancements in robots for human welfare: A review. *International Journal of Research in Management, Science & Technology, 1*(2), December 2013. E-ISSN: 2321-3264. Available at www.ijrmstorg.

Hanchuan, P., Fuhui, L., & Chris, D. (2005). Feature selection based on mutual information: criteria of max-dependency, max-relevance and min-redundancy. *IEEE Transactions on Pattern Analysis and Machine Intelligence, 27*(8), 1226–1238.

Hansen, P. G. (2014). *Nudge and libertarian paternalism: Does the hand fit the glove?* (p. 23). Copenhagen: Roskilde University.

Hao, M., Liu, G., Gokhale, A., Xu, Y., & Chen, R. (2019). Detecting happiness using hyperspectral imaging technology. *Computational Intelligence and Neuroscience, 2019.*

Hasan, M. F. (2019). *Sustainability of Bata as market leader in footwear industry in Bangladesh.*

Hewett, K., Rand, W., Rust, R. T., & Van Heerde, H. J. (2016). Brand buzz in the echoverse. *Journal of Marketing, 80*(3), 1–24.

House of Lords. (2011). *Behaviour change, the house of lords* (p. 111).

Infantolino, Z., & Miller, G. A. (2017). Psychophysiological methods in neuroscience. *Noba Project*.

Jones, C., & Deeming, A. (2008). Affective human-robotic interaction. In *Affect and emotion in human-computer interaction* (pp. 175–185). Berlin, Heidelberg: Springer.

Jyoti, J., & Sharma, J. (2012). Impact of market orientation on business performance: Role of employee satisfaction and customer satisfaction. *Vision, 16*(4), 297–313.

Kiesler, S., & Goodrich, M. A. (2018). The science of human-robot interaction.

Kim, L. y., Mažar, N., Zhao, M., & Soman, D. (2013). *A practitioner's guide to nudging.* Rotman School of Management University of Toronto.

Klatt, L.-I., Schneider, D., Schubert, A.-L., Hanenberg, C., Lewald, J., Wascher, E., Getzmann, S. (2020). Unraveling the relation between EEG correlates of attentional orienting and sound localization performance: a diffusion model approach. *Journal of Cognitive Neuroscience, 32*(5), 945–962. https://doi.org/101162/jocn_a_01525.

Kringelbach, M. L., & Berridge, K. C. (2009). Towards a functional neuroanatomy of pleasure and happiness. *Trends in Cognitive Sciences, 13*(11), 479–487.

Kringelbach, M. L., & Berridge, K. C. (2010a). The functional neuroanatomy of pleasure and happiness. *Discovery Medicine, 9*(49), 579.

Kringelbach, M. L., & Berridge, K. C. (2010b). The neuroscience of happiness and pleasure. *Social Research, 77*(2), 659.

Lakens, D. (2013). Using a smartphone to measure heart rate changes during relived happiness and anger. *IEEE Transactions on Affective Computing, 4*(2).

Lamichhanea, B., Westbrookbc, A., Coled, M. W., & Bravera, T. S. (2020). Exploring brain-behavior relationships in the N-back task. *NeuroImage, 212*, 116683, 15 May, 2020. https://doi.org/101016/jneuroimage2020116683

Lum, H. C. (2020). The role of consumer robots in our everyday lives. In *Living with robots* (pp. 141–152). Academic Press.

Mayer, R. N. (2012). The US consumer movement: a new era amid old challenges. *The Journal of Consumer Affairs, 46*(2), 171–189.

McColl, D., & Nejat, G. (2014). Recognizing emotional body language displayed by a human-like social robot. *International Journal of Social Robotics, 6*(2), 261–280.

Mohanty, M. S. (2014). What determines happiness? Income or attitude: Evidence from the US longitudinal data. *Journal of Neuroscience, Psychology, and Economics, 7*(2), 80.

Oliver, A. (2013a). From nudging to budging: Using behavioural economics to inform public sector policy. *Journal of Social Policy, 42*, 685–700. https://doi.org/10.1017/S0047279413000299.

Oliver, A. (2013b). From nudging to budging: using behavioural economics to inform public sector policy. *Journal of Social Policy, 42*(4), 685–700.

Oliver, R. L. (2014). *Satisfaction: A behavioral perspective on the consumer: A behavioral perspective on the consumer*. Routledge.

Oliver, A. (2015). Nudging, shoving, and budging: Behavioural economic-informed policy. *Public Administration, 93*(3), 700–714.

Osawa, H., Akiya, N., Koyama, T., Ema, A., Kanzaki, N., Ichise, R., Hattori, H., & Kubo, A. (2017). What is real risk and benefit on work with robots? From the analysis of a robot hotel. In: *HRI2017 Companion* ACM, Vienna.

Pakrasi, I. L. (2018). *Towards expressive mobile robots* (Doctoral dissertation).

Pappas, I. O., Pateli, A. G., Giannakos, M. N., & Chrissikopoulos, V. (2014). Moderating effects of online shopping experience on customer satisfaction and repurchase intentions. *International Journal of Retail & Distribution Management*.

Parasuraman, R., & Rizzo, M. (2007). *Neuroergonomics at work: The brain at work*. Oxford: Oxford University Press.

Parasuraman, R., & Rizzo, M. (2008). *Neuroergonomics—The brain at work*. Oxford University Press: Oxford.

Petersen, F. E., Dretsch, H. J., & Loureiro, Y. K. (2018). Who needs a reason to indulge? Happiness following reason-based indulgent consumption. *International Journal of Research in Marketing, 35*(1), 170–184.

Pham, M. T. (2015). On consumption happiness: A research dialogue. *Journal of Consumer Psychology, 25*(1), 150–151.

Piçarra, N., & Giger, J. C. (2018). Predicting intention to work with social robots at anticipation stage: Assessing the role of behavioral desire and anticipated emotions. *Computers in Human Behavior, 86*, 129–146.

Piçarra, N., Giger, J. C., Pochwatko, G., & Gonçalves, G. (2016). Making sense of social robots: A structural analysis of the layperson's social representation of robots. *European Review of Applied Psychology, 66*(6), 277–289.

Pozharliev, R., Verbeke, W. J., & Bagozzi, R. P. (2017). Social consumer neuroscience: Neurophysiological measures of advertising effectiveness in a social context. *Journal of Advertising, 46*(3), 351–362.

Rouhiainen, L. (2016). Facebook marketing tips and strategies for small businesses. *Lasse Rouhiainen Lexington USA, 101*.

Saleem, A., Ghafar, A., Ibrahim, M., Yousuf, M., & Ahmed, N. (2015). Product perceived quality and purchase intention with consumer satisfaction. *Global Journal of Management and Business Research*.

Saphire-Bernstein, S., & Taylor, S. E. (2013). Close relationships and happiness. In *Oxford handbook of happiness, psychology, social psychology, clinical psychology*. Online Publication. doi: 101093/oxfordhb/9780199557257.013.0060.

Schellong, M., Kraiczy, N. D., Malär, L., & Hack, A. (2019). Family firm brands, perceptions of doing good, and consumer happiness. *Entrepreneurship Theory and Practice, 43*(5), 921–946.

Shaffer, J. (2016). *Neuroplasticity and Clinical Practice: Building Brain Power for Health Front Psychol, 7,* 1118. https://doi.org/10.3389/fpsyg201601118.

Simon, C., & Tagliabue, M. (2018). Feeding the behavioral revolution: Contributions of behavior analysis to nudging and vice versa. *Journal of Behavioral Economics for Policy, 2*(1), 91–97.

Sinek, S. (2009). *Start with why: How great leaders inspire everyone to take action.* Penguin.

Smaby, K. (2011). Being human is good business. *A List Apart.*

Sporns, O. (2014). Contributions and challenges for network models in cognitive neuroscience. *Nature Neuroscience, 17*(5), 652.

Stahl, B. C., McBride, N., Wakunuma, K., & Flick, C. (2014). The empathic care robot: A prototype of responsible research and innovation. *Technological Forecasting and Social Change, 84,* 74–85.

Strulika, H., & Trimbornb, T. (2018). Hyperbolic discounting can be good for your health. *Journal of Economic Psychology, 69,* 44–57. December 2018. https://doi.org/101016/jjoep201809007

Suardi, A., Sotgiu, I., Costa, T., Cauda, F., & Rusconi, M. (2016). The neural correlates of happiness: A review of PET and fMRI studies using autobiographical recall methods. *Cognitive, Affective, & Behavioral Neuroscience, 16*(3), 383–392.

Sugden, R. (2009a). On Nudging: A review of nudge. Improving decisions about health, wealth and happiness by Richard H Thaler and Cass R Sunstein. *International Journal of the Economics of Business, 16*(3), 365–373.

Sugden, R. (2009). *On nudging: A review of nudge: Improving decisions about health, wealth and happiness by Richard H. Thaler and Cass R. Sunstein.*

Thaler, R. H., & Sunstein, C. R. (2008). *Nudge: Improving decisions about health, wealth, and happiness.* New Haven, CT: Yale University Press.

Tussyadiah, I. P., & Park, S. (2018). Consumer evaluation of hotel service robots. In *Information and Communication Technologies in Tourism 2018.* Springer International Publishing AG 2018. https://doi.org/101007/978-3-319-72923-7_24.

Tussyadiah, I. P., & Park, S. (2018). Consumer evaluation of hotel service robots. In *Information and communication technologies in tourism 2018* (pp. 308–320). Cham: Springer.

van Pinxteren, M. M. E., Wetzels, R. W., Rüger, J., Pluymaekers, M., & Wetzels, M. (2019). Trust in humanoid robots: implications for services marketing. *Journal of Services Marketing.*

Vaufreydaz, D., Johal, W., & Combe, C. (2015). Starting engagement detection toward a companion robot using multimodal features. *Robotics and Autonomous Systems.* http://doi.org/101016/jrobot 201501004

Vaufreydaz, D., Johal, W., & Combe, C. (2016). Starting engagement detection towards a companion robot using multimodal features. *Robotics and Autonomous Systems, 75,* 4–16.

Viale, R. (2019). Architecture of the mind and libertarian paternalism: Is the reversibility of system 1 nudges likely to happen?. In *Mind & SOciety.* Germany: Springer GmbH. Part of Springer Nature. https://doi.org/101007/s11299–019-00218-z

Wang, L., Luo, X. R., & Xue, B. (2018). Too good to be true? Understanding how blockchain revolutionizes loyalty programs 24th Americas conference on information systems, New Orleans, pp 1–10

Warriner, A. B., Shore, D. I., Schmidt, L. A., Imbault, C. L., & Kuperman, V. (2017). Sliding into happiness: A new tool for measuring affective responses to words. *Canadian Journal of Experimental Psychology/Revue Canadienne de Psychologie Expérimentale, 71*(1), 71.

Weinmann, M., Schneider, C., Brocke, J. V. (2016a). Digital nudging. *Business & Information Systems Engineering,* 58, 433–436. https://doi.org/101007/s12599-016-0453-1.

Weinmann, M., Schneider, C., & vom Brocke, J. (2016b). Digital nudging. *Business Information Systems Engineering, 58*(6), 433–436.

Weiss, L. A., Westerhof, G. J., & Bohlmeijer, E. T. (2013). Nudging socially isolated people towards well-being with the 'Happiness Route': Design of a randomized controlled trial for the evaluation of a happiness-based intervention. *Health and Quality of Life Outcomes, 11*(1), 159.

Whitehead, M., Jones, R., Howell, R., Lilley, R., & Pykett, J. (2014). *Nudging all over the world. ESRC Report*. Swindon and Edinburgh: Economic and Social Research Council.

Wu, Y. H., Cristancho-Lacroix, V., Fassert, C., Faucounau, V., de Rotrou, J., & Rigaud, A. S. (2016). The attitudes and perceptions of older adults with mild cognitive impairment toward an assistive robot. *Journal of Applied Gerontology, 35*(1), 3–17.

Wyatt, J., Browne, W. N., Gasson, M. N., & Warwick, K. (2008). Consumer robotic products. *IEEE Robotics & Automation Magazine, 15*(1), 71–79.

Yamazaki, Y., Vu, H. A., Le, P. Q., Liu, Z., Fatichah, C., Dai, M., … Nagashima, N. (2010, July). Gesture recognition using combination of acceleration sensor and images for casual communication between robots and humans. In *IEEE Congress on Evolutionary Computation* (pp. 1–7). IEEE.

Young, J. E., Sung, J., Voida, A., Sharlin, E., Igarashi, T., Christensen, H. I., et al. (2011). Evaluating human-robot interaction. *International Journal of Social Robotics, 3*(1), 53–67.

# Author Index

**A**
Ahn, Paul Hangsan, 179
Azambuja de, Gina Pipoli, 3

**B**
Bucea-Manea-Țoniş, Rocsana, 231, 255

**D**
Dominic, Elizabeth, 163
Dutta, Tanusree, 153

**E**
Ercan, Özge, 89
Eternod, Vivian, 23

**F**
Felfernig, Alexander, 203

**G**
Gurgu, Elena, 231, 255

**K**
Kazemekaityte, Austeja, 69
Krajnović, Aleksandra, 127

**L**
Le, Viet-Man, 203

**M**
Mandal, Manas Kumar, 153
Martins, Oliva Maria Dourado, 231
Mochón, Francisco, 43

**P**
Pillai, Anil V., 117

**R**
Rodríguez-Peña, Gustavo, 3

**S**
Savadori, Lucia, 69
Simion, Violeta Elena, 231
Sur, Ratul, 101

**T**
Tran, Thi Ngoc Trang, 203

**V**
Van Swol, Lyn M., 179
Victor, Vijay, 163